Ore Geology, Economic Minerals and Mineral Economics

VOLUME 1

S. K. TIWARI

ATLANTIC
PUBLISHERS & DISTRIBUTORS (P) LTD

ATLANTIC
PUBLISHERS & DISTRIBUTORS (P) LTD
7/22, Ansari Road, Darya Ganj, New Delhi
Tel.: +91-11-4077 5252, 2327 3880
E-mail: orders@atlanticbooks.com
Web: www.atlanticbooks.com

Reprint 2026

Published by Atlantic Publishers & Distributors (P) Ltd.

Printed & bound in India by Atlantic Print Services

Preface

This textbook is based on various economic minerals and deals with Ore Geology and Economic Minerals as well as Mineral Economics. Section A "Resources/Mineral Resources" provides general information about our mineral resources, while Section B "Ore Geology" describes in sufficient detail ore formation periods, ore genesis, ore fluids, forms, modes, structures, tectonics and mineralization, secondary sulphide deposits, etc. Section C "Economic Minerals" comprises discussion on deposits of diamond, gold, iron, manganese, tin, tungsten, copper, lead, zinc, bauxite, mica, limestone and phosphorite.

Keeping in view the development in marine science as well as increasing concerns for environment, ocean resources, manganese nodules and forest resources have been specially considered in Section D "Ocean and Forest Resources". Section E is based on Mineral Economics. It encompasses classification norms for mineral deposits, geo-statistics, criteria for mineral exploration and investment, ore microscopy, processing, etc. Moreover, legal provisions, national mineral policy and mineral conservation have been included along with ocean/sea policy and forest policy. Section F "Mineral Based Industries" describes in detail iron and steel, aluminum, cement, fertilizer, refractory and abrasive industry need.

It will be pertinent to point out that such type of work can only be a cohesion of appropriate information gathered from various sources, resources and persons of eminence, and all those have been duly mentioned in the text as well as in the list of references. However, those missing are due to oversight and not at all deliberate.

For their valuable works, I feel indebted to Prof. B.B. Dhar, former Head, Department of Mining Engineering, IT-BHU and former Director, CMRS, Dhanbad; Prof. Rekha Ghose, ISM, Dhanbad; Prof. D.P. Singh, former Vice Chancellor, Rewa University, M.P., Lucknow University, Lucknow and U.P. Rajarshi Tandon Open University, Allahabad; and Prof. G. Kumar, BIT, Sindri (Dhanbad). My gratitude is due to Prof. R.M. Singh, Prof. R.N. Tiwari, Prof. G.C. Choudhary, Prof. M.S. Srinivasan and Prof. M.P. Singh of Department of Geology, BHU. They have always been a source of inspiration and motivation.

My acknowledgements are due to all those who have directly or indirectly helped and encouraged me in completing this work. I am grateful to my family members who helped me a lot without demanding my time and attention to them. I am also thankful to the team of Atlantic Publishers and Distributors (P) Ltd., who did their best to give a perfect shape to my vision.

I hope the book will be of equal interest to academicians, students, those preparing for various competitive examinations, environmentalists and general readers as well. I will always welcome reactions, suggestions and recommendations.

Banaras Hindu University **S.K. Tiwari**
Varanasi-221005

List of Elements

Symbol	*Element*	*Symbol derived from*
Ac	Actinium	Greek *aktis* = ray
Ag	Silver	Latin *argentum*
Al	Aluminium	Latin *alumen* = alum
Am	Americium	America
Ar	Argon	Greek *argos* = inactive
As	Arsenic	Latin *arsenicum*
At	Astatine	Greek *astatos* = unstable
Au	Gold	Latin *aurum*
B	Boron	Persian *burah*
Ba	Barium	Greek *barys* = heavy
Be	Beryllium	Greek *beryliion* = beryl
Bi	Bismuth	German (Origin unknown)
Bk	Berkelium	Berkeley, California
Br	Bromine	Greek *bromos* = stench
C	Carbon	Latin *Carbo* = Charcoal
Ca	Calcium	Latin *calx* = lime
Cd	Cadmium	Greek *Kadmeia* = Calamine
Ce	Cerium	asteroid ceres
Cf	Californium	California
Cl	Chlorine	Green *Chloros* = green
Cm	Curium	Marie and Pierre Curie
Co	Cobalt	German *Kobold* = goblin
Cr	Cromium	Greek *Chroma* = Colour
Cs	Caesium	Latin *Caesium* = bluish-grey
Cu	Copper	Cyprus
Dy	Dysprosium	Greek *dysprositas*
Er	Erbium	Ytterby, a Swedish town
Es	Einsteinium	Einstein
Eu	Europium	Europe
F	Fluorine	Latin *fluor* = flow
Fe	Iron	Latin *ferrum*

Fm	Fermium	Fermi
Fr	Francium	France
Ga	Gallium	Latin *Gallux* = Cock
Gd	Gadolinium	Gadolin, a Finnish Chemist
Ge	Germanium	Latin *Germania* = Germany
H	Hydrogen	Greek *hydro* = water
He	Helium	Greek *helios* = sun
Hf	Hafnium	*Hafnia* = Copenhagen
Hg	Mercury	Latin *hydrargyrum* = mercury
Ho	Holmium	*Holmia* = Stockholm
I	Iodine	Greek *ioeides* = violet
Il	Indium	its indigo spectrum
Ir	Iridium	Latin *iris* = a rainbow
K	Potassium	Latin *Kalium*
Kr	Krypton	Greek *Kryptos* = hidden
La	Lanthanum	Greek *lanthanein* = Conceal
Li	Lithium	Greek *lithos* = stone
Lu	Lutetium	*Lutetia,* ancient name of Paris
Lw	Lawrencium	Lawrence, US Physicist
Md	Mendelevium	Mendeleyev, Russian chemist
Mg	Magnesium	Magnesia, district in Thessaly
Mn	Manganese	Magnesia (as above)
Mo	Molybdenum	Greek *molybdos* = lead
N	Nitrogen	Greek *nitron* = saltpetre
Na	Sodium	Latin *natrium*
Nb	Niobium	Niobe (Greek myth)
Nd	Neodymium	Greek *neos* = new+didymos = twin
Ne	Neon	Greek *neos* = new
Ni	Nickel	German, abbreviation of *küpfernickel* = Copper devil
No	Nobelium	Nobel
Np	Neptunium	Planet Neptune
O	Oxygen	Greek *oxys* = acid
Os	Osmium	Greek *osme* = odour
P	Phosphorus	Greek *phosphoros* = light bearing
Pa	Protactinium	Greek *protos* = first + *actinium*
Pb	Lead	Latin *plumbum*
Pd	Palladium	asteroid pallas
Pm	Promethium	Prometheus, stealer of fire from heaven (Greek myth)

Po	Polonium	Poland
Pr	Praseodymium	Greek *prasios* = green + didymos = twin
Pt	Platinum	Spanish *plata* = silver
Pu	Plutonium	Planet Pluto
Ra	Radium	Latin *radius* = ray
Rb	Rubidium	Latin *rubidus* = red
Re	Rhenium	Latin *rhenus* = river Rhine
Rh	Rhodium	Greek *rhodon* = rose
Rn	Radon	Radium emanation
Ru	Ruthenium	*Ruthenia* = Russia
S	Sulpher	Latin *sulfur*
Sb	Antimony	Latin *Stibium*
Sc	Scandium	Scandinavia
Se	Selenium	Greek *selene* = moon
Si	Silicon	Latin *silex* = flint
Sm	Samarium	Samarski, a Russian savant
Sn	Tin	Latin *stannum*
Sr	Strontium	Strontian, Scottish village
Ta	Tantalum	Tantalus (Greek myth)
Tb	Terbium	Ytterby, a swedish town
Tc	Technetium	Greek *technetos* = artificial
Te	Tellurium	Latin *tellus* = earth
Th	Thorium	Norse God Thor
Ti	Titanium	Latin *titanes* = suns of earth
Tl	Thallium	Greek *thallos* = green shoot
Tm	Thulium	Greek and Latin *Thule* = Northland
U	Uranium	Planet Uranus
Une	Unnilennium	Greek *ennea* = nine
Unh	Unnilhexium	Greek *hex* = six
Uno	Unniloctium	Greek *octo* = eight
Unp	Unnilpentium	Greek *pente* = five
Unq	Unnilquadium	Latin Quattuor = four
Uns	Unnilseptium	Latin *septem* = seven
V	Vanadium	Norse Goddess Vanadis (Freya)
W	Tungsten	Wolfram
Xe	Xenon	Greek *xenos* = stranger
Y	Yttrium	Ytterby, a Swedish town
Yb	Ytterbium	Ytterby, a Swedish town
Zn	Zinc	German *Zink*
Zr	Zirconium	Persian *Zargun* = gold coloured

List of Organisations and Acronyms

AIEC	—	Association of Iron Ore Exporting Countries
AMD	—	Atomic Mineral Division
AOC	—	Assam Oil Company
APEC	—	Association of Petroleum Exporting Countries
ASEAN	—	Association of South-East Asian Nations
BCCL	—	Bharat Coking Coal Limited
BOC	—	Burma Oil Company
BPCL	—	Bharat Petroleum Corporation Limited
BSI	—	Botanical Survey of India
CBD	—	Convention on Biodiversity
CCL	—	Central Coalfield Limited
CFRI	—	Central Fuel Research Institute
CGWB	—	Central Ground Water Board
CIL	—	Coal India Limited
CITES	—	Conventional International Trade in Endangered Species
CMPDIL	—	Central Mineral Production Division of India Ltd.
CMRS	—	Central Mining Research Station
COM	—	Commission on Ore Microscopy
CPCB	—	Central Pollution Control Board
CRI	—	Cement Research Institute
CSIR	—	Council of Scientific and Industrial Research
DGMS	—	Director General of Mines Safety
DNES	—	Department of Non-Conventional Energy Sources
DOD	—	Department of Ocean Development
DRDO	—	Defence Research & Development Organisation
ECL	—	Eastern Coalfields Limited
ECO	—	Economic Cooperation Organisation

EIA	—	Environmental Impact Assessment
EIL	—	Engineers India Ltd.
EMS	—	Environmental Management System
EPA	—	Environmental Protection Agencies
FAO	—	Food and Agriculture Organisation
FCCC	—	Frame-work Convention on Climate Change
GAIL	—	Gas Authority of India Limited
GATT	—	General Agreement on Tariff and Trade
GEF	—	Global Environmental Facility
GSI	—	Geological Survey of India
HCL	—	Hindustan Copper Limited
HPCL	—	Hindustan Petroleum Corporation Limited
HZL	—	Hindustan Zinc Limited
IBM	—	Indian Bureau of Mines
ICAR	—	Indian Council for Agriculture Research
ICON	—	Indian Consortium of Oil and Natural Gas
IEDP	—	Indian Ecology Development Project
IGC	—	International Geological Congress
IIT	—	Indian Institute of Technology
IMA	—	International Mineral Association
IOC	—	Indian Oil Corporation
IPCC	—	Inter-governmental Panel on Climate Change
ISM	—	Indian School of Mines
ITC	—	International Tin Council
ITRC	—	Indian Toxicological Research Centre
IUCN	—	International Union for Conservation of Natural Resources
IUGS	—	International Union of Geological Sciences
MEC	—	Mineral Export Corporation
MoEF	—	Ministry of Environment and Forest
MSY	—	Maximum Sustainable Yield
MTO	—	Mineral Trade Organization
NABE	—	National Advisory Board on Energy
NBG	—	National Botanical Garden
NCA	—	National Commission on Agrriculture
NCDC	—	National Coal Development Corporation
NCERT	—	National Council for Educational Research and Training

NCL	—	Northern Coalfields Limited
NCWC	—	National Landuse and Wasteland Development Council
NEERI	—	National Environmental Engineering Research Institute
NGO	—	Non Governmental Organisation
NIO	—	National Institute of Oceanography
NMDC	—	National Mineral Development Corporation
NML	—	National Minerals Ltd.
NWDB	—	National Water Development Board
NWP	—	National Water Policy
OCC	—	Oil Coordinating Committee
OECD	—	Organisation of Economic Cooperation and Development
ONGC	—	Oil and Natural Gas Commission
OPEC	—	Organisation of Petroleum Exporting Countries
SAARC	—	South Asian Association for Regional Cooperation
SPCB	—	State Pollution Control Board
UGC	—	University Grants Commissions
UN	—	United Nations
UNCED	—	United Nations Conference on Environment and Development
UNDP	—	United National Development Program/Projects
UNEP	—	United Nations Environment Program
UNESCO	—	United Nations Educational Scientific and Cultural Organization
UNO	—	United Nations Organisations
USA	—	United States of America
USGS	—	United States Geological Survey
WAPCOS	—	Water Protection and Conservation Society
WHO	—	World Health Organization
WMO	—	World Meteorological Organisation
WSSD	—	World Summit on Sustainable Development
WTO	—	World Trade Organisation
WWF	—	World Wildlife Federation
ZSI	—	Zoological Survey of India

List of Abbreviations

3 D	Three dimensional
AD	Anno Domini = in the year of christ
AP	Aerial Photography
APs	Aerial Photographs
B	Billion
B & W	Black and White
B.a./B.y.a.	Billion years ago
B.y.	Billion years
BC	Before christ
BIF	Banded Iron Formation
BOD	Biological Oxygen Demand
BOPD	Barrels of oil per day
BT	Billion tons
C	Circa = about
CCD	Carbonate Compensation Depth
Cf	Confer = compare
CIOB	Central Indian Ocean Basin
cm	Centimetre
CNG	Compressed Natural Gas
COD	Chemical Oxygen Demand
CPC	Civil Procedure Code
dB	Decibel
DNA	Deoxyribbose Nuclic Acid
DO	Dissolved Oxygen
DU	Dobson unit
e.g.	exempli gratia = for example
EEZ	Exclusive Economic Zone
EM	Electro Magnetic
EMS	Electro Magnetic Spectrum
et al.	et alia and other
Fm	Formation
ft	Feet
GDP	Gross Domestic Product

GHG(s)	Greenhouse Gas(s)
GMT	Greenwich Mean Time
GW	Ground Water
HSD	High Speed Diesel
i.e.	idest = that is to say
IAQ	Indoor Air Quality
IPC	Indian Penal Code
IR	Infra red
K cal	Kilo Calories
Kg	Kilogram
Kgoe	Kilogram of oil equivalent
Km	Kilometre
LOS	Law of Sea
Lr/L	Lower
LVZ	Low Velocity Zone
M	Million
m	Metre
M.y. /M.y.a.	Million years ago
m^2	Square metre
m^3	Cubic metre
MBF	Main Boundary Fault
Mcmd	Million Cubic metres per day
MCT	Main Central Thrust
Mha	Million hectares
Mid/M	Middle
Mkl	Million kilolitre
MOR	Mid Oceanic Ridge
MSS	Multi Spectral Scanner
MT	Million tons
Mtoe	Million tons of oil equivalent
MTPA	Million tons per annum
MW	Mega watt
N, S, E, W	North, South, East, West
N-, S-, E-, W-	Northern, Southern, Eastern, Western
NB	nota bene = note well
NELP	New Exploration Licencing policy
NMP	National Mineral Policy
NNMP	New National Mineral Policy
NRSE	Non-conventional Renewable Sources of Energy
OB	Over Burden

OC	Open Cast
OEG	Oil Equivalent of Gas
op cit	opere citato = just quoted (in the book/chapter)
P & T	Pressure & Temperature
P	Pressure
pH	Hydrogen Ion Concentration
pNdB	Perceived noise level decibel
ppm	Part per million
R & D	Research and Development
Rad	Radiation absorbed dosage
Rem	Roentgen equivalent man
RS	Remote Sensing
S & T	Science and Technology
S.G.	Super Group
Sp. gr	Specific Gravity
SPM	Suspended Particulate Matter
SW	Surface Water
T	Temperature
t	Ton/tonne
t/a, tpa	Tons per annum
t/d, tpd	Tons per day
TDS	Total Dissolved Solids
TNT	Tri nitro toluene
TPS	Super Thermal Power Station
UG	Under Ground
Up/U	Upper
UV	Ultra-violet
viz.	Videlicet = namely
Vs	Versus = against
WT	Water Table
>	More than
<	Less than
%	Percent
°C	Degree celcius
°F	Degree fahrenheit
α	Alpha
β	Beta
γ	Gamma
µg	Micro gram
µm	Micro metre

List of Compounds

MIC	Methyle Isocynate
CS_2	Carbon Sulphide
C_6H_6	Benzene
CNG	Compressed Natural Gas
H_2SO_4	Sulphuric Acid
HF	Hydrofloric Acid
H_2S	Hydrogen Sulphide
SiO_2	Silica
NPK	Nitrogen-Phosphorus-Potassium
HNO_3	Nitric Acid
TEL	Tetra Ethyl Lead
DDT	Dichloro-diphynyl-trichloro-ethane
BHC	Benzene hexachloride
$FeSO_4$	Ferrous Sulphate
$Fe_2(SO_4)_3$	Ferric Sulphate
$FeCl_3$	Ferric Chloride
N_2O	Nitrous Oxide
PCB	Polychlorinated biphenyl
CFC	Chloro-fluoro Carbon
H_2O	Water
CH_4	Methane
NH_3	Ammonia
$CaCO_3$	Calcium carbonate
CO_2	Carbon dioxide
PO_4	Phosphate
SO_4	Sulphate
FeO	Iron oxide
PAN	Per-oxyacetyl nitrate
HC	Hydrocarbon
H_2O_2	Hydrogen peroxide
O_2	Oxygen
O_3	Ozone
SO_x	Sulpher oxides
SO_2	Sulpher dioxide
SO_3	Sulpher trioxide
NO_x	Nitrogen oxides
NO_2	Nitrogen dioxide
NO	Nitric oxide
NO_3	Nitrate
CO	Carbon monooxide
CO_3	Carbonate
HCO_3	Bicarbonate
$CaSO_4$	Calcium sulphate
$CuSO_4$	Copper sulphate
$(NH_4)_2SO_4$	Amonium sulphate
CoHB	Carboxyhaemoglobin

Contents

Preface *iii*
List of Elements *v*
List of Organisations and Acronyms *viii*
List of Abbreviations *xi*
List of Compounds *xiv*

VOLUME - I

Section A — Resources/Mineral Resources
1. Resources: Survey, Conservation and Utilization 3
2. Mineral Resources 34

Section B — Ore Geology
3. Periods of Formation of Mineral Deposits 55
4. Ore Genesis and Mineral Exploration 69
5. Ore Fluids 84
6. Classification of Deposits 108
7. Forms and Modes of Occurrence of Ore Deposits 120
8. Structures, Tectonics and Mineralization 133
9. Supergene Sulfide Enrichment 141

Section C — Economic Minerals
10. Minerals and Metals 154
11. Diamond Deposits 174
12. Gold Deposits 210
13. Iron Deposits 233
14. Manganese Deposits 265
15. Tin Deposits 284
16. Tungsten Deposits 287
17. Copper Deposits 291
18. Lead Deposits 301

19. Zinc Deposits 308
20. Bauxite Deposits 314
21. Mica Deposits 325
22. Limestone Deposits 335
23. Phosphorite Deposits 345

VOLUME - II

24. Precambrian Phosphate Deposits 381

Section D — Ocean and Forest Resources

25. Oceans: Resource Reservoir 405
26. Ocean Policy 431
27. Manganese Nodules 441
28. Forest Resources 498
29. Forest Policy and Future Prospects 510

Section E — Mineral Economics

30. Mineral Deposits and their Classification Systems 521
31. Requirements for Mineral Exploration 538
32. Geostatistics and Mineral Resources 548
33. Classical Statistics in Mineral Exploration 559
34. Economic Criteria for Investment Decisions 566
35. Ore Microscopy 577
36. Mineral Processing 610
37. Constraints in Mineral Exploration 616
38. Mineral Exploration—Acts, Rules, Regulations and Policies 631
39. National Mineral Policy 646
40. Mineral Conservation 663

Section F — Mineral based Industries

41. Iron and Steel Industry 681
42. Aluminium Industry 688
43. Cement Industry 699
44. Refractory Industry 712
45. Fertilizer Industry 719
46. Abrasive Industry 732

Bibliography 743

SECTION-A
RESOURCES/MINERALS RESOURCES

1

Resources : Survey, Conservation and Utilisation

The strength of nations, social or political, is determined mainly by the resources they have and their capacity to utilise and conserve these resources. The evolution of civilisation is normally to think in terms of the advances made in their capacity to utilise resources and the induced social, economic, cultural and political changes. Sumerian, Mesopotamian, Babylonian and Gangetic river valley civilisations flourished because they could make use of the water resources available to them. Some countries have been less developed because they could not adequately develop the capacity to exploit their resources. The social and economic institutions which evolved in the framework of inadequate utilisation of resources, in turn, became obstacles in the path of development of underdeveloped countries. Today, most of the countries are paying serious attention to the problem of resources, their assessment, conservation and utilisation. The reasons for this are different for developing countries from those of developed countries. For developing countries it is the problem of more efficient and higher utilisation of resources and their distribution amongst the masses of their population. So it is high time to focus our attention on the survey, conservation and utilisation of resources.

Factors Affecting the Development & Utilisation of Resources

We need to understand the factors which affect the development and utilisation of resources and which, in turn are affected by the existing level of resource utilisation in any society. In past, the older social, economic and political institutions could not provide sufficient scope for the development of new modes of resource utilisation. These had to give way to a new order of society which could provide nourishment for the growth and maturisation of the new forces of production. In other words, a change in the old order of society is a prerequisite for a more efficient and higher utilisation of resources at a pace which will provide adequate solutions to the fast multiplying problems. This implies that we can not characterise the problem of resource conservation and utilisation as a technological problem even in the broadest sense of expression. We have to recognise that the attitude of our people towards life and work, and the existing social, economic and political institutions determine, to a large extent, the ways and scope for resource utilisation. Superstitious beliefs and irrational outlook, lack of alertness and adaptability, general resistance to change and experiment, submissiveness to exploitation by vested interests, low standards of health and productivity, low levels of work discipline and existence of a land tenure system detrimental to agricultural advance, under-developed institutions for enterprise and employment, weak infrastructure of voluntary organisations, and at the root of all these deficiencies, a low degree of popular participation, and a rigid, inegalitarian social stratification have, time and again, frustrated our attempts to successfully undertake the task of national development. Attitudes have generally supported the institutions and at the same time are supported by them. Again, these attitudes and institutions are effects of some more primary causes. Low level of national income, its uneven distribution, and detrimental effect of such inequalities on education and health of our increasing population are the fundamental causes which have adversely affected the development of our country.

To find satisfactory solutions to the number of our unsolved problems, we need to adopt an integrated approach to the task of higher and efficient utilisation of resources. Finding solutions to merely technical problems will not suffice as we can not assume that the existing social structure is conducive for the implementation of those solutions, nor can we assume that the social system will adjust itself to become receptive to the introduction of new doses of technology. The two have to go together; the answers must consist of an integrated solution to both-technical and social problems. Thus it is time to analyse the important resources of the nation, and to explore some avenues of thought and action and the steps to be taken to conserve and efficiently utilise our resources.

Global Context

Developing countries produce a large fraction of world's raw materials and control an insignificant fraction of international trade in finished goods. Their resources have been constantly exploited by developed countries. The Industrial Revolution could not 'trickle down' among developing countries due to colonial rule and they were left far behind in the race for development. With the dawn of independence, most of the developing countries, inherited a large and relatively uneducated population, scarce capital, nascent industry, a rudimentary infrastructure and a primitive agricultural system. The problems were gigantic, particularly in the absence of any technological base which could have provided an impetus to the rate of resource utilisation. Though we did make significant progress after independence in developing our scientific and technical capabilities and in bringing about social reforms, it was not adequate in its pace and magnitude to meet the revolution of rising expectations amongst the large masses of our population.

What is more striking is the fact that while over-utilisation of natural resources by developed nations got reflected in the adverse shift of the ecological balance of world, the less developed countries are still living in poverty because of under-utilisation of their resources and are compelled to live in an environment polluted by the waste generated by industrialised nations. If present trends of industrial and agricultural production, population growth and exploitation of non-renewable natural resources continue, there could be so much damage done to the ecology that Earth's carrying capacity of human beings may drastically decline. Also, the oil crisis which continues to be a significant event in the world economic scene, lends support to the grim realisation on the part of raw-material producing countries of the prospects of dwindling reserves of their natural resources largely due to over-exploitation by developed nations. Recent trends in oil production and pricing have added new problem and the countries have learnt their first major lesson on the need for conservation and efficient utilisation of resources, specially the non-renewable natural resources. This is a time for all developing nations to join hands in their attempt to protect their declining natural resources, which are being over-exploited to suit the needs of developed nations. International prices of raw materials have been artificially pegged at a lower level whereas those of finished goods have not only been inflated but their demand and usage have also been artificially created to spread the phenomena of "limitless consumerism." These prices, in most cases, do not reflect the true economic value of the products. In the present scenario self-reliance is the paramount goal of developing countries and only a rational policy of international trade can help in removing the obstacles on the path of development. Developing countries will have to take the responsibility of jointly safeguarding their natural resources and have to overcome the unequal terms of trade. This is the only way in which they can pull themselves out of the poverty and its consequences.

Indian Context

Availability of facilities like housing, medical care, schools, water supply, transport, per capita consumption of food, clothing, energy, minerals and other natural resources are the principal indicators of levels of living of any country. Per capita consumption of natural resources and hence the levels of living in India are miserably low as compared to advanced countries. With >15% of world's population, India consumes only 1.5% of world's energy output. Even the techniques of agriculture are largely primitive and modern Science & Technology (S&T) have made little inroads in the countryside. About 2/3 of our energy consumption is accounted by human and animal labour and non-commercial fuels like cow-dung, firewood, vegetable wastes. There is an acute scarcity of drinking water and sanitation facilities in our country resulting in high risk of contracting diseases who are already weak and vulnerable to incipient diseases on account of poor health and lack of nutrition. Added to this are the problems of rising population and large scale illiteracy.

Thus our total efforts for exploration, development and conservation of natural resources have been meagre. An integrated approach to long term resource planning has been practically absent. It was only in a crisis situation that we paid adequate attention to our resources and this necessarily resulted in short term resource planning which did not have any corresponding long term plans in which it could fit in. Lack of education and adequate health care, and presence of irrational attitude towards life and work, and traditional social institutions detrimental to advancement, have constantly threatened to thwart the forces of modernisation. Absence of fuller knowledge, development, utilisation and conservation of our natural resources, in turn, is the cause of our inability to generate powerful forces of modernisation under whose thrust alone the existing blockades can be torn apart. It is not only that the effect is the cause of its cause, but it is 'circular causation with cumulative effect.'

Yet we have to break the ice, and that too at its weakest spot if we are allowed to try our hands once only. To mobilise sufficient strength of modernisation which emanates from the quantum and state of development of resources, we have to answer the questions raised at present time. We have to simultaneously attack the traditional institutions, which have determined the rhythm of our life for centuries and which can be changed only through education of our people and introduction of a series of reforms. It has been realised that it is not possible to bring about technological reforms unless accompanied by institutional and attitudinal reforms, education useful in the immediate working environment and assured supply of critical inputs like water, power, fertilisers, pesticides, and modern agricultural implements. A series of integrated reforms need to be worked-out for each region and probably for each district at the micro level. Obviously, there is no general panacea which would fit the needs for each region of the country. The predominant forces responsible for a certain state of the society differ from region to region and the relative importance which will have to be given to different aspects of modernisation will, therefore, be correspondingly different. This is particularly true in the context of uneven distribution of coal, oil and natural gas, forests, water, iron ore and other natural resources in the country which also has to put up with significantly different climates, rainfall, productivity of the soil, cultures and languages in its different regions. Paucity of capital makes it imperative to rely on those resources which are abundantly available in country such as labour, land and other natural resources. Central focus of this complex situation is that unless there is mass participation in the endeavours for advancement, there could be no prospects of rapid progress along a wide front.

Concept of Regional Economics

A broad assessment of the problems and opportunities related to development and utilisation of resources underlines the immensity and complex nature of tasks which we need to tackle within a definite time frame. The most encouraging features of the landscape which inspire optimism in the face of these gigantic tasks are the assets we have in the form of intellectual potential of our society embracing various types of scientific, technical and managerial, skills, tools, techniques and temper provided by the recent advances in S&T and a wealth of resources endowed by Nature. The most urgent is to reorient our modes of thinking and action in evolving a multi-pronged strategy to mobilise our resources. Secondly, we have to select a line of attack which would pay immediate dividends without sacrificing long-term interests or more lasting social values. One of the alternatives on the horizon is the concept of regional economics. This concept aims at a fully integrated development of national resources in a region which would cut across state boundaries and would be chiefly focussed on a comprehensive survey of new resources and maximum exploitation of the existing natural assets in the area with a balanced policy of conservation. Regional economics can grow to a point where several disciplines including technology, geology, economics and sociology are synthesised to obtain maximum benefits from the results of scientific and technological progress. The main objective is to create conditions under which the efficiency of capital investment is increased, the utilisation of raw materials and energy improved and benefits of interlinked and balanced development realised. The other aim of this strategy is to create supplementary opportunities for the growth of social and cultural values. Regional economics would cover the integrated development of the productive forces and of concrete social processes to ensure that every region operates under the most favourable conditions for growth. This idea can be given concrete shape by the combined efforts of various Ministries and Departments and non-governmental agencies. We have, no doubt, made significant advances in the planned management of our economy. But we are now confronted with the task of increasing the sophistication of our planning to match it to the new dimensions and complexities in the national scenario. We should formulate plans for geographical complexes which would be built around sources of power, water and raw materials while, in areas distant from coal and oil fields, other power stations can serve as focal points in view of the flexibility in their siting. These energy centres can secure the most efficient clustering of productive forces and fulfil the need for developing new regions on techno-economic considerations, particularly regions with large resources of primary materials and fuels. The surplus of goods and services in one region would, on a scientific basis, meet the needs of less developed regions which could be linked with the developed regions in securing the needed inputs in terms of men, money or materials. Perhaps a chain reaction which would be generated by the success of a regional complex will transform our economy within the shortest possible time and will assist in 'leap frogging' into the future by blending the achievements of our scientific and technological revolution with the objective of a socialist order of society.

Planning for Development

We need to state the objectives for which we plan to make use of our resources. These could be rationality, development and development planning, rise of productivity, rise of levels of living, social and economic equalisation, improved institutions and attitudes, national self reliance, social discipline and participation of the people in the task of nation-building. Though it may be difficult to attain all these goals for any nation what needed is a change of direction towards the desired goals of development planning. It is in this direction that we have to reorient our thinking and plans of action. And as we approach the desired goals, the changed social order will, in turn, accelerate the rate at which resources can be utilised or conserved.

The type of resources which need to and can be conserved depends on the availability of other resources, the final demand for them and the level of technology. For example, coal can be utilised for the production of power, fuel and has a potential for conversion into oil and other chemicals. Yet, to meet the demand for power, we have to use our coal reserves for quite sometime to come. But on account of its exhaustible nature, and its other uses, alternatives have to be found for the production of power. These could be through nuclear energy, solar energy or geothermal energy. And till a certain point of time, which we can call the threshold point, we must make the maximum use of our coal resources and after this point we must conserve them. The period between the present point of time and the threshold point is the time available to us for the development of alternative methods of power production which are not dependent on non-renewable natural resources. Other modes of conservation and efficient utilisation of resources could be through appropriate technological reforms, recycling of waste, matching of the objectives of organisations engaged in resources exploitation with the national objectives, reform of the social institutions and education of our people in new techniques of resource development and utilisation. In fact, survey, conservation and utilisation are three interlinked aspects of resources and cannot be considered in isolation. Geological and geophysical surveys identify the available resources, their concentration and their geographical distribution. This combined with the knowledge of the distribution of demand for these resources across the country can help us in working out the economics of their extraction, establishing the need for conservation and developing the technology for their optimum utilisation. It may be prudent to develop a technology of extraction which simultaneously tackles the problems of utilisation and conservation of resources. This requires an integrated approach to resource planning. For, otherwise there is a risk of unintelligent overexploitation of resource as happened in advanced countries or under-utilisation of resources as in developing countries. Economic and social advancement left to "natural forces" is highly unlikely to follow the desired direction. On the contrary, we may end up in a crisis situation—as happened during the oil crisis (1973) from which it may be difficult and costly to retrieve lost ground and which certainly will impede our advancement. Certain specific resources vital to our national development and the issues stemming from the broad spectrum of the resources in the order of their importance are discussed here.

Human Resources

Human beings are at centre of two different equations. First, it is for them that other resources are being utilised and second, they themselves are resources and if adequately developed are capable of unleashing an unprecedented scale of productive forces. Yet this probably is the most underutilised resource in our country. A sizable part of our labour force does not engage in any form of work and most of those who do work, work only for a fraction of a day, week, month and year, and that too not efficiently. Such a state of affairs has drained our agriculture, except in plantations and a few market-oriented sub-sector, of much of its vitality. These attitudes towards work could survive over a long period of history only because of the traditional institutional framework of our society. The situation has become more and more serious as these attitudes and institutions are handed down to an ever-increasing population from their predecessors. Though more determined steps have been taken in the field of family planning in the last couple of years, the labour force will continue to increase somewhere between 2-3% in coming years. As in past most of these newcomers to the labour force will have to earn their living in the village economy, particularly in agriculture. Large or medium scale industries may not be able to absorb even a small fraction of the underutilised or unutilised labour force of the country even if industrialisation proceeds at a more rapid rate. The reasons for this are, first, the relative smallness of non-agricultural sector or conversely, the vast size of agricultural sector which sustains >70% of the country's population and second, the absorptive capacity of non-agricultural sector is very limited. In fact, in the initial stages industrialisation may leave a backwash effect on traditional industries, and thus reduce the total employment in the manufacturing industry as a whole. Small scale industries have a good scope for growth and though capable of generating higher employment opportunities per unit of capital employed for which reason they must be encouraged, it is doubtful if the problems of unemployment can be adequately tackled through setting up of small scale units alone. There is enormous scope for the creation and expansion of cottage industries in rural areas but they will have to face the competitive threat posed by the modern small-scale and large-scale industries which in longer run will intensify inspite of protection and will thus compel the cottage industry to gradually go through the process of labour saving modernisation. Thus, we come to the conclusion that even if more employment is generated in small-scale and cottage industries, they are unlikely to absorb a sizable portion of our underutilised or unutilised labour forces which must, therefore, find work in agricultural sector. Fortunately, there are certain important features of agricultural sector which keep our hopes alive and permit us to state with certainty that a large part of our labour force can be productively absorbed in this sector. First, unlike traditional crafts agricultural sector is not vulnerable to the backwash effects of industrialisation and also unlike cottage industry agriculture does not face any market competition. Food will be required in larger and larger quantities in future, as population increases.

On the subject of human resources development, emphasis on the concept of 'Investment in Man' has to be given due recognition both in theory and practice, particularly in developing countries. There is hardly any doubt that the interdependent and self-supporting system of attitudes and social institutions plays a dominant role in utilisation of human resources, and in fact, all resources. Yet, our traditional social institutions and attitudes are in no way conducive for rapid development and utilisation of our resources. This undesirable system which probably is the single most important obstacle on the path of our development, can be removed only through formulation and implementation of more rational plans which give much deserved importance to health and education of our people. Our national planning models are still based on the traditional concepts, many of which were borrowed from western economics and are inadequate to explain or predict the reality in a developing country. Traditional economic and fiscal opinions continue to ignore the drift of current development and significance to public policy of the new insight which is emerging in the field of social sciences. So exclusively preoccupied are we with assumptions of physical investment that we continue to build models of economic growth on strictly materialist assumptions without any cognisance of the role of human skills and creativity. Our major concern seems to be with elegant economic models which tend to ignore whatever is difficult to quantify even if these unquantifiable factors happen to be more important than those which are quantifiable. Investment in Man is one such area which is unable to find its due place in our economic models.

The developing countries have given too little importance to the development of their human skills and knowledge which could be of direct use in their immediate environment and have overemphasised on the physical investment, installation of plants and equipments. This has led to an imbalance between human resources and physical resources and has consequently led to the failure of the countries to attain their optimum rate of economic growth. Yet it is remarkable that the theory of economics or in other words the theory of optimum resources allocation was built, based solely on physical investment. One wonders how the models of economics could be developed without incorporating the social structure and attitudes supported by that structure which are responsible for the widespread existence of absentee land ownership and tenancy and the consequent low productivity of land and labour.

It is true that our Plan documents have emphasised the importance of health and education or in other words of 'Investment in Man' but that can only be regarded as an appreciation of the concept and are certainly vague and incomplete in the sense that they do not answer the questions of where should health and educational programs be directed, how far should they be pushed, what means they should employ and what are other policy measures needed. The theory of 'Investment in Man', though of great importance, is not sufficiently developed so as to enable the decision makers in the optimum utilisation of our resources.

Our political leadership had been aware of this problem from the pre-independence days and Mahatma Gandhi and Pandit Nehru had emphasised the importance of education and health which only has the capacity to transform our society into the new era. To quote Pandit Nehru : "It is clear that in the final analysis it is the quality of human beings that counts. It is man who builds up the wealth of a nations as well as its cultural progress. Hence education and health are of high importance so as to produce that quality in human being.... We have to remember always that it is right education and good health that will lay the foundation for economic as well as cultural and spiritual progress."

In late 1970s we took certain crucial and decisive steps towards the reformation of our social institutions under the inspired leadership of our then Prime Minister Mrs. Indira Gandhi and we have been able to usher into a new era under the guidance of her 20 point Program. These are relevant even today and for a speedy implementation of the program, we have to increase the level of education and health care. One way which needs a large scale expansion is the use of mass media provided by TV, radio and films for education of our people. We cannot wait for the availability of trained teachers in large numbers and will have to utilise the mass media to broadcast programs on education tailored to the needs of each region or each district.

Unfortunately 'Investment in Man', has not been recognised in our policies and programs. There is a need to define the problem more explicitly, to assess the needs for investment in human beings for each region and to develop appropriate theory based on empirical studies and to set up institutions which could implement our education and health plans. This is the single most important area which will have far reaching consequences on the conservation and utilisation of our resources.

Agricultural Sector

It is in the field of agriculture that battle for development will be won or lost. There is an urgent need to evolve strategies for exploration, development and conservation of our agricultural resources whose potential has hardly been tapped. Except probably for plantation and a few market-oriented sub-sectors, we have been unable to attain the maximum possible output feasible from optimum combination of land, labour, water and such other locally available natural resources which do not necessarily involve any fresh input of capital or even technology. Additional inputs of capital and technology will further raise the potential of the sector.

The low yields per acre are particularly remarkable when we consider that about 70% of the population of India is agricultural and has not been consistently able to produce the country's entire food needs till recent. The labour input per worker is generally low in terms of man hours and is of low efficiency. This means that not only the productivity of land but also of labour is low. And this, in turn, implies that agriculture in our country is not labour-intensive but, instead, labour-extensive. These work-patterns are supported by institutions of economic and social stratification and are related to land ownership and tenure which determine the use of land. These work patterns are also supported by the low nutritional levels caused by poverty which in turn impairs the willingness and ability to work, thus keeping the yields down and causing poverty. This process of circular causation leads us to believe that if we have to succeed in exploration, development, utilisation and conservation of our natural resources, we have to understand reality in its totality and determine solutions which are all-encompassing and are of necessity inter-disciplinary in nature. There could be nothing short of an integrated approach which can solve our problems. When we observe that in only 15% of the gross cropped area, the output is only 12% ha. Most of our districts have achieved a growth rate of 5% in agricultural production, we realise that urgent steps are required for improving agricultural productivity through irrigation, the adoption of improved technology and a speedy implementation of land reform legislations.

Emphasis on the social aspects to the problem is essential not because S&T are secondary in nature, but because the parameters which are commonly not understood or are given relatively low importance are often social parameters. A major emphasis on these aspects is warranted because we can not conceive of bringing about important technological changes in our country unless accompanied by no less-important social changes. Even without any increase in capital or technological inputs, substantial increase in the output can be achieved merely by increasing work in terms of number of man hours/day and of higher efficiency. Increasing number of farm-management studies point out that there are significant differences in agricultural productivity not only from region to region, but also within a region, within a district, and a village and, in fact, adjacent farms. The yield per acre of the best farmers have been found to be several times higher than those produced by average farmer. The idle manpower can be used in such essential and labour intensive works as the construction of roads, bridges, schools, irrigation canals, warehouses for storing crops and farm supplies, in digging wells to provide drinking water, for soil conservation, and labouring on afforestation and pasture improvement. Idle time can be used in keeping the villages clean, and in the manufacture of agro-industrial products. But the difficulties in undertaking these works are enormous. The success of these ventures is dependent on cooperation, collective action, and hence organisation, that is often absent in a faction-ridden village. The one question which will be of paramount importance to each individual is: who is going to benefit from such ventures of cooperation? Our social institutions are such that a large fraction of the benefit of any development work accrues to a small but dominant minority and there is a sense of suspicion in the minds of the larger majority. This situation can be mended through legislative and administrative actions directed towards transforming the existing order of society. We have taken a number of such steps since independence in this direction with varying degrees of success and we are fully aware that much more is required if we do not want to permanently lag behind on the path towards progress and prosperity.

Historically, few nations have managed to industrialise without first becoming self-sufficient with respect to agricultural products; hence the oft-cited generalisation that agricultural revolution is a prerequisite for a successful industrial revolution. This is particularly true in the case of densely populated agrarian economics with a low technological base. As a large part of the labour force is, and will continue to be, in the field of agriculture, it is very difficult to generate sufficient demand for industrial goods unless the agricultural productivity, level of employment and utilisation of the labour force are considerably improved to raise the existing low levels of income. In conjunction with attitudinal and social changes and increased work input, we have to bring about such technological changes which are not capital intensive but, instead, labour intensive, which can make use of locally available resources; which can be mastered by the less educated; and which are capable of substituting primitive modes of production. Technological reforms are not necessarily labour saving if intelligently selected. Appropriately chosen techniques lead to more abundant crops requiring more work at the time of preparation of soil, sowing, weeding and harvesting. It is not possible to transplant western agricultural technology which is oriented to save labour inputs into Indian agricultural scene. In fact, while implementing our technological reforms, we have to restrict the availability of machines which just substitute labour.

To be successful in bringing about agricultural revolution we have to intensify our work on a comprehensive survey of climate, soil, crops, breeds of livestock, etc. which are necessarily localised in nature and hold large dividends in store. In particular, soil surveys must be carried out in different parts of the country to identify the correct type and dosage of fertilisers that can be prescribed. Urgent steps are necessary to raise the quality of soil and for increasing its potential for production through irrigation, soil conservation, chemical and biological control of the soil and use of fertilisers and manures. About 150 Mha out of 300 Mha are estimated to require soil conservation measures.

It is essential to increase the land forested area not only to supply timber, pulp wood, and fuel but also to provide soil protection and moisture control. Soil experts recommend that 30-33% of the total area be forested with due consideration to their geographical distribution. In India, forests account for only 22% of the total area. Despite all plans a steady destruction of forests seems to have continued for a long time. Yet the fact that more forests are needed does not in itself demonstrate that potential for cultivated area will be reduced. Much land suitable for reforestation is either remote from existing villages and put to no worthwhile use or is used only intermittently in shifting cultivation or grazing. Reclamation of large parts from 33% of the total land area which is put to no use at present, is possible through better drainage and soil conservation system.

One of the most promising ways of increasing the acreage available for cultivation and labour utilisation is through a reduction of claims on space now made by animals. The same number of animals could be much better nourished on tilled fodder crops requiring only a fraction of the acreage now under pasture. It is not right to say that all the land now used for grazing is waste land and could not be tilled in any case. A good fraction of this 'waste land' can be used for food production by the application of modern agricultural methods. On the other hand, India commands a large number of domestic animals many of which are stray, and useless and those which are in use are of poor quality and we have not been successful in their full utilisation. There is a need to control the unchecked increase in the animal population and to raise good quality cattle through cross-breeding and proper nourishment. If appropriate steps are taken to improve the quality of cattle, it is not only possible to reduce the cattle population but also to effect a substantial increase in their total output.

During preparing our perspective plans for agriculture for the future to ensure supply of food for the increasing population, we have to adopt modern methods of cultivation. This would imply increased supply of irrigation water, fertilisers, improved seeds and other technological inputs. Through appropriate technological improvements, we have to increase the quality of organic manures abundantly available in the countryside that can be utilised along with the inorganic fertilisers supplied by the industrial sector.

Water Resources

Our water requirement for irrigation, drinking, industrial use and power generation is mainly met by rainfall and ground water (GW) natural reserves. The criticality of water resources underscores the fact that if immediate action at the national level for conservation and efficient utilisation of water resource is paralysed due to whatever reason, a delay of decades may be caused in the economic development of nation. We need to intensify explorations to identify potential GW sources and take urgent steps for conservation of water which is lost through seepage, evaporation and evapo-transpiration. Average annual rainfall all over India amounts to > 3,000M acre ft water. Out of this, 1000M acre ft are immediately lost due to evaporation and roughly 650M acre ft seep into soil, leaving 1350M acre ft to flow into the river systems. One-third of the river flow, *i.e.*, about 450M acre ft is considered utilisable for irrigation. Use of chemical monolayer films need to be tested and extensively employed to conserve water in reservoirs. It is estimated that abnormal evaporation can be reduced by 33% through efficient use of such films.

At the end of 4th Plan (1969-1974) there was 175 Mha of arable area out of which 142 Mha was under cultivation; 107 Mha of land was irrigated both from surface and GW sources. A potential of 44.9 Mha and utilisation of 43.1 Mha was achieved at the end of the plan. In 5th plan (1974-79) irrigation potential of 13.1 Mha was created, thus increasing the total land under irrigation to 58 Mha.

More accurate meteorological forecasts are necessary to enable the agricultural sector to plan its harvest. Training in modern techniques of utilisation of irrigation water in combination with fertilisers, pesticides, high yielding seeds, *etc.* appropriate to the local soil and climatic conditions is a prerequisite for efficient utilistaion of irrigation water. Results, particularly of major irrigation works, have not been encouraging due to the inability of the farmers in making a quick and full use of the facilities offered. In the absence of adequate funds and the capital intensive nature of irrigation projects, it is essential that a reasonable contribution from the local manpower is sought in the setting up of irrigation projects. Planning Commission observes that participation of local manpower in construction considerably enhances the prospects that the irrigation water, once available, will be intelligently and efficiently used.

Because of the regional imbalances in water availability due to uneven precipitation and routes taken by our perennial and seasonal rivers, we have not been able to tap the full potential of water in our agricultural sector. In fact, a portion of our water resources becomes detrimental to our agricultural system in the form of floods and accumulation due to inefficient drainage system. The sediment and detritus transported by the rivers whether in suspension or along their beds have an important bearing on storage, flood control, soil conservation, etc. There is an urgent need for a critical study of the behaviour of flowing water under different conditions and design of structures for the control and conservation of water. Construction of multipurpose dams should be planned in such a way that water can be fully used for the purpose of irrigation down-stream land besides flood control. Work on the construction of distributories water courses, and field channels should proceed in step with the construction of dam and the reservoir. Many of our projects in the past had to face long delays sometimes through inadequate planning, sometimes because of budgetary and political constraints.

Though, rivers are now considered as national assets, studies need to be conducted for an optimum apportionment of water depending upon irrigation needs and intra-basin and inter-basin needs. Immediate steps are required for the protection of water from pollution and for monitoring environmental characteristics of inland water.

Only about 37% of Surface Water (SW) was utilised in early 1970s and much less of GW potential. The efficiency of utilisation is, of course, highly questionable. According to some experts, the actual cultivator of soil, has frequently viewed irrigation more as an insurance against monsoon failure than as a device to raise his yield and possibly engage in multiple cropping. This could be due to the inherited institutional systems as well as inefficient administration of the distribution of water. In past, irrigation has not been thought of as the focal point around which coordinated programs with respect to simultaneous introduction of technological inputs, supplies, credits, marketing, better communication, land reforms could have been introduced. Hence, combined action on the part of a number of government agencies as well as action at the community level becomes crucial to the speedy utilisation of new irrigation facilities.

Energy

One of our foremost concerns today is the energy crisis. This crisis is being faced by most of the countries of world and it highlights the central role of energy in our civilisation. The magnitude, type and the nature of use of energy inputs throughout the ages have determined the life styles, economic prosperity, military power and the social modes of various countries at different points of time. Energy can no longer be considered as an isolated economic and social problem. The generation of electricity from very large stations has resulted in many interdependencies among areas and regions. Any disruption in the supply of fuel or energy in an advanced society today, will result in a serious dislocation of economic activities and life styles of people. The impact of oil crisis is an example of our dependence on energy.

An energy deficient society is weak and is not adaptable to too fast changing environmental conditions in comparison to a high energy society. Such a weak society can not take advantage of expanding opportunities for economic development generated by advances in S&T.

Ours is not high energy society. However, political compulsion and social aspirations of our people demand that we produce and consume more energy in future to improve the quality of life of the large masses of our population. There is fragmentation and inefficiency in the production, distribution and consumption of energy in our country. Urban areas consume energy in a large measure. Distribution of energy consumption like the distribution of income and wealth is largely skewed in favour of urban areas. To promote development and social equality in future, energy policy appears to be a potent instrument. The policy options open to us are *inter-alia*, in the choice of technologies, inter-fuel substitution, pricing, energy allocation, and environmental quality. While choice of social goals is a political process in our country, we find that this choice takes place at various points of time by the force of events in an unarticulated manner and not by conscious planning. Even if the goals are defined, the alternative strategies that are open, are not often explicit.

Strategies of Energy Planning

(i) Since the design and construction of a power station takes 7-12 years, energy planning necessarily requires, a long term integrated approach. Energy planning in India, in past, has been largely a choice of series of projects with different technologies. As no clear understanding of the nature and components of demand for electric power exists, capacity addition and pricing of power are usually done in adhoc manner. Current methods of cost-benefit analysis are somewhat inadequate to evaluate power system alternatives. These methods take into account only monetary costs and benefits and not social costs and benefits. However, questions like who pays and who benefits are not satisfactorily answered by cost-benefit analysis. Current pricing policy of electricity undertakings, largely based on accounting data, does not reflect social priorities for the use of energy.

(ii) If we make an assumption that with the beginning of 21st century our per capita electricity consumption should be about half of the per capita consumption in W-Europe, we need a total installed capacity of about 180,000 MW (at 60% load factor). Assuming that our hydro-electricity potential of about 40,000 MW is fully tapped, there will still remain a gap of 140,000 MW which has to be filled by coal or some other source of energy. In the event of coal having to fill the gap of 140,000 MW, total coal production including coal required by industries would have to be over 1,000 MT. Production and distribution of such an amount of coal would pose formidable financial and technical problems. Other sources of energy like geothermal, wind, tidal or solar power are either very limited or localised and seasonal or may involve considerable amount of R&D over a long period of time. Thus it is inevitable that in the long run, nuclear power will have to substantially contribute much towards electricity generation in India. The intrinsic compactness of nuclear fuel as well as of the effluent of a nuclear power plant make them eminently portable, and will reduce not only transport cost and hence the price of electrical energy, but also the strain on our already over-worked transportation network. In 1970s installed capacity of nuclear power was 600 MW. With the available uranium (U) resources and reactor technology, the total potential for nuclear power was about 5,000 MW. An active atomic mineral exploration program is under way and it is expected that more U-resources would become commercially exploitable. Simultaneously with the development of breeder technology more power can be generated from the existing U and thorium (Th) resources. In addition to fission energy, generation of fusion energy will ensure an almost inexhaustible source of energy.

(iii) Most of the state electricity grids have been formed by interconnection of power systems so that integration and optimisation of power generation and distribution can be achieved at state level. A few zonal grids are also in operation with some exchange of power between the states. Formating of a national grid has ensured an overall optimisation. Distribution and utilisation of various types of fuels can be done in the best possible manner with necessary interfuel substitution by the interconnected grids. Significant improvements in energy generation and distribution systems can be brought about by using operations, research and computers, better training of personnel and maintenance of equipments, a rigorous estimation of sectoral demand of energy so that allocation and price could be in accordance with the priority of activities, reduction of construction time, and cost of power stations. These tasks can be carried out by a national energy authority such as NTPC which could plan, finance, co-ordinate and regulate energy generation, distribution and utilisation. Or, in other words this agency can coordinate and regulate the activities of various organisations like Coal Mines Authority, NCDC, BCCL, ECL, GSI, Mineral Exploration Corporation of India (MEC), ONGC, Railways, State Electricity Boards and some municipal corporations which are engaged with energy systems in some form or other.

(iv) Since losses due to use of low quality coal and poor transmission of power are very high, R&D programs have to be accelerated to solve this problem as soon as possible. Research also needs to be done in improving conversion efficiency and thermal exchange of different types of generating plants. Many of our electric appliances are very inefficient in the use of power. These have to be redesigned so that they use power more efficiently. Fuel efficiency in many of our industrial enterprises can also be improved significantly by redesigning industrial equipments like boilers.

(v) In rural areas we need to make use of locally available sources of energy. There is great potential in non-commercial fuels like vegetable wastes, firewood, and animal wastes. According to certain estimates "every year some 500 MT of cow dung are burnt in India providing 70% of household fuel requirements. This dung contain 1.38 MT of nitrogen and 0.69 MT of phosphates. If this were applied to fields instead of going smoke, it could produce an extra 14 MT of food grains. The investment needed to mine and distribution of 46 MT of coal to replace dung completely would be prohibitive."

Towards utilising this important resource we had talked about bio-gas plants and more generally about appropriate technology. The idea was to ferment the dung anaerobically in a pit, the gas being collected in a tub and transferred by pipe to cookers and gaslights, and to use the residue as a manure. In a specific case, plants were developed for commercial scale operation and installed. On follow up it was found that only 1 in 5 of the plants was in use. The moisture in the gas had clogged the pipes and spoiled the gaslight mantles. The flame was uneven and took a long time to cook food. Farmers were still compelled to burn dung because the volume of gas generated was not enough to meet the family cooking needs. Leaving apart-socio-economic resources like the cost of the plant and need for a minimum of five cows which only the richer farmers could afford, the principal failing was due to lack of research. The conservation efficiency of the plant turned out to be only half of what was expected; because the average temperature of the plant in operation was 10°C lower than that claimed for the pilot plant.

If we had given more attention to the problem, we would have developed not only less costly plants but also directed our efforts towards breeding a new strain of bacteria capable of multiplying much faster and at lower temperatures to achieve the needed gas output with the available dung input.

While energy crisis is a global phenomenon, its solution will continue to be national in a very large measure for a long time to come. Also, in India we are, to some extent dependent on other countries for supply of fuels like oil and enriched uranium. We therefore, have to look towards increasing our coal production, tapping hydroelectric potential and intensifying oil exploration. Commendable efforts have been made by ONGC in this direction. Department of Atomic Energy (DAE) is concentrating on the use of natural uranium based nuclear power to eliminate our dependence on imports. We recognise the important role of atomic energy in meeting the future power requirement and we have been concentrating our efforts on the development of fast breeder technology which would generate more fuel than what is consumed in the reactor.

Science and Technology

To achieve the goal of self-reliance, urgent steps are needed to streamline the efforts of various public and private institutions towards establishing the most efficient and effective methods of mobilising indigenous resources in tune with our needs of industrial and agricultural development. Most of the developing countries today, are seized with the chronic problem of sub-optimal utilisation of available manpower, machinery and natural resources. This is due to indiscriminate adoption of foreign technology and inadequate orientation of indigenous efforts. We are aware of the social problems which arose from direct transplantation of imported technology. We are also aware that many times we ended up with imported plant and machinery which could not efficiently process local raw materials. We had the painful experience of capital-intensive technologies which did not create sufficient employment opportunities. Nor did they achieve the results in terms of desired products and scale of operation in relation to the limited size of the domestic market. We did not give due recognition to the fact that optimum size of an enterprise is determined not only by technological factors, but is also governed by several other considerations like economics in marketing and finance. Even India, which is relatively more fortunate in the field of S&T among developing countries, is faced with wide and substantial gaps on the frontiers of its R&D. More than 50% of country's land surface had not been mapped geologically on 1:1 scale by 1970s and a total of only 500 odd geophysical investigations were carried out for minerals till then, only 20% of the land surface had been surveyed by reconnaissance geology and hardly 100 geophysical studies were conducted for GW in that period since independence. Our success ratio for exploring GW tube-wells was at a low figure of about 35% whereas in some countries it is between 82-85%. On one hand, we are heavily dependent on the import of metals, minerals, petroleum etc., while on the other hand we allow a time lag of 5-7 years between surveys and production of maps thus affecting the timely execution of natural resource development programs.

It is particularly important for a country like India to know the nature and extent of its physical resources through effective surveying. It is a global phenomenon that many of the earlier geological surveys in many countries did not disclose the mineral wealth which has been discovered in more recent decades making use of more sophisticated technologies. Likewise our country may be much richer than we thought a few years ago. The most arid or unprofitable deserts are now found in many cases to be natural geological basins containing oil, minerals and vast quantities of underground (UG) water which are capable of being exploited and could more than adequately pay for the capital and efforts involved. There seems to be boundless possibilities in an integrated survey of physical resources, study of rocks, minerals, soil, rainfall and river-flows, subsoil water, fisheries, forests, and what is more important, ocean bed resources. The agencies which make these studies often suffer from lack of trained personnel, shortage of suitable equipments and want of multidisciplinary effort which alone can match the magnitude of the task and its varying complexities.

Next in importance is a survey of our capabilities in the field of biological sciences for adapting new technologies for better breeding of plants and animals, control of pests and diseases, evolution of new food materials and plant nutrients. Most of these problems can be solved by the application of existing knowledge and skills to concrete conditions in a particular region. The main thrust of our effort has to be in the direction of stimulating development quickly through inexpensive methods and simple technologies so that existing manpower would become more efficient and the surplus utilised in more productive vocations.

All these facts focus our attention on the need for accelerated indigenous R&D. Successful development and implementation of S&T plans which must be accompanied with social reforms, necessarily depend upon local parameters like terrain, quality of soil, climate, distribution of water resource, irrigation network, available infrastructure of transport, power and other utilities and geographical distribution of natural resources. We can not depend upon the developed countries to help us out in the field of S&T because their social setting and physical system provided by nature is very different from ours and hence their modes of exploration, development and conservation are unlikely to be applicable in our setting. This, of course, is not to imply that we should not go in for foreign collaboration in very high technology fields where we may not have adequate know-how and the output of which is of utmost importance to the economy. In fact, there is a strong case for utilising the most modern technology among developing countries in certain selected areas. Use of satellite for surveys, use of Remote Sensing (RS) technology for assessing mineral resources as well as standing crops, use of radioisotopes for establishing the characteristics of flowing water, use of atomic energy for the generation of power, use of the most modern equipments for more reliable meteorological forecasts, use of mass media for education and also for rapid information flow, are just a few of the illustrations. Modern technology does not presuppose existence of sound industrial base and is not necessarily costlier than the traditional technology. In fact, the scope for using modern technology provides a very effective way for leap-frogging into the modern technology era and thus avoiding the long and torturous path of development followed by today's developed countries.

For a smooth and rapid progress in the field of indigenous S&T what is most important is to forge links amongst research and educational institutions, national laboratories, government agencies in the fields of S&T, industry and agriculture. Considerable efforts have already gone in this direction, yet we cannot say that we are anywhere near the optimum point of inter-organisation cooperation and collective action. We have to speed up the communication between organisations, set up inter-organisation teams for research projects, decentralise authority and take concrete steps towards motivating scientists and technicians. These are probably the only ways of creating a favourable atmosphere for a rapid advancement of S&T in country.

Non-renewable Natural Resources and Ecology

In context of the energy sector we had discussed the exhaustible nature of fossil fuels and its adverse impact on environment. The discussion needs to be extended to minerals, metals and such other non-renewable natural resources which we cannot indiscriminately exploit, nor can we keep them importing, nor create ecological complications.

Of the 3.28 Mkm^2 area of the country, 32% is accounted by sedimentary areas favourable for petroleum, natural gas and GW resources, 16% is covered with the Deccan Lava flows and the remaining 52% needs to be explored for minerals. This would involve much intensified geological and geophysical surveys for effective action to quickly meet the requirements of non-renewable natural resources. Optimum exploration, utilisation and conservation of non-renewable natural resources can be effected only through organised machinery whose actions can be subjected to co-ordination and which is capable of developing an integrated approach to non-renewable natural resource management. We have to implement our ideas on the export of higher value added products instead of just ores. To counter the machinery of international trade which is heavily biased against the raw material producing developing countries, we have to reduce our emphasis on export of raw materials. Countries are realising in an increasing manner the importance of conservation of non-renewable natural resources. Recent trends in the developing countries indicate that steps are being taken to cut down the exploitation of their natural resources not only to conserve them for future generations but also to obtain a better price per ton of export. As we are aware, the cost of mining sharply increases as we go deeper and deeper, and this fact needs to be kept in mind while exploiting, selling and pricing such minerals. Plans have been formulated to supply the growing needs of non-renewable natural resources at minimum social cost and to utilise nation's non-renewable natural resources at optimal depletion rate. Government rightly intervened to nationalise coal mines in 1973 and thus save them from the ruthless exploitation by vested interests whose profit motivation predominated national objectives.

We have to undertake social cost benefit analysis before embarking on exploitation of non-renewable natural resources.

The parameters which will go into this will be much more than simple financial and economic analysis. Analysis must take into account who is going to pay for the investment and who is going to benefit from it, what are the ecological consequences, and questions raised by resources balancing. Many of these decisions are political in nature and can be brought into effect after creating a deeper awareness and understanding of underlying factors on a technological and social plane.

Nature in the process of its evolution, has possibly experimented with almost all permutations and combinations of simple and complex molecules. Whatever was found to be functional for life was retained and whatever was disfunctional was rejected by it. There is nothing like waste in the delicately balanced kingdom of Nature. The waste of one species is food for another. The balance is maintained through the control of population operating in the cycles of Nature. Human beings by their ability to control some of the natural forces, of course without much understanding the complex cybernetic system of Nature, disturbed the balance by creating linear chains of resource exploitation which are characterised by rapid generation of waste for which there was no readily available mechanism at the disposal of Nature for adequate reprocessing. Nor was the disposal of waste taken care of by the people who created it. We are aware of how the rivers and consequently harvest and fish were affected by pollution generated by chemical plants. Unlike advanced countries, pollution in India is fortunately confined to certain pockets only. Yet, if we want to make use of the experience of advanced countries, along with industrialisation we should undertake simultaneous preventive measures to control pollution. This will not only avert crisis at a later point of time but it is also socially and economically more prudent. Waste can be treated chemically and can be used as a raw material. Ash generated in Thermal Power Stations (TPS) can be used for making bricks and urban sewage can be treated and used as fertiliser. DAE has been successful in setting up plants for the manufacture of heavy water based on NH_3 produced as a by-product in fertiliser plants. A good deal of R&D is required to identify ways of using waste as raw material not only for economic reasons but also for ecological protection.

Industrialisation Issues

In developing countries characterised by high man:land ratio, a low technological and industrial base and a large underutilised labour force, the agricultural sector has to play a prime role for some time to come in the upliftment of the levels of living of the weaker sections of society. Yet, in view of the rapid and accelerating increase in labour force there is no prospect that incomes and levels of living can be substantially and permanently improved in the longer run, unless a much larger proportion of labour force can be effectively utilised outside agriculture and especially in modern industry. Though in the early phases of industrialisation the absolute size of labour force engaged in all types of manufacturing may happen to fall because of the risk of backwash on traditional manufacturing, industrialisation will have an important employment-creating influence once it has reached a higher level. And that precisely is the reason that industrialisation should forge ahead at a much faster rate. As a matter of fact, our government has vigorously promoted industries, oriented towards export or import substitution not only to achieve the goal of self-reliance but also to minimise the adverse effects on traditional industries. Through the instrument of industrial licensing we have also taken important steps towards protecting the interests of small scale industries and village craftsmen. Steps are being taken to modernise small scale industries and village crafts through introduction of technological changes which do not decrease the demand on labour and which attempt to utilise the local resources to the maximum possible extent.

Generally speaking, the extent to which industrialisation can be brought about through assimilation of stimuli for change and improvement is a function of levels of mass education and ability to change attitudes and social institutions. Improvements in education and health and change of attitudes and institutions are necessary not only to compensate for the relative absence of automatic spread effects but also to support the industrialisation drive itself. To promote the growth of modern small scale industry and craft industry in villages which are responsible not only for utilisation of local resources but also for generating employment, there is an urgent need for guaranteeing market protection to them and efforts to increase their productivity. Development and application of appropriate technology not only in small scale sector but also among the large scale industries will have the most telling effect on the success of our industrialisation efforts.

Increased efforts are required for maximum utilisation of already established capacity in industrial sector. This can be achieved through increased R&D, better maintenance of our plants, efficient management, change of product mix and possible change of designs.

Urgent steps are needed for developing indigenous industrial designs which have a profound effect not only on the utilisation of country's resources, but also on the quality and scale of whatever is produced. Good designs should make it possible to produce the articles actually needed by people, thus ensuring a stable domestic market for the local industry. This would require co-operation and collective action on the part of various public and private research and educational institutions engaged in the task of developing indigenous know-how and scaling it up to the commercial level. What is presumed, of course, is the existence of an integrated plan and the administrative machinery through which such efforts can give fruitful results.

Here, it has been tried to outline the important factors which affect the survey, conservation and utilisation of our resources. Certain suggestions have been made in relation to improving the level of mass education and health as well as changing the existing system of attitudes and social institutions which are mutually supporting and have become a major obstacle on the path of our development. This, of course, is not to imply that modern S&T has a secondary role to play in our developmental efforts. On the contrary, we must give urgent attention to the development and application of appropriate S&T to achieve the goal of self-reliance, but such efforts must give due cognisance to the sociological factors right from the conception of an idea to its implementation. This would require an integrated approach to resource planning on the part of the state agencies and agriculture sector, steps need to be undertaken to tap the potential provided by Nature and technology. This can be achieved only through a series of reforms effectively administered.

Though number of studies are being made to obtain a better grasp of the reality in its totality, there is an urgent need for intensifying efforts towards obtaining further insight through empirical studies. This would involve not only the analysis of presently available data with various institutions, but also conducting further surveys at the regional, district or even block level and putting them together for arriving at important generalisations.

It has become necessary that from among the experts within the country, interdisciplinary teams consisting of specialists from social sciences, economics, agriculture, geology, geography, environmental SC statistics, *etc.* are formed to undertake further studies on survey, conservation and utilisation of resources. Such studies should be formalised under the existing educational and research institutional and budget grants made for the purpose. The teams could be formed around each important resource with an organisation to ensure inter-team discussions as the resources themselves are closely interlined.

Results of these studies will be useful not only in obtaining a deeper insight into the existing pattern of utilisation and conservation of resources, but can also be used in arriving at an integrated approach to exploration, conservation and utilisation of our resources.

2

Mineral Resources

It has become essential to highlight and bring to focus the current problems regarding exploration of our minerals, existing gaps in our knowledge and R&D needed for an integrated development of our resources in view of its great bearing in our National Planning and rapid development with the prime object of attaining self-sufficiency in industrial sector. Though it would be impossible to deal with all the aspects of mineral exploration, R&D programs as well as personnel management here, the purpose is to describe some of the significant topics that need our immediate attention.

Important Groups of Minerals

Following are some important groups of minerals which have an important role to play in Indian economy.

Ferrous Group

First in importance are the ferrous group of minerals, namely—iron ore, manganese and chromite.

(a) **Magnetite:** Magnetite deposits have been established at Kudremukha (Karnataka), Kanjamalai and Salem (T.N.) and Kerala. The reserves at Kudremukha are exported through Manglore port. These also meet the requirements of the coast based steel plant at Manglore. Kanjamalai deposits meet the requirement of Salem plant.

Magnetite deposits of Kudremukha and Kodachadri are close to W-Coast and those of Bababudans are slightly to East. These three adjoining deposits contain considerable reserves of Banded Magnetite Quartzite (BMQ) and need to be explored for an integrated development in near future. Estimated reserves of magnetite are 34.08 MT (as in 1999).

(b) Hematite: Jharkhand-Orissa deposits alone account for nearly 50% of the known reserves of hematite Fe-ore. These are located quite close to the coal and limestone belt and also close to the ports of Haldia and Paradeep and offer scope both for the new as well as expansion of the existing steel plants and for export.

The reserves have mainly been established in Rowghat and Bailadila region and adjoining areas. Rowghat deposit is closer to the coal belt while Bailadila ones are in the interior and the Fe-ore from this area is being exported through Visakhapatnam. Development of Bailadila deposits is directly linked with the haulage capacity and cost and the capacity of Visakhapatnam port. It also supplies ores to the steel plant at Visakhapatnam and Bailadia.

Bellary-Hospet deposits and other areas of Karnataka are also located in the interior but being equidistant from the E and W-coasts, the ore is exported through Chennai, Goa and other harbour like Karwar on W-Coast. These deposits also sustain the steel plant at Vijayanagaram and its expansion.

Goa-Ratnagiri deposits are located on W-Coast itself but their reserves are only meagre. If a coast based steel plant is to be set up, there will have to be a reduction in export.

The belief that we have inexhaustible Fe-ore reserves, is no longer valid. Though we have comfortable surplus at present the need to assess area-wise measured reserves of lump and fines is essential to meet the requirement of steel plants. Estimated reserve of hematite is 14,630 MT and of magnetite is 10,619 MT (as in 2007). Production status of Fe-ore, Fe and steel is given in Table 2.1.

An examination of the inventory of Fe-ores prepared by GSI and IBM reveals a noticeable gap. Hence, we have to convert the indicated and inferred category to measured category and a few deposits have to be proved to meet the requirements of the steel plants and increase export. Our knowledge about the existence or absence of blue dust in various deposits, its extent and composition have to be made up by our future programs of exploration.

Table 2.1: Production of Iron-ore, Iron & Iron Products (Steel)

	1994-95	*95-96*	*96-97*	*97-98*	*98-99*	*99-2000*
1. Iron	53.4	65.9	66.1	75.7	70.7	—
2. Hot Metal (includes pig-iron)	17.1	17.8	18.8	19.2	18.2	—
3. Crude Steel (ingots + concast + blooms + slabs)	15.9	22.4	23.5	24.7	23.1	—
4. Finished Steel (includes secondary products)	17.8	21.4	22.7	23.4	23.8	27.2
5. Steel Casting	249.4	303	393.4	392.5	373.1	386.9

Major problem in the case of iron have been the utilisation of large quantities of fines generated in the process of mining Fe-ore and natural blue dust. The fines so generated are much larger than the projected demand pattern in country. It is, therefore, essential that steps are taken for their proper utilisation by sintering/pelletsing or manufacture of sponge iron.

In the context of rapid expansion of steel industry all over the world and the great strides made in the steel technology, it is also essential to consider the production of pre-reduced pellets which will incidentally case the shortage of coking coal in the country to some extent. The other problem related to Fe-ores is the high alumina content particularly in Goa, Orissa-Jharkhand ores.

Enough attention is needed to be paid to the theoretical aspects of Fe-ore sedimentation in Precambrians of India. Some distinctive geo-chemical association can be discerned in Fe and Mn bearing formations; (1) a dominant Fe-ore in Kudremukha-Bababudan (Karnataka), Bailadila (M.P.) and Badampahar-Gorumahisani (Orissa); (2) a lime, Mn, iron in Karnataka (Mysore-Karwar) Goa, Ratnagiri region and in Keonjhar-Singhbhum region; and (3) a lime and manganese in Ganpur region of Orissa, Sausar formation of M.P. and Maharashtra. In M.P. and Maharashtra, the lime, Mn, Fe-ore sequence has not been established.

For a better understanding of the character of various Fe-ore deposits, detailed stratigraphic, palaeo-geographical and geochemical studies of various Fe-ore basins are needed.

(c) **Manganese:** The *in-situ* reserves of Mn-ores of all categories are located in Maharashtra, M.P., Karnataka, Orissa, and A.P. These reserves are of a rather moderate order considered in the background of the envisaged expansion in indigenous steel production. At present only little are in measured category and the rest are in inferred and indicated categories.

Estimated reserve (extractable) has been estimated to be 167 MT out of which 40 MT, 49 MT and 78 MT are under proved, inferred and indicated categories respectively (as in 1999).

The other problems include the beneficiation of low grade ores and reduction of the phosphorus content in some of M.P., Maharashtra and A.P. ores. So the future studies should be intensified and a solution found. Export of Mn also needs to be reviewed so that the country may reduce import of Mn-ores.

(d) **Chromite:** Total reserves of chromite are located in Orissa, Karnataka, T.N., A.P., Manipur, Maharashtra and Jharkhand. About 95% of this comes from Orissa. Of total resources of 213 MT, 31% (66MT) is under reserve category while 69% (147 MT) is remaining resource. Proved lump ore of metallurgical grade and fines are considerably less. It is obvious that we have to explore and prove gradewise reserves by drilling all the deposits. Since much of the reserves are of powdery nature, it is necessary to carry out tests for their agglomeration and bonding characteristics.

Base Metals

Importance of base metals (Cu, Pb and Zn), in which we have limited resources, needs special emphasis in the context of self-sufficiency and national interests. Fully conscious of this fact, GSI embarked on a large scale exploration of base metals for 4 decades and covered extensive area in various mineralised belts by large scale mapping, extensive ground geo-chemical and geo-physical prospecting. More than 1Mm of drilling and 50,000 m of exploratory mining has been carried out. As a result of early exploration programs the reserves position has changed in a phenomenal way. Estimated reserves of Cu which were placed at 54 MT in 1961, rose to 1.39 BT in 2006-07. Similarly Pb-Zn ore reserves also showed an increase from 21 MT in 1961 to 522.58 MT in 2006-07. Due to this intensive work on base metals, admittedly we have more data available on fundamental aspects than on ferrous group of minerals. In 1999, estimated reserve of Cu-Pb-Zn as shown in Table 2.2.

Table 2.2: Reserve Status: Cu-Pb-Zn

Type of Reserve	*Reserve in MT*		*Metal Content*		
	Cu ore	*Cu metal Content*	*Pb-Zn ore*	*Pb*	*Zn*
Quarriable	416.8	4.37	1791.3	2.32	1.0
Restricted	647.1	1.89t	203.87	—	—
Indicated	769.9	—	150.00	—	—

During early years of Independence >500 prospects spread over a wide area ranging from Central Himalaya in North to T.N. in South, were examined in a systematic fashion. On the basis of the data collected the following generalisations have been made about the distribution pattern of base metal occurrences in India.

In Peninsular Precambrian shield areas, economic ore deposits are confined to Archaean to Lower Proterozoic Bhilwara-Aravalli-Raialo-Delhi Groups of Rajasthan; Lower Proterozoic Singhbhum-Gangpur zone in Jharkhand-Orissa; Nellore Group of Metamorphites and Lower Proterozoic Cuddapahs of A.P.; Archaean metavolcanics-greywacke assemblage of Karnataka; Lower Proterozoic granites of M.P. and Sakolis of Maharashtra.

No workable Cu-ore deposit could be located within Charnockite-Khondalite province, extending from Orissa down South to T.N. and Kerala. Similarly, the extensive Tertiary tholeitic magma province, spread over a total area of 3,20,000 km^2 in Western and Central India, is almost barren of Cu-mineralisation except for a few local occurrences of academic importance.

The bulk of ore tonnage is found in meta-sedimentary rocks. Porphyry Cu-deposits in association with acid to intermediate volcanic rocks, which account for a sizeable part of the world Cu-production, could not be discovered in India in early years. Similarly, mineralisation, in granitic rocks is insignificant except for the ore deposits at Mosabani, (Jharkhand) and Malanjkhand (M.P.). The ultramafic bodies in Peninsular Precambrians are also devoid of significant Cu-mineralisation.

Sulphide metallogeny in Rajasthan metamorphites is distinctly different from other shield areas of Indian Peninsula. There is a broad change from dominantly Cu-Pb-Zn metallisation in pre-Aravallis to Pb-Zn metallisation in Aravalli Group and Cu-mineralisation in Raialos and Delhis. Such a stratigraphic specialisation is not recognisable in other parts of Indian Shield. Strangely, the bulk stratigraphic characters of this metallogenetically 'specialised' ore province in Rajasthan-Gujarat have no comparable counterparts in Southern, Central or E-Indian shield areas.

In Extra-Peninsular tract feeble and sporadic base metals mineralisation is confined to 3 different domains : (a) impersistent pyrite-chalcopyrite or sphalerite-galena pods and disseminations in pre-Silurian meta-sedimentaries; (b) small lenticular ore shoots within dolomite-orthoquartzite assemblage of Jammu, Kumaon and Garhwal Himalya and (c) around and within Cretaceous and Tertiary intrusive. Even where ore shoots of high grade and width have been found, their persistence is very small and variable.

Although available radiometric and stratigraphic data on various ore deposits are not adequate, there are still discernible major epochs in the formation of base metal ore deposits in Indian sub-continent (Table 2.3).

Most of the deposits discovered are moderate to steeply dipping tabular to lenticular ore bodies, essentially composed of a complex sulphide or sulphide-sulfosalt assemblage, except for strata-bound deposits, *e.g.*, Agnigundala, where comparatively simple mineralogy and insignificant wall rock alteration favour the theory of syngenetic or diagenetic origin.

Table 2.3: Major Epochs of Base Metal Formations

Age/Epochs	*Deposits*
Tertiary metallogenesis	Pb-Zn-Sb in Spiti-Lahaul granite
Cretaceous-Eocene-	(i) Native Cu localities in Deccan Trap;
	(ii) Native Cu in Dras Volcanics.
Hercynian metallogenesis	Stray Cu/Pb/Zn pockets in the Kashmir Himalaya.
Pre-Silurian (? Caledonian)	
metallogenesis (Perhaps including parts of Proterozoic)	Dispersed Cu/Pb/Zn in lower Himalaya metamorphites
Upper Proterozoic metallogenesis (650-1100 M.y.).	Pb in Vindhyans
Lower Proterozoic metallogenesis	
(1100-1900 M.y.)	(i) Cu in Delhi group
	(ii) Pb-Zn in Aravalli group
	(iii) Cu-Pb-Zb in Singhbhum-Gangpur
	(iv) Cu-Pb pockets in granitised charnockites
	(v) Pb-Cu in Agnigundala
	(vi) Cu in Malanjkhand granite and quartz veins
	(vii) Cu in Mailaram metamorphites
Archaean metallogenesis	
(1900-2700 M.y.)	(i) Complex Pb-Zn-Cu ores in Dariba-Rajpura.
	(ii) Cu-Pb-Sb in Dharwarian Volcano-sedimentary pile.

Lower Proterozoic era have yielded the bulk of ore tonnage of base metals so far

Exploration by drilling has generally indicated that ore deposits in major mineralised zones have >500m of depth continuity, wherever the strike length of mineralisation exceeds 1,000m. But this general behaviour of depth persistence in major mineralised belts like Singhbhum, Khetri, Agnigundala etc., is not applicable to the isolated ore occurrences, where ore bottoming takes place at any level. For instance, Turamdih Cu-deposits in Singhbhum Cu-belt with a strike length of 1,100m have been found to continue beyond a vertical depth of 250m from the surface, whereas ore shoots in Baraganda Cu-deposits in Hazaribagh spread over a strike length of

1100m do not generally continue below 200m from the surface. On the other hand, Mailaram Cu-deposit with a strike length of only 750m has been proved to continue beyond 240m depth from the surface.

Diverse structural control have been observed in the localisation of replacement sulphide bodies. Dariba-Rajpura-Bethumni Pb-Zn mineralisation is controlled by a steep dipping (65°-80°) fault zone oriented parallel to the regional trend of litho-units. In Ambamata Pb-Zn field also, ore localisation appears to be controlled by fault zone intersections with suitable host rocks. Shear zone control has also been invoked for Saladipura pyrite-pyrrhotite deposit, notwithstanding S-ward convergence of the mineralised zone and of the associated marble bands. In Khetri and Singhbhum Cu-belts, ore mineralisation appears to be controlled both by lithology and by closely spaced en echelon movement zones affording space for ore deposition. Individual shoots in these two ore fields are broadly conformable with the associated metasedimentaries in strike and dip, and in some sections rich ore shoots defined by assay contacts plunge either parallel to the fold axis or in the direction of maximum tectonic transport. Transverse faults in Singhbhum and Khetri Cu-belts are apparently post-mineralisation displacements. They cannot obviously be considered as providing channels for movement of ore solution. Even in strata-bound sulphide deposits, *e.g.* Bandalamottu Pb-ore body thickening of the ore zones along the hinges of folds and thinning (with localised disruption) on the limbs indicates ore flowage during folding.

Role of intrusives as a source for ore is somewhat ambiguous. From Mosabani and Dhobani to Turamdih, Singhbhum Cu-belt mineralsation cuts across, on a regional scale, various igneous, sedimentary and metasomatised rocks. In Khetri again, some of the blocks (*e.g.*, Kolihan) display a low angle discordance between the attitudes of ore body and host rocks, although pre-mineralisation granitic bodies in the region do not show any obvious spatial association with sulphide mineralisation. Malanjkhand Cu-mineralisation in granites and quartz veins affords refreshing contrast to the above uncertain picture. In Sargipali, Orissa, granitic mass has engulfed and disrupted the Pb-ore shots in the central section, and is obviously post-mineralisation; while in Dariba-Rajpura, the only intrusives are thin quartz veins, away from the mineralised zones. The range of diversity is so wide that no simple formula can be evolved for further localisation of ore search.

Another interesting point about sulphide deposits is their occasional association with carbonaceous or graphitic bands, which introduce a measure of uncertainty in surface geophysical exploration. In Dariba-Rajpura the richest mineralistaion is localised along the contact between the calc-silicate band and the overlying graphitic schist. A pronounced S.P. anomaly axis with the values on the negative centres ranging from nearly 100-500 m.v. was mapped over the sulphides; but to what extent the graphitic schists present along the fault zone accentuated the S.P. anomaly is not clearly known. Again at Akwali, Khetri Cu-belt, Cu-mineralisation is partly localised within carbonaceous phyllites whose geophysical noise might be louder than the response over the sulphide ore body. It is in this domain that application of statistical geophysics can be a useful exercise.

Widely varying types of wall rock alteration have been noticed around replacement sulphide deposits. But no direct relationship between the diversity of wall rock alteration and the tonnage of ore bodies could be established. *Prima facie* this general observation appears to run counter to mass balance considerations in epigenetic hydrothermal systems, although it is recognised that an interplay of tectonic and geochemical forces might produce such apparent anomalies in space even within broadly closed systems. Here is scope for a statistical geochemical study of this problem over a number of ore fields.

Our country's annual demand for base metals is increasing day by day and there is a huge shortfall between demand and supply. Substitution of Cu by Al can at best partially reduce the gap. Therefore, the available options for re-orientation of base metals exploration can be brought under two heads:

Empirical Options

1. Saturation exploration of all the known localities of base metal occurrences which have not yet been explored in detail;
2. RS Survey (including aerial survey) of the whole Peninsular India and Inner Himalaya, North of Tertiaries using instruments of maximum depth penetrability;
3. Regional geochemical survey of the whole Lower Proterozoic and Archaean (Fe-ore-Dharwar-Bhilwara group and above) particularly the metasediments and volcanics and including all the known major mineralised districts;
4. Detailed geological and geochemical survey of Himalaya, North of Tertiary sedimentaries, to locate ore occurrences in presently inaccessible areas;
5. Rapid geological survey of all the unmapped areas in Rajasthan and Gujarat for gossan shows.

Saturation exploration is usually a wasteful practice and can only be recommended for terrains about which no previous geological knowledge is available and where operations have to be completed within a very limited period, irrespective of the cost involved. India can ill afford the venture on such a costly hobby.

Less costly is the program of regional geochemical survey of all the known mineralised districts and of Lower Proterozoic and Archaean stratigraphy in terms of input and output equations. One positive advantage of this approach would be a greater understanding of the geological processes in Lower Proterozoic and Archaean sequences. Discovery of a single major Cu-deposit by this program may bring in adequate return for this investment.

The 4th empirical option given above involving a massive geological and geochemical survey program of Himalaya, North of Tertiaries, will be equally costly but more uncertain because of its inaccessibility, limited size, and dislocation and disruption undergone by the ore bodies. However, a long-term program can be planned. Preliminary geological survey of the country for ore indication, including gossan shows, was outlined by GSI as early as in 1970s and the target date for completing the first phase of the work was set at 1985.

Conceptual Options

1. Detailed stratigraphic tectonic studies for locating Agniundala type ore bodies in other parts of Cuddapah basin of A.P.;
2. Critical stratigraphic and geochemical studies of Vindhyan basins of Sone Valley and Rajasthan for locating Katanga type base metals mineralisation;
3. Extensive geochemical studies of Punjab and Dras volcanics, particularly near their contacts with younger granitic and monzonitic bodies for locating porphyry-type Cu-mineralisation;
4. Extensive geochemical studies of Dongargarh, Dhanjori and Dalma volcanics in Jharkhand, Orissa and M.P. and of Upper Dharwar volcanics in Karnataka for base metal mineralisation;
5. Extensive geochemical study of Delhi Group in the stretch between Ajmer and Abu for locating Khetri type Cu-mineralisation.

Thus, it is evident that the task confronting the exploration geologists is vast and manifold. It calls for a good theoretical understanding and unraveling the total geological set-up in order to solve many problems that exist. Even in areas where comprehensive exploration has been done a clear exposition of geological processes involved and their sequences are not fully established. Here is a field which requires our immediate attention, if we are to implement a program of conceptual exploration in future. In this effort, the resources of GSI, research institutes in geology and earth science faculties in different universities will have to be pooled. Since this would involve a detailed stratigraphic-structural-geochemical re-examination of many proved deposits on sound theoretical models, an inter-disciplinary approach appears to be the only answer.

Bauxite (Al-ore)

Exploration for bauxite received priority since 1960s, both for establishing adequate captive reserves for public sector Al-plants and for locating new deposits. Major achievements were recorded from Orissa, A.P., M.P., Maharashtra and the total metal grade bauxite reserves which were estimated at 43 MT in 1961 increased to 230 MT in 1971. Later the deposits of Visakhapatnam, A.P., Gujarat, Bihar augmented the reserves further. With such new findings, there has been a considerable upgradation in the reserves (Table 2.4).

Table 2.4: Reserve Status of Bauxite (GSI, 1999)

Type of Reserve	*Reserve (MT)*
Estimated	3037.0
Quarriable	2525.0
Restricted	0.5
Others	90.0

This led to increase in production/extraction of aluminium (Table 2.5).

Table 2.5: Production of Aluminium

Year	*Production (MT)*
1994-95	467.2
1995-96	527
1996-97	509.6
1997-98	550
1998-99	536.8
1999-2000	497.9

In view of the fact that reserves of high grade bauxite deposits in the country are just adequate for meeting the modest indigenous demand of next 2-3 decades, leaving very little excess reserve for export, it is imperative that in the case of future expansion programs of existing plants as well as for all the proposed plants, the flow sheet for alumina extraction should be so designed that lower grade bauxite (with alumina between 40-45%) can be used within an economically permissible framework. Beneficiation of low grade, high-silica and high-iron bauxite needs more and more attention, and ways and means for optimum recycling of the caustic soda may be devised and introduced in the new plants so that lower grade bauxite could be utilised in these units.

It is to be stressed here that a break-through in the process technology is essential so that sub-marginal and low grade deposits could also be explored and utilised. That seems to be the only alternative to augment our resources very substantially and also go in for export of coastal bauxite deposits in Gujarat, Goa and A.P.

Water

Of all the natural resources, water constitutes the most important mineral since there can be no life without it. In order to understand its origin, occurrence, movement, development and utilisation an integrated study of hydrology and geology (hydrogeology) is needed for

1. Geologic and geomorphic evaluation of GW regimen,
2. GW resources, availability and hydrologic control,
3. Geochemical evaluation of GW including its role in mineral exploration,
4. Geothermal potential of GW,
5. Geotechnical evaluation of GW,
6. GW and its relation to applied civil engineering, mining and environmental control including agriculture and forestry, and,
7. GW utilisation for domestic and irrigational requirements.

Comprehensive evaluation of GW resources is possible only by integrating the varied scientific data collected on the above lines. Data collected thus, can be subjected to system analyses to solve GW regimen problems. In more complicated cases, where such an application is not possible, analytical models may prove helpful.

For a complete understanding GW and SW should be studied as a whole. This then calls for an evaluation of inter-state basins of India. Well known are the existing controversies in regard to inter-state river basins and sharing of waters there from, which is a very delicate situation full of controversies, pulls and pressures of all kinds. Recent controversy over sharing of convey water between Karnataka & T.N. gives a testimony to this fact (2002). In this case even Supreme Court had to intervene to direct Karnataka to release allocated share of river water to T.N. The only rational answer to such situation is scientific evaluation by nonpartisan national survey organisations like GSI and CGWB.

GSI till 1960's has been responsible for GW resource evaluation. Having the built-in expertise and the necessary infrastructure, it has been the best suited Organisation to carry out the systematic survey of GW on a continuing basis throughout India without which short term plans could not be dovetailed into the ultimate master plan, be it state level or inter-state level. In order to integrate SW studies, other organisations like CGWB and state departments, are needed to be brought together under a suitable machinery but transferring the function of all GW to CGWB perhaps has not been a correct step. It is a negation of scientific approach based on the concept that the work will progress faster by a Department having greater resources. Integrated and multidisciplinary approach is very much necessary for an overall development of GW. Policy makers should allow GSI also to continue with basic systematic GW studies along with their regional basin-wise mapping programs.

Integrated Surveys

Any program of mineral resources evaluation requires, as an essential pre-requisite, a geological map, showing the rock/soil types in their compositional, genetic, stratigraphic and tectonic relationship, mineral deposits occurring in the area and geological and terrain parameters affecting the stability of man-made structures. The whole of India has been covered on ¼" sheets and out of 32.7 lakh kms^2 about 20 lakh km^2 on 1 : 63,360 and 1 : 50,000 scales. The geological map of India, has been revised many times. Thus the broad frame work of the country as a whole is fairly well known. Deccan traps and Indo-gangetic alluvium occupy about 9 lakh km^2. Plans have now been drawn up to map all the portions of the country so far not covered on the standard scale, by ground surveys and photogeological mapping. Though geological maps on 1 : 63,360 scale are not available for some parts of the country; the priorities and needs of various development programs have not suffered in any way as such areas have been covered on the scales required for mineral exploration, river valley, GW and other projects.

Use of modern techniques such as Aerial Photography (AP) and RS has speed up many mapping and mineral exploration programs and cut down considerably the time and cost involved in the preparation of large scale and detailed maps necessary for various projects, as well as the expenditure of ground surveying. In any program of delineation of known mineralised zones and their possible extensions, based on extrapolation of major controls of mineralisations, following fields offer scope for use of these new techniques:

Large Scale Photography for Detailed Mapping

Scale of geological mapping commonly preferred at the time of detailed mineral assessment is of the order of 1 : 2,000-1:1,000 or even larger. Contoured maps on 1: 2,000 scale can easily be made from aerial photos (APs) on twice or thrice the smaller scales. For plotting of ground data, prints enlarged as much as four times can be very effective. Since such photographs will have to be obtained by special flying tasks, there is ample provision for obtaining colour and colour IR-photography of such regions whenever necessary.

Prospecting with γ-ray and U-V Detectors

Many economic mineral deposits contain traces of redioactive or fluorescent material which can be detected with γ-ray or UV sensors. Airborne radiometric surveys carried out by DAE : and with those EM-methods by Airborne Mineral Survey and Exploration Wing of GSI, are instances in this context. UV-photography is being used, in the search for calcite, phosphate deposits and petroleum seepages etc.

Photogeochemical Prospecting

Analogous to biogeocemica and geobotanical prospecting is the field of photogeochemical prospecting. If concentration of certain elements takes place in the foliage, they can be detected by air-borne instruments sensitive to narrow-band EM radiation. Relationships have been established between the content of certain metals in soils and directional reflectance of vegetation growing over them. Examples are the anomalous reflectance, observed of balsam fir trees growing on Cu and Mo-bearing soils and of Pinus ponderosa growing on Cu-rich soil are from U.S.A. Our environmental conditions deserve serious consideration in this regard.

Thermal Mapping for Sources of Geothermal Energy

Sources of geothermal energy such as hot springs and fumaroles are often located along zones of volcanism of deep seated faults, which are favourable locii for deposition of certain types of epigenetic mineral deposits. A direct way of locating such areas is with thermal IR imagery wherein ground temperature variations are recorded. Thermograms have even been utilised for geological type discrimination.

Side-looking Radar (SLR) Imagery

On a radar image one can easily distinguish major structures such as zones of faulting and fracturing, which often contain associated ore deposits. An advantage of SLR is its ability to penetrate and 'strip away' trees and bushes, thereby revealing the underlying topography.

Ground stereo-photography

At the stage of exploratory mining terrestrial stereo-photography can be resorted for updating the survey work. This can also be utilised for geological mapping of mine faces as soon as blasting is over.

Other Investigations

Among other fields of specific studies/investigations are hydrogeological investigations for assessment of water resources; geotectonic studies for various engineering projects; and environmental studies for detecting levels and types of pollution, etc.

(I) Hydrogeological Investigations. In the inventory and assessment of quality of water resources following aspects may be highlighted :

(a) Sequential AP both on panchromatic and IR film, can be used to obtain valuable data on the nature of surface run-off, delineation of catchment areas and optimum location of gauging stations. In addition, they also provide up-to-date information on changes in stream and river courses.

(b) Multi-spectral photography, thermograms and multi-spectral scanner systems (MSS) can be used to delineate indirect hydrogeological evidences such as vegetation guides, permeable zones along structurally weak locii and degree of surface moisture.

(c) In case of shallow deltaic areas like Sunderbans, the limits of tidal incursions and salinity levels have considerable effect on over all hydrologic regimen. Distinguishing benthic vegetation in such an environment and thereby plotting the changing fresh-water/salt water interface, from air borne MSS data, has been demonstrated to be feasible in Florida Keys.

(II) Geotechnical Studies. Panchromatic APs are being increasingly utilised for feasibility studies of major engineering projects. Their use can be extended in the matter of site selection as well as over all planning in relation to economic potential of the control area, distribution of natural resources and availability of construction material. Unconventional photography like IR, B&W, colour and colour IR as well as multi-spectral B&W aerial photos are of high potential usefulness in engineering, soil mapping and stability evaluation.

For rapidly evaluating a poorly mapped or previously unmapped area, SLR imagery can be used as a primary sources of preliminary data.

(III) Environmental Studies. In a developing country like India, the problem of pollution both in the air and water is not as yet that acute as experienced in the highly industrialised and developed countries. If we are to benefit from mistakes committed by the more affluent nations, it is time to give more consideration to this problem. It should form part of our integrated surveys. In this regard the need is to refer to the immediate problem relating to pollution of water, both SW and GW, about which adequate information is available.

Salinity problem and high incidence of fluoride in A.P. and Rajasthan deserves further study. GSI has patented processes for defluoridisation by the use of (a) MgO and CO_2 and (b) serpentine.

For a successful implementation and completion on schedule of any plan, importance of training and personnel management can never be underestimated. Therefore, any massive mapping or exploration program requires a large number of trained scientific and technical personnel. Fortunately Personnel management is not unknown in our Government departments now.

SECTION-B
ORE GEOLOGY

3

Periods of Formations of Mineral Deposits

Pre geological history of Earth comprises the sequence of events through which it passed before the time when physical condition of the surface became much as it is today, *i.e.*, a surface partly of rocks, partly of water, with a mean temperature determined by solar radiation. This is the zero datum for geological time, a datum following which Earth's surface has been subjected to normal processes of weathering and erosion. Pre geological history began with Earth originated as an individual body within the universe. We do not know the length of time covered by this period with any degree of precision. Earth's crust probably dates back to a maximum age of about 4200 M.y. though the age of Earth as an individual body is probably of the order of approximately 5,000 M.y.

It is common belief that Earth has been formed from the same material that gave rise to the formation of Sun and other planets. However, the controversies arise regarding the mode of origin. One theory propogates that it has been formed because of the condensation of incandescent gases, whereas the another theory suggests that it has been formed by gradual accretion of solid particles in a cosmic dust cloud. When we consider the solar composition with that of the Earth, the major point of difference is the great abundance of H & He and to a lesser extent C & N in Sun. These differences can be adequately explained on either hypothesis, since the elements are either gaseous or from stable gaseous compounds. If Earth grew by the accretion of solid particles, it is readily seen that these elements have always been of low abundance on Earth. On the other hand, if Earth has been formed by the condensation of a mass of incandescent material, the lighter gases would have tended to escape from Earth's gravitational field during cooling from high temperatures. The logical evidence favours the accretion hypothesis, because of the fact that not only the lighter gases are in low abundance on Earth's crust but Xenon, a million times less abundant than their immediate neighbours in the periodic table, a definite deficiency compared to their probable abundance throughout the universe. If Earth condensed from the incandescent material, these gases should be more abundant than they are and if Earth was formed by the accretion of solid particles, then they were never present in any amount. About the accretion hypothesis Earth's atmosphere was formed by the release of gases occluded and chemically combined in solid particles.

Whichever theory of origin we accept, the internal structure of Earth with its marked density stratification seems to demand that at an early period in its history, it was sufficiently hot for metallic iron to liquify and undergo gravitational accumulation to form core (Mason, 1966). If, as appears probable, Earth formed by the accretion of planetesimals resembling chondritic meteorites, the initial state of our planet was an intimate mixture of Ni-Fe, troilite, and silicate minerals, broadly homogeneous throughout. The time of formation of Earth by this accretion planetesimals is usually placed 4.5×10^9 years ago, the figure obtained for the time when certain stony meteorites were enriched in and certain Fe-meteorites were depleted in U & Th. However, it is not obvious that the latter event and formation of Earth were necessarily contemporaneous, and there has been a tendency to push back the time of accretion of Earth to about 5×10^9 years.

The next stage after accretion of Earth was one of heating, mainly by radioactivity. Here it is important to note that during that time there were many short lived radioactive isotopes, effects of which were tremendous as shown in Fig. 3.1 (Vinogradov, 1961), which is based on the known heat production by radioactive nuclides and their estimated abundances in Earth. This fig. shows that at 5×10^9 years ago, radiogenic heart production in Earth would have been approximately 6 times greater than today, and most of this heat was contributed by K^{40} and U^{238}, whereas today these nuclides are greatly diminished in amount and the principal sources of radiogenic heat are U^{238} and Th^{232}.

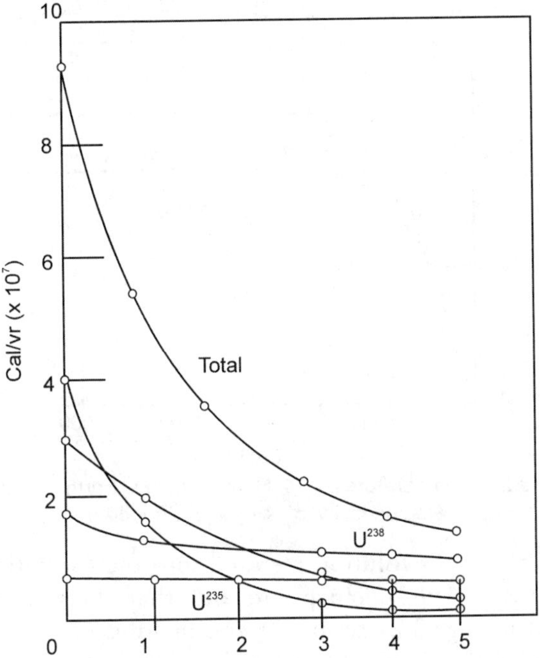

Fig. 3.1: Radiogenic Heat Formed in Earth by U, Th and K^{40} Separately and Plotted Against Time.

Under these circumstances, roughly homogeneous undifferentiated Earth would heat up comparatively rapidly. With the chondritic values for radioactive elements, temperatures within primitive Earth would increase as shown in Fig. 3.2 (MacDonald, 1964). On this basis, melting temperature of iron reached in about 600 M.y. at depth of a few hundred kilometers. At this stage a regime set in and the curves for later times are realised.

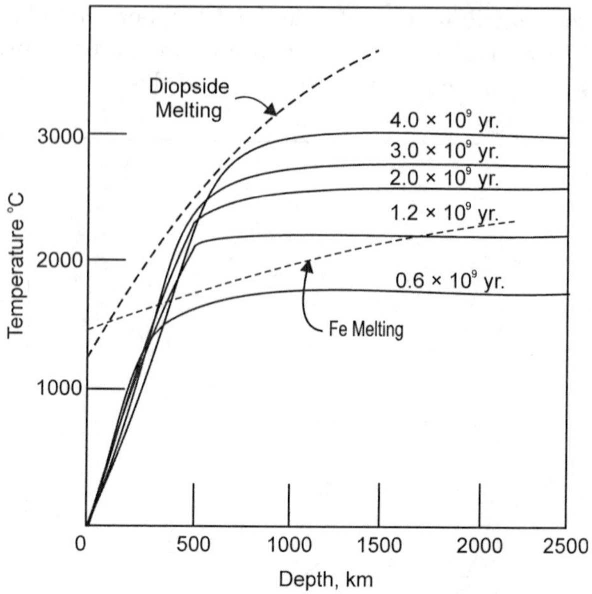

Fig. 3.2: Temperatures in a Homogeneous Earth Heated by Radioactivity of Average Chondrite

This thermal evolution resulted into the formation of Fe-core (Elasser, 1963). He points out that thermal evolution (depicted in Fig. 3.2) results in the melting of free iron in a zone at a depth of several hundred kilometers and accumulation of this metal into a layer of molten iron. A coherent layer of molten iron within a predominantly silicate mantle is clearly unstable, because material of higher density overlies that of lower density. This results into development of a large 'drop' (Fig. 3.3), which then sinks towards the center of Earth, thereby displacing silicates. Sinking of this large mass of molten iron to the center would transform a considerable amount of gravitational energy into heat, sufficient to raise the temperature of interior of Earth by some 2000°C. Thus, process of core formation was strongly exothermic and self-accelerating. It is presumed that the time required for such a 'drop' to form and fall was of the order of 100,000 years.

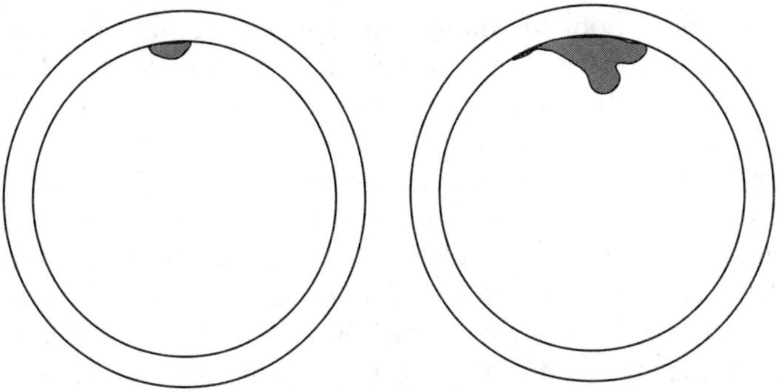

Fig. 3.3: Stages of Drop Formation from a Heavy Liquid Layer

With the movement of free iron, approximately 1/3 of the total mass of Earth was confined in towards the center, the whole planet was profoundly reorganized. While iron was asymmetrically sinking, lighter fractions of silicate material were asymmetrically rising, undergoing partial fusion, mixing and in-mixing. The original surface was probably engulfed undigested. Under these circumstances it is not surprising that crustal rocks older than 3.5×10^9 years have not been found. Development of thick, stable crystal blocks had to await the decay of radioactivity and solidification of most of the mantle.

Development of crust is a further problem of special importance. Considerable diversity of opinion exists. One school of thought pictures the entire crust of Earth solidifying as basalt. No differentiation into continents and ocean basins would then exist. Localised fractures in this initially thin basaltic crust would channel lava, hot gases and solutions from hotter sub-crustal layers and subsequently, these fractures would provide nuclei (resembling present day island arcs) for nascent continents. Action of solutions carrying SiO_2 and alkalis would produce less basic crustal rocks. Erosion and sedimentation at first on a small scale, would accelerate chemical differentiation of crustal material. This concept pictures the continents growing through pre-geological (and geological) time by successive welding of sedimentary deposits and new island arcs to the initial continental nuclei.

Second school of thought on the other hand, pictures the initial continents as the final product of solidification of crust. It carries the differentiation by crystallisation, already exemplified by the separation of basaltic crust from periodotitic mantle and a step further considers the continental nuclei as being raft-like masses essentially of quartz and feldspar, *i.e.*, of granitic or granodioritic composition. Elasser, 1963 suggests that asymmetrical distribution of continents and oceans, the land areas being concentrated in N-Hemisphere, may have been produced by asymmetrical production of large iron 'drops'. The first protocontinent would form in the region above the first large falling drop, and the antipodal region would remain oceanic.

Present evidence suggests that development of a stable crust was not a rapid, once-for-all process. The interval between the age of Earth and that of the oldest dated rocks is over a billion years. Great crustal instability during this interval is indicated. Original continental nuclei may have been partly remelted and rebuilt many times before they grew to a sufficient size and thickness to resist further engulfment. However, by about 3.5×10^9 years ago, pattern of continental nuclei and oxygen basins was probably established, since rocks approaching this age are known in most of the continents.

Now, the next step in Earth's geological history was the origin of atmosphere. Growth of Earth by accretion in a solar dust cloud suggests an initial atmosphere rich in H and lacking free O_2. It possibly consisted largely of H, He, N, H_2O vapour, CH_4 and NH_3 (CH_4 and NH_3 present in the atmosphere of larger planets). Gravitational attraction of Earth is insufficient to retain H and He and they would gradually diffuse into outer space. Evolution of primeval atmosphere into the present one, which consists essentially of N_2 and O_2 is because of the photochemical dissociation of H_2O vapour by solar radiation, thereby producing free O_2. CH_4 would then oxidize to CO_2 & H_2O. Other photochemical reactions would produce more complex organic compounds, ultimately leading to self-reproductory systems, in effect living matter. The probability of such reactions has been experimented by Miller, 1957, who passed electrical discharges through a mixture of CH_4, NH_3, H_2O vapour and H_2 and was able to demonstrate the presence of several carboxylic and aminoacids in the products.

Various suggestions as to how life originated have been explained by Oparin (1957) and Barghoorn (1957). They explained that initial development of living matter was dependent on pre-existence of organic molecules such as amino acids. However, once cells of a kind were established, photosynthesis became possible. Photosynthesis was a revolutionary event in Earth's history. It enabled organisms to collect and use solar radiation directly and thereby utilise CO_2 to synthesise more complex compounds, it was also responsible for the gradual release of O_2 into atmosphere. Photosynthesis may have originated through development of certain coloured compounds capable of acting as catalysts in photochemical reactions. These compounds were probably porphyrins, which are easily synthesised and are extremely stable for organic substances. Porphyrins are formed by condensation of four pyrrole derivatives around a metal atom. Pyrrole is a five membered ring compound with the empirical formula (CH_4NH). It is extremely suggestive that porphyrins act as light receivers in chlorophyll of plants, as O_2 carrier in many animals and as respiratory pigments in cells. Once photosynthesis was established, a vastly greater supply of energy was available for organism, whether they used light themselves or are others that did so.

The length of period between establishment of a stable crust and appearance of life is difficult to evaluate. The oldest dated fossils are the microflora in Gunflint chert, a Precambrian formation on North shore of Lake Superior, whose age is fairly well established at 1.9×10^9 years. Stromatolites, laminated calcareous structures interpreted as fossil algae, are known from Bulawayan System of S-Rhodesia, believed to be older than 2.6×10^9 years, but neither the interpretation nor the age is firmly established. Most of the oldest sedimentary rocks are conglomerates, sandstones and shales showing few features indicating that conditions in atmosphere and hydrosphere were very different from those today. Cloud,

1965 pointed out that the oldest formations, are devoid of limestone. However, limestone first became abundant in younger Precambrian deposits, with ages $<1.2 \times 10^9$ years. This indicates a rather high partial pressure of CO_2 in atmosphere and hydrosphere, acting to keep $CaCO_3$ in solution. Red beds, coloured by ferric oxides (Fe_2O_3) produced by oxidative weathering, appeared about the same time, suggesting an increase in atmospheric O_2 over that present earlier. The most characteristic of Precambrian sediments are banded iron formations (BIF) all over the world, and not younger than 1.7×10^9 years. Combined evidences of geochemistry, palaeontology and stratigraphy indicate that photosynthesis by green plants was probably established at least 2.0×10^9 years ago, but that atmospheric O_2 was first available in relatively large quantities about 1.2×10^9 years ago. By the end of Precambrian, organic evolution had produced metazoa (multicellular organisms), and probably by that time atmosphere and hydrosphere were notably different geochemically from their present state.

It is, thus, obvious that Earth had to undergo a long journey since its formation so as to acquire its present status. This journey is characterised by distinct geo-chemical differentiation of matter in stages so that finally the crust, mantle, core along with hydrosphere, biosphere and atmosphere constitute Earth of today.

Table 3.1 to 3.5 illustrate the different features of Earth.

Table 3.1: Rocks Expressed as percent Volume of Crust

	Rock Types	*%*
(A)	**Igneous Rocks :**	
	Basalts, etc.	42.5
	Granites	10.4
	Granodiorites & Diorites	11.2
	Syenites	0.4
	Peridotites	0.2
	Total	**64.7**
(B)	**Sedimentary Rocks :**	
	Sandstones	1.7
	Clays & Shales	4.2
	Carbonates & Evaporates	2.0
	Total	**7.9**

(C) Metamorphic Rocks :	
Gneisses	21.4
Schists	5.1
Marbles	0.9
Total	**27.4**
Grand Total (A+B+C)	**100.0**

Earth's crust includes both oceanic & continental crusts and its total mass has been estimated by Evickson (1973) as 24×10^{28}t.

Table 3.2: Crustal Abundance of Elements

Element	*Symbol*	*Crustal abundance gm/t*	*Total in 10^{12} t*
Aluminum	Al	83,000	1,990,000
Antimony	Sb	0.62	14.9
Barium	Ba	390	9,400
Beryllium	Be	1.3	31.2
Bismuth	Bi	0.0043	0.1
Chromium	Cr	110	2,600
Cobalt	Co	25	600
Copper	Cu	63	1,510
Fluorine	F	450	10,800
Gold	Au	0.0035	0.084
Iron	Fe	58,000	1,392,00
Lead	Pb	12	290
Lithium	Li	21	500
Manganese	Mn	1,300	31,200
Mercury	Hg	0.089	2.1
Molybdenum	Mo	1.3	31.2
Nickel	Ni	89	2,130
Niobium	Nb	19	460
Phosphorus	P	1,200	28,800
Platinum	Pt	0.046	1.1
Selenium	Se	0.075	1.8
Silver	Ag	0.075	1.8
Tantalum	Ta	1.6	38.4
Tellurium	Te	0.00055	0.013
Thorium	Th	5.8	140
Tin	Sn	1.7	40.8
Titanium	Ti	6,400	153,600
Tungsten	W	1.1	26.4
Uranium	U	1.7	40.8
Vanadium	V	140	3,360
Zinc	Zn	94	2,250

Source: Coates (1981)

Table 3.3: Common Chemical Elements in Earth's Crust

Elements	*Wt. %*	*Atom %*	*Radius (Å)*	*Volume %*
O	46.60	62.55	1.40	93.77
Si	27.72	21.22	0.42	0.86
Al	8.13	6.47	0.51	0.47
Fe	5.00	1.92	0.74	0.43
Mg	2.09	1.84	0.60	0.29
Ca	3.63	1.94	0.99	1.03
Na	2.83	2.64	0.97	1.32
K	2.59	1.42	1.33	1.83
Total	98.59			100.00

Table 3.4 & 3.5 show crustal abundance of elements and abundance of minerals in Earth's crust.

Table 3.4: Crustal Abundance

Elements/Oxides	*Wt. %*
SiO_2	57.60
TiO_2	0.84
Al_2O_3	15.30
Fe_2O_3	2.53
FeO	4.27
MnO	0.16
MgO	3.88
CaO	6.99
Na_2O	2.88
K_2O	2.34
P_2O_5	0.22
C	0.14
CO_2	1.40
S	0.04
Cl	0.05
H_2O	1.37
Total	**100.01**

Table 3.5: Abundance of Minerals in the Earths Crust

Minerals	*Wt. %*
Plagioclase	39.0
K-Feldspar	12.0
Quartz	12.0
Pyroxene	11.0
Mica	5.0
Amphibole	5.0
Clay & Chlorite	4.6
Olivine	3.0
Calcite and Aragonite	1.5
Dolomite	0.5
Magnetite	1.5
Others (*e.g.*, Garnet, Kyanite, Apatite)	4.9
Total	**100.00**

History of Mineral Operations

Evolution of man is the most fascinating phenomena of Earth's history. Even since the earliest period, dated back to some 10,000 years ago, man was instrumental for the growth of present day civilisation. However, systematic trend of civilisation is marked by systematic trend of mineral utilisation techniques adopted by them. Table 3.6 gives an idea of the known mineral operation activities by man.

Table 3.6: Mineral Operation Activities since Pre-historic Time

Mineral Operations	*In use Prior to*
Au washed from alluvial deposits	Recorded civilisation
Cu reduced from ores by smelting	Recorded civilisation
Bitumen mined and used	Recorded civilisation
Sn Reduced from ores by smelting	3500 B.C.
Bronze Made	3500 B.C.
Fe reduced from ores by smelting	3500 B.C.
Au reduced from ores by concentration	2500 B.C.
Ag Reduced from ores by smelting	2000 B.C.
Pb reduced from ores by smelting	2000 B.C.
Ag Parted from Pb by cupellation	2000 B.C.
Bellows used in furnaces	1500 B.C.
Steel Produced	1000 B.C.
Base metals separated from ores by water conc.	500 B.C.
Au refined by cupellation	500 B.C.
Sulfide ores smelted for Pb	500 B.C.
Mineral Operations	In use Prior to
Hg Reduced from ores	400 B.C.
Touchstone known for determining Au & Ag Fineness	300 B.C.
Quicksilver Reduced from ore by distillation	Christian Era
Ag Reduced from Au by cementation with salt	Christian Era
Brass made by cementation of Cu and Calamine	Christian Era
Zn-oxides obtained from furnace fumes by construction of dust chambers	Christian Era
Sb Reduced from ores by smelting	Christian Era
Au recovered by amalgamation	Christian Era
Refining of Cu by repeated fusion	Christian Era
Sulfide ores smelted for Cu	Christian Era
Alum made	Christian Era
Cu refined by oxidation and poling	1200 A.D.
Au parted from Cu by cupelling with Pb	1200 A.D.
Au parted from Ag by fusion with S	1200 A.D.
Au parted from Ag by HNO_3	1400 A.D.
Au parted from Ag with Sb sulfide	1500 A.D.
Au parted from Cu with S	1500 A.D.
Ag parted from Fe with Sb sulfide	1500 A.D.
Ag recovered from ores by amalgamation	1500 A.D.
Separation of Ag from Cu by Liquefaction	1540 A.D.
Co and Mn used for pigments	1540 A.D.
Roasting Cu ores prior to smelting	1550 A.D.
Stamp-mill used	1550 A.D.
Bi reduced from ore	1550 A.D.
Zn reduced from ore	1500 A.D.

Formation of Mineral Deposits

A mineral deposit is always associated with a certain geological formations of a particular age. Magmatic, contact metasomatic, pegmatitic and hydrothermal mineral deposits of a region are usually related to one or the other period of igneous intrusion. An igneous activity itself is generally associated with a period of crustal movements. Similarly, many of the extensive sedimentary deposits have been formed during periods of quiet sedimentation when there was very little crustal disturbance. Thus, there have been certain specific time during Earth's history in which many useful minerals would have been formed over an extensive area of Earth crust. This specific period of mineral formation is known as Metallogenetic epochs. In the context of India's mineral deposits 8 such epochs have been recognized associated with various primary and metamorphic mineral deposits closely connected with basic or acidic igneous activity in organic belts of large scale crustal disturbances. These are :

1. Middle Dharwar
2. Upper Dharwar
3. Post Dharwar
4. Singhbhum-Granite
5. Post Delhi
6. Post Kaimur
7. Early Palaeozoic
8. Cretaceous-Eocene

On the other hand, important epochs of formation of various secondary mineral deposits, more or less unconnected with igneous activity and deposited in quiet conditions are :

1. Dharwar (Pre-Cambrian)
2. Cuddapah (Pre-Cambrian)
3. Vindhyan (Pre-Cambrian)
4. Cambrian or Middle Carboniferous
5. Upper Carboniferous to Cretaceous
6. Eocene to Pliocene
7. Pliocene to Recent.

Pre-Cambrian Epoch seem to have been a period of great and varied mineral formations and to this period belongs world's leading deposits of Au, Mn, Cu & Fe. More than 75% of world production of important metals belong to this period. In fact, barring a few mineral deposits such as Coal, Petroleum and Bauxite, all other minerals have been formed during this period. In general, it may be stated that ancient Pre-Cambrian stable land masses are the scenes of the greatest mineral deposits of world.

Table 3.7 gives the list of deposits formed during different periods of Indian stratigraphy. Whereas Table 3.8 and 3.9 illustrate the concentration clarkes for bodies of common metals and annual world consumption of the elements respectively.

Table 3.7: Formation of Mineral Deposits during Different Periods of Indian History

Formation	*Age (M.y.)*	*Mineral Deposits*
Recent	0-5	Agricultural soil, Brick earth, Sand, Gravel, Salt, Salt peter, Placers (Ilmenite, Monazite, Zircon, Garnet, Diamond etc.)
Pleistocene	2-3	Brick earth, Sand, Gravel, Boulder, Kankar, Laterite, Bauxite, Fe and Mn Ores etc.
Pliocene	12	Sand, Sandstone, Petroleum, Lignite etc.
Miocene	25	Petroleum, Limestone, Gypsum
Oligocene	40	Coal, Petroleum, Limestone
Eocene	60	Coal, Petroleum, Limestone, Lignite, Gypsum
Cretaceous	135	Road Metal, Building Stones, Chromite etc.
Jurassic	180	Thin Coal Seams, Limestone, Gypsum
Triassic	225	Sandstone, Limestone
Permian	270	Coal, Fire Clay, Cu-Pb-Zn Ore
Carboniferous	350	Quartzite, Limestone
Devonian	400	Quartzite, Limestone
Silurian	440	Quartzite, Limestone
Ordovician	500	Quartzite, Limestone
Cambrian	600	Quartzite, Limestone, Sandstone
Proterozoic	2500	I. Building Stones (Sandstone, Limestone, Dolomite, Marble, Quartzite) II. Non Metallic Minerals (Asbestos, Apatite, Feldspar Mica, Pyrite, Quartz) III. Precious and Semi Precious Stones (Diamond, Ruby, Sapphire, Emerald, Chrysoberyl, Garnet) IV. Metallic Ores (Au, Cu, Pb, Ag, Zn, Fe, Mn, Cr, Ni, Co, Ilmenite etc.)

Table 3.8: Concentration Clarkes for Ore Bodies of Common Metals

Metal	*Clarke*	*Minimum % Profitably Extracted an ore body*	*Conc. clarke Necessary for*
Al	8.13	30	4
Fe	5.00	30	6
Mn	0.10	35	350
Cr	0.02	30	1500
Cu	0.007	1	140
Ni	0.008	1.5	175
Zn	0.013	4	300
Sn	0.004	1	250
Pb	0.0016	4	2500
U	0.0002	0.01	500

Table 3.9: Annual World Consumption of Elements (and their compounds) in Tons, in Powers of 10

Consumption in t^{10}	*Elements*
10^9-10^{10}	C
10^8-10^9	Na, Fe
10^7-10^8	N, O, S, K, Ca
10^6-10^7	H, F, Mg, Al, P, Cl, Cr, Mn, Cu, Zn, Ba, Pb
10^5-10^6	B, Ti, Ni, Zr, Sn
10^4-10^5	Ar, Co, As, Mo, Sb, W, U
10^3-10^4	Li, V, Se, Sr, Nb, Ag, Cd, I, Rare Earths, Au, Hg, Bi
10^2-10^3	He, Be, Te, Ta

4

Ore Genesis and Mineral Exploration

The basic knowledge on genesis of ore deposits has developed along with the evolution of human race on Earth. Ores, metals and non-metals today occupy a dominant position in the industrial set up of a country. It was the utility of different minerals that became pivotal in acquiring knowledge on minerals. Greek philosophers have laid first foundation for making an intellectual attempt on occurrence and origin of mineral deposits.

Before Civilization

Since time immemorial mankind has been using stones, precious stones and later on metals as implements for protection, and also for decoration purposes (4000 B.C.). First scientific study of ores and minerals could be ascribed to the ancient Greek and Roman philosophy of Plinny, Aristotle, Ptolemy, and Avicenna propounded more or less scientific theories to distinguish precious stones and metallic deposits from the ordinary stones in the ancient Hindu writings. References to gemstones are found in the Valmiki Ramayana. During the entire period following the prehistoric civilization, there is no record of the contribution of mankind to the study of stones and ores, except in the Hindu scripts i.e. *Arthashastra* and in various treatise dealing with gems and gemstones.

The early man was acquainted with ore deposits and worked many of them. He used cinnabar and soft red hematite as pigments long before smelting. Gem stones and native Cu, Ag and Au were highly valued. Essentially nothing was understood of their origin till Greek and Roman civilization. The hypotheses about ore genesis were postulated by philosophers, none of whom has first-hand information of ores. The 10th century is called the dark age of the science. From the time of Greeco-Roman supremacy until 18th century many imaginative theories were advanced. These are:

(1) Some philosophers envisioned and animated Earth that breathed ore or gave off metallic exhalations as a regular function of metabolism.

(2) Others considered ores themselves to be alive, growing from seeds or ripening from base metals into precious metals.

(3) A popular conception among such "*naturalists*" was that ore existed in the form of subterranean "*gold tree*" whose twigs and branches were various kinds of metals and whose roots led down to the centre of Earth.

Early hypotheses are not based upon field evidences and thorough field knowledge. Indians, Greeks, Egyptians were first in the field of development of mineral deposits particularly gold and gem stones and search for them on a more scientific scale has been noticed around 3400 BC. Egyptians sank number of shafts over 250 m along Red Sea coast for gem stones in which about 400 men worked during 1925 B.C.

Metallurgical utility of different metals particularly Cu and Fe during Bronze Age have led to the identification of distinct metallogenous deposits and their exploitation and refinement. For example :

Au	...	18000 years BC
Cu	...	12000 years (widely used since 4000 BC)

The ancient mines of Cassandra, Greece, where during 2500-365 BC extraction of Au-As ores based on their knowledge of localization of fissures and identification of availability of native metals along zone of oxidation. Understanding of complicated fault pattern, bears an ample testimony to the zeal with which during ancient times Greeks were involved in the development of science on occurrence of ores.

Herodotus (484-425 BC), a Greek and Roman philosopher, was first to indicate in his writings on "Genesis and Occurrence of Ores" (Gold-Quartz-vein). Theophrastas (372-287 BC) wrote a book entitled *Book of Stones*, the first mineralogy text book on metals-stones-earth. Alluvial placer gold was also reported and mined from Spain and Strabo in 19th AD made a reference of the same.

Ancient Greeks have sunk >2000 shafts all over the Mediterranean region. Their locations disclose accurate knowledge of occurrence of ores. Many modern theories can be traced to them.

The first modern treatise in ore geology was published in 1556 by George Bauer also known as Georgius Agricola (1494-1555) who in his book *De Re metallica* made first scientific observation on the classification and mode of occurrence of ore deposits. He distinguished the deposits as transported or alluvial from the deposit which is *in situ*. The *in-situ* deposits are classified into embedded deposits as veins, as stringers, as impregnations etc. During Agricola's time the prospecting for water and minerals was done by non-scientific methods using divination rod or hazel fork or twig. It was Agricola' who very forcibly argued against the use of such non-scientific method in the search for ore deposits. He can be called the father of modern Ore Geology. In the same century one more Danish scientist came into prominence. Nicholas Steno correctly identified minerals and their crystalline form. He can be called as father of modern mineralogy and crystallography.

Alchemy flourished during 16th and 17th centuries and they believed that ore deposits were generated by such celestial powers as Sun-rays or planetary influence. These fanciful ideas were not restricted to alchemists. Their proponents included some of the world's most renowned thinkers like Aristotle, Plinny, Avicenna, etc. Avicenna, an Arabian translator (980-1037 AD) was first to group sulphide minerals, while translating the works of Aristotle.

In view of the religious attitude and rampant superstitions of early days of civilization mysterious forces were called upon to explain ore deposits and to detect buried minerals. Even today the ancient practice of dousing, or witching for ore deposits is employed by many of our less sophisticated prospectors. Ancient Egyptians and Greeks used divination rods; and until 15th century; forked twigs were used to locate mineral deposits. Credit for the adoption of a divination rod to prospecting has been given to the Germans of Hay Mining Region. As the theories of ore genesis improved so did prospecting techniques with the result that advanced geophysical and geochemical methods have nearly replaced the "hazel stick."

In 16th Century

A solid base for modern theories of ore deposits was formulated in 16th Century by Gorgeous Agricola. He is considered as the father of economic geology. He lived in Erzgeberge Region of Saxony and gained an intimate first hand knowledge of the mines in that district. He wrote many treatises on geology and mining, the most significant of which was his comprehensive and pioneering work, entitled "*De Re Metallica*" (1556). His writings have a significant stimulus to Earth science. Agricola's principal contribution was his attempt to classify ores. His classification was based upon genesis—Whether the deposits were (i) alluvial or *in-situ* and upon form (ii) bedded or vein type.

Hoover and Hoover translated Agricola's work and attributed two fundamental principles in addition to his classification of ore deposits. These principles are : (1) ore channels are secondary features, younger than the country rocks, and (2) ores are deposited from solutions circulating in these channels. This great geologist argued for such a prospecting technique like a 20th Century scientist. Agricola described that 'lodes' are originated by deposition of minerals in "canals" (fissures) from circulating underground waters, largely of surface origin that had become heated within Earth and had dissolved the minerals from rocks. He made a clear distinction between "homogenous minerals" (minerals) and "heterogeneous minerals" (rocks). He also classified ore deposits genetically into : (1) Vein, (2) Beds, and (3) Stringers. Agricola's contributions made an indelible mark on the scientific thought of later workers who made significant development on the genesis of ore deposits.

In 17th & 18th Century

During the next two centuries, i.e. 17th and 18th, the European geologists took the lead in the development of theories of the origin of the metallic as well as non-metallic minerals regarding the true nature of the mineral deposits. Following Agricola's generation, Nicolus Steno, who worked in Italy, stands out in 17th Century as a scientist of great perception and argued that ores are a product of condensation from vapors ascending through open fissures (Steno, 1669), a remarkably advanced concept and one that has been retained in modern theory too.

Desecrates wrote *Principia Philosophia* in 1664 and observed that ore minerals were driven upward from a deep metalliferous shell by interior heat in the form of exhalations and resurgent surface waters to be deposited as lodes in the fissures of outer stony crust. This concept is a forerunner even to the present day ideas.

German geoscientists have contributed most in the 18th Century. Of course, Swedes also contributed to some extent.

Bacher (1703) and Henkel (1725) argued "origin of ore veins to the action of vapours over study material—fermentation in the bowels of earth." According to Henkel, idea of transmutation is similar to metasomatism. In 18th Century Henkel (1725, 1727) and Zimmermann (1746) acknowledged the importance of hydrothermal solutions or vapours of deep-seated origin, which, they correctly reasoned should contain dissolved rock material. They even recognised the products of ore deposition by metasomatism (replacement).

Delius (1770, 1773) was first to recognize superficial alterations of ores by atmospheric agents like Sun. He also observed the development of secondary minerals and zone of supergene enrichment.

Charpenter (1778, 1799), a professor at Mining Academy at Freiberg. (Germany), believed that the veins in these mines were formed by alteration of the country rock, and he showed as supporting evidence how some of them grade into wall rock.

Gerhard (1781), a German author, considered the veins to the open fissures filled by minerals leached from the country rock. From Charpenter's and Gerhard's writings has come the 'theory of lateral secretion', which considered that the contents of an ore deposit are leached from the adjacent wall rock by water that are usually but not necessarily of meteoric origin.

The theory of lateral secretion in the genesis of ore deposits was advocated during 18th century by German scientist Sandberger (1882). This theory of lateral secretion was neglected for almost 100 years, until he (1882) published his "*Investigations Concerning Mineral Veins.*" He attempted to show that most mineral deposits were formed according to this theory.

The German contributors Von Oppel (1749) and Zimmermann (1749) advocated the magmatic nature of many ore deposits. Zimmerman (1749), anticipated the idea of metamorphism—origin of lodes and transformation of rocks into metallic minerals and vein stones by action of solutions transmitted through inter-granular openings in rocks. Oppel (1749), slightly modified the ideas of earlier workers (Agricola) and made important distinction between 'veins' and bedded deposits.

According to Lehman (1753), depositions from exhalations and vapours emanated from earths' interior.

In 19th Century

Towards the end of 18th century two prominent schools of thoughts came up in the geological field which influenced the course of the geological action and geological knowledge for nearly 100 years. These scientists were Abraham Gottlob Warner and James Hutton (1788).

Hutton and Warner, with diametrically opposed views dominated the geological thought. Their ideas have had a far-reaching effect on the science of ore deposits. The first *neptunistic* views were postulated by Warner (1749-1807). Warner propounded universal theory of the formation of mineral deposits in a primeval ocean whereas Hutton in the dynamically opposite manner advocated a universal magmatic origin of ore deposits.

Hutton (1788) in his "*Theory of the Earth*" was first to advocate plutonist view (ore magma) and applied magmatic origin of igneous rocks to all mineral deposits. He postulated, "The materials which fill the mineral veins were melted by heat and forcibly injected into the clefts and fissures of the strata." Huttons ideas were supported by Fournet (1856), Bell (1861) and Spurr (1923). Warner and his followers are known as "Neptunist" school of geology whereas Hutton and his followers are known as "Plutonist" school of geology. These two extreme and widely diverted views of origin of rocks and mineral deposits led to controversy in the theory of genesis of ore deposits and dominated the geological thought well up the middle of 19th century. The Plutonists won out with respect to rocks. The Neptunists prevailed with respect to dominance of water in the formation of mineral veins.

The suggestion that ores are direct magmatic products or ore formed as product of differentiation was also advocated by Joseph Bruner (1801). Breislak (1811) also called upon the process of magmatic segregation to explain how ore minerals became concentrated in definite layers within igneous rocks. This theory was changed by Fournet, who was an able advocate of the theory of "ore magmas" and his ideas, with little modification, were revived in times by Spurr (1923). Today ore magmas are widely accepted as the origin of certain types of ore bodies.

Warner argued that all rocks and ore deposits were formed as sediments in a primeval ocean. He pictured veins, as originating as cracks formed on the bottom of a primeval ocean by slumping or earthquakes, the cracks were subsequently filled with chemical precipitations. Play fair (1802) for Hutton and Charles Anderson (1809) for Warner popularised their theories.

Heated arguments between plutonists and neptunists have continued for years. It was readily shown that lavas are not sedimentary formations. Even today plutonists and neptunists or as "magmatists" and "syngeneticists" have not resolved.

Recognition of certain ore deposits, intrusive rocks and their association have helped in establishing the foundation of modern theories concerning the origin of hydrothermal ores. During this period, the geologists began to distinguish between ore deposits of igneous affiliation and those of sedimentary origin.

French school of economic geologists during 17th and 18th century produced a dozen eminent economic geologists, outstanding among them being Daubree, Lassius, Von-Cotta, Bischoff, Fourenet, Charles de Launay (1893). French school gave the most convincing arguments in favour of magmatic origin of ore deposits regarding the theories of magmatic differentiation and the formation of mineral deposits. They also demonstrated the theories of the weathering of ore deposits and showed the supergene enrichment phenomenon of sulphide deposits. Theory of lateral secretion of ore genesis also received a strong support in French school of economic geologists.

The connection of mineral forming solutions with magmas was given emphasis by French geologists—Daubree, Scheerer, Elie de Beaumont. Daubree was first to introduce experimental methods into the study of mineral deposits in 1841 and produced artificial mineral cassiterite. The 19th century witnessed a pronounced development of theories of ore genesis not only in Europe but also in USA. During this century metasomatism and metasomatic ore deposits received great attention among the economic geologist and in USA the leading school believed in epigenetic or hydrothermal model of the ore deposition. Some of the United States early geologists such as Beck classified deposits in syngenetic and epigenetic categories. The work of other mining geologist in USA like Penfield, Kemp gave rise to more analytical description of the characters of the ore deposits.

The field data accumulated in past showed that many large ore deposits are localized in areas of structural complexity and igneous activity, a fact that is used extensively in exploration. Most evidences cited to show that certain types of ores are so commonly found in and around igneous masses. For example, Ni ores with norite and peridotites, disseminated Cu deposits with monzonite or quartz monzonite stocks. This association strengthened the assumption that a genetic relationship exists between igneous rocks and many ores. Zonal arrangements of minerals suggest the spreading of ore solution from channels of deep seated resources (magma chamber). As the fluids rise and spread throughout the rocks, they deposit their minerals in favourable structural and stratigraphic environments. Much of the evidence supported the idea of relationship between ore deposition and igneous activity.

Beaumont, the father of modern thought on formation of mineral deposits observed that mineral deposits are the result of one phase of igneous activity. Water vapour was essential feature in volcanic activity and metallifereous veins were formed as encrustations upon walls of fissures from hot waters of igneous origin. In the middle of 19th century Beaumont made several contribution still generally accepted. He thought that hydrothermal solutions have played a role in the formation of ore deposits. Besides hydrothermal deposits, he recognized and described magmatic segregation and replacement ores in the contact metamorphic zones of intrusive igneous rocks. During this period the chemical principle began to influence ore genesis.

The latter half of 19th century found many eminent scientists contributing to theories of ore transport and deposition. With enough field data collected by observers they were able to understand clearly that no single theory will explain the genesis of all ore deposits (Zirkel & Rosenbusch).

Beaumont recognised many types of mineral deposits with igneous affiliations. (Chromites-Magmatic concentration). Similar views are even held today on some of the deposits.

Von Cotta (1870), in treatise on "Ore Deposits" presented concise information on the content, character and structural features of mineral deposits. He observed that "no single theory was applicable to all deposits. Because formations of ores are many fold the whole process is not confined to any particular geologic period or locality". Such statements are part of present day modern text books also.

Mineral deposits are products of deep seated igneous action and recognized zonal arrangement of minerals based on temperature and pressure of deposition and the after effects and modifications/alterations, imparted due to weathering. Von Cotta's observations exerted profound influence on Economic Geology.

Bischoff (1847), while following Neptunists ideas, posed the "Lateral secretion theory" of meteoric origin. The theory was supported by data regarding dissemination of lode minerals in superficial rocks and their significance in ore genesis. Similar ideas were floated by T. Sterry Hunt (1881, USA); JA Phillips (1875, England), Sand Berger (1882, Germany), SF Emmons (1886), Van Hise (1901).

Broad theories of Ore genesis are following :

(a) Descensionists Theories
(b) Ascensionists Theories
(c) Lateral secretionists Theories

Stelzner (1879), Patera (1888), Posepny (1894), Launay (1893) strongly supported the views of Beaumont and advocated hot water "accessionists" theories, further supported by Kemp (1901), W. Lindgren (1911) and Weed (1903).

Presently, these ideas are accepted but modified by $\delta 0^{18}$ analyses that indicate the importance of meteoric water drain into igneous magmas.

J.H.L. Vogt (1893) strongly advocated the influence of magmatic differentiation and formation of certain mineral deposits during the course of crystallization of magmas *i.e.*, magmatic segregations, etc.

Loius de Launay, J.H.L. Vogt, Waldemar Lindgren, R. Beck, C.R. Van Hise, F.L. Ransome, W.H. Fmmons, J.F. Kemp, W.H. Weed and other scientists formulated and stated many of our modern theories early in 20th century.

In 20th Century

Lindgren (1911), world's most regarded economic geologist, wrote "Mineral Deposits" which is termed as the best text book on mineral genesis even today. In a symposium at Rickard (1903), Lingren gave a popular genetic classification of mineral deposits as given below :

(A) ***Pyrometasomatic Deposits***: High temperature replacement bodies near the border zone of igneous intrusive.

(B) *Hydrotherrrmal Deposits*: The ores formed by hot, aqueous solutions were further sub-divided according to temperature and depth of formation.

(a) *Hypothermal*: great depth (300°-500°C)

(b) *Mesothermal* : intermediate deposits (150°-300 °C)

(c) *Epithermal* : shallow depths (50°-150 °C)

Graton (1933) introduced the term "*telethermal*" *i.e.*, deposits formed at shallow depths from nearly cooled hydrothermal solutions *i.e.*, deposits formed at long distance from the igneous source.

Buddington (1935) introduced the term "*xenothermal*" zone to deposits formed at high temperature and shallow depths. Granitization-metallic elements sweated out and migrated into fissures and channels of low pressure and temperature. Other deposits are :

Sedimentary deposits
Residual deposits
Igneous deposits
Metamorphic deposits

The recent trend on metallogenesis is towards the variations in lateral secretion theory which considers that certain mineral deposits have originated by mobilization of metals from underlying and adjacent rocks which later migrated upward and deposited their content in the existing sites of deposition.

In Submarine exhalative process accompanying submarine volcanic eruptions massive sulphide deposits are recorded. It was observed in many massive sulphide deposits that were sandwitched between older andesitic rocks and overlying rhyolitic rocks or their metamorphic equivalents. Biogenic processes have also been recognized as a factor in the formation of mineral deposits e.g. some sulphide deposits, Fe deposits etc.

Advances in the study of mineral deposits in respect of character, distribution and localization has immensely facilitated by the advance techniques of fluid inclusions, stable isotope studies and thermodynamic considerations.

In recent years, majority of the workers are carefully considering the localization of deposits in their geological setup and with application of recent techniques of deciphering the various phases of deposition. All possible attempts are being made to evolve a genetic model.

The studies clearly suggest that the deposits have been formed by diverse processes of ore formation and it is essential to have a pragmatic scientific approach to decipher the ore genesis which will have far reaching implications in the development of natural resources.

It was only in 20th century that the more development in the theories of ore genesis and the idea regarding the formation of ores took place particularly in the development of the new technique of the investigations namely (i) microscopy (ii) Geochemical technique (iii) Sophisticated instrumental technique like mass spectrometer, X-rays diffraction, electron microscope (TEM and SEM), Electron microprobe (EMP).

Some of the concept which has been introduced during the century can be put as follows:

(1) Concept of ore magma (Spurr, 1923): Experimentally proved by Fischer (1950).
(2) Concept of an exhalative theory of ore deposition, (Oftedehl, 1954).
(3) A source bed concept in ore genesis (Gray, 1954): Concept successfully employed to explain genesis of many low temperature alpine type of sulphide deposits.
(4) Colloid diffusion theory of ore genesis (Hill, 1948): Explain many of the banded character of sulphide deposits.
(5) Concept of remobilization or secondary regeneration of ore fluid- Schneider Hohn (1932) and reported by Fermor: This concept is now accepted worldwide and find place in modern economic geology book of Stanton.
(6) Concept of biogenetic mineralization—(Contribution of organism in formation of mineral deposits, Harder): He described Fe and Mn bacteria. This had been followed by Jensen to explain genesis of biogenic sulphide deposits. Algae have been today found effective in precipitation of many metallic & nonmetallic deposits.

PLATE TECTONICS AND MINERAL DEPOSITION

Plate tectonics provide a basis for understanding the distribution and origin of mineral and energy resources in space and time (Rone, 1977). The occurrence of energy and mineral deposits can be related to plate tectonics in three ways:

(a) geological process driven by energy liberated at plate boundaries control the formation of energy and mineral deposition,
(b) such deposits form in specific tectonic settings, which are in turn controlled by plate tectonics, and
(c) reconstruction of fragmented continents can be used in exploration for new mineral and energy deposits.

Deposits at Oceanic Ridges

Discoveries have shown that oceanic ridges are important sites of metallic mineral deposition (Corllis *et al.*, 1979). Hydrothermal activity at ridges is responsible for the formation of sulfide deposits and metalliferous sediments on the flanks of the ridges. Important metals are Fe, Zn, Cu, Pb, Au and Ag. Sampling of the upper 10 m of metalliferous sediments in Red Sea indicates the presence of about 8×10^6t of ore comprised chiefly of Fe, Zn and Cu. Mn oxide deposits are also important at some ridge locations. Deposits formed at oceanic ridges are accessible in ophiolite complexes in Phanerozoic orogenic belt. Ultramafic rocks in ophiolites contain asbestos, chromite and Ni-ores. Podiform chromite deposits are commonly associated with serpentinised ultramafics. Cyprus-type massive sulfides (Cu-Fe rich) also occur in association with ophiolites and appear to represent hydrothermal deposits formed at ocean ridges.

Deposits at Convergent Plate Boundaries

Metalliferous mineral deposits are important at both continental and convergent plate boundaries. A major part of known base-metal (Zn, Cu, Mo, Pb), precious-metal (Ag, Au, Pt) and other metal deposits (Sn, W, Sb, Hg) are found associated with present or former convergent margins. In circum-pacific area, for example, major metal deposits occur along convergent margins in W N&S America, Japan, Philippines, New Zealand and Indonesia. More than half of world's Cu-production comes from Cu-porphyries (large-volume, low-grade deposits and disseminated Cu-sulfide) in this region, the most important occurrences of which are in SW-USA and in Andes. Red-bed U-deposits are also found associated with convergent margins, for example in SW-USA. Zonation of metallic mineral deposits has been reported in late tertiary rocks from Andes (Sillitoe, 1976).

Going east from the coast towards greater depths to subduction zone, the major zones encountered are : contact metasomatic Fe-deposits, Cu-Au and Ag veins, porphyry Cu-Mo deposits, Pb-Zn-Ag vein and contact metasomatic deposits and Sn-deposits. Belts are believed to be caused by progressive liberation of metals from the descending slab, with Sn coming from an extreme depth of about 300 Km. Geochemical and isotopic data support the general concept and that metal deposits associated with subduction are derived from some combination of descending slab and overlying mantle wedge. Metals move upward in magmas or in fluids and are concentrated in late hydrothermal or magmatic phases.

Deposits at Collision Boundaries

Metalliferous deposits are also abundant at collision boundaries where a variety of local tectonic settings exist, depending on distance from the boundary zone and stage of evolution. Mineral deposits associated with oceanic ridges (ophiolites), convergent plate margins and cratonic assemblages and continental rifts may all occur in collision zones.

Deposits in Cratonic Rift Systems

As previously mentioned, crustal doming precedes rifting and appears to be related to hot spot activity in the upper mantle. Granites intruded at this stage have associated Sn and fluorite deposits. At more advanced stages of cratonic rifting evaporites accumulate in this rifts, and eventually oceanic metalliferous deposits are formed as the rifts open into small ocean basins. Aborted rifts are characterized by fluorite and barite deposits, carbonatites (with associated Nb, P, REE, U, Th, etc.) and Sn-bearing granites. Pb-Zn-Agdeposits occur in limestones during early and middle-stages of cratonic rifting.

The high heat flow from upwelling asthenosphere or mantle plumes provides a source for geothermal heat in cratonic rifts. Origin and occurrence of metallic mineral deposits, oil and gas, and geothermal energy are related to plate tectonics. Metalliferrous sediments, massive sulfides and Ni and Cr deposits form at oceanic ridges. A variety of metallic mineral deposits and geothermal energy occur in the over riding plate at convergent boundaries, and oil and gas are formed in many back-arc basins; cratonic rift systems contain geothermal energy, Sn, U, Th, REE and other rare element deposits, collision boundaries are characterized by a variety of metallic mineral deposits and by oil and gas occurrences; and cratonic basins are sites of major HC accumulation.

5

Ore Fluids

An ore is a naturally occurring mineral substance or a geologic body from which metals can be obtained after treatment. The ores are divided in two categories (a) Metallic ore; (b) Non-metallic deposit.

Obviously the metallic ores are source of metals. Sometimes it can be monometallic but mostly the ores are polymetallic. Non-metallic ores are not ores in the true sense of the term. They are useful not for their metallic content but for their application as industrial raw material for example—limestone, clay, feldspar, sand etc. These are also known as industrial minerals or rocks.

Any ore deposit consists of two types of minerals (Table 5.1-5.2).

(i) Ore minerals (ii) Gangue minerals.

Table 5.1: Common Ore Minerals

Metal	*Ore Mineral*	*Composition*	*Percent Metal*	*Primary*	*Super-gene*
1	2	3	4	5	6
Gold	Native gold	Au	100	X	X
	Calaverite	$AuTe_2$	39	X	
	Sylvanite	$(Au,Ag)Te_2$	..	X	
Silver	Native silver	Ag	100	X	X
	Argentite	Ag_2S	87	X	X
	Cerargyrite	$AgCl$	75		X
Iron	Magnetite	$FeO.Fe_2O_3$	72	X	
	Hematite	Fe_2CO_3	70	X	X
	Limonite	$Fe_2O_3.H_2O$	60		X
	Siderite	$FeCO_3$	48	X	X
Copper	Native copper	Cu	100	X	X
	Bornite	Cu_5FeS_4	63	X	X
	Brochantite	$CuSO_4.3Cu(OH)_2$	62		X
	Chalcocite	Cu_2S	80	X	X

1	2	3	4	5	6
	Chalcopyrite	$CuFeS_2$	34	X	X
	Covellite	CuS	66	X	X
	Cuprite	Cu_2O	89		X
	Enargite	$3Cu_2S.As_2S_5$	48	X	
	Malachite	$CuCO_3.Cu(OH)_2$	57		X
	Azurite	$2CuCO_3.Cu(OH)_2$	55		X
	Chrysocolla	$CuSiO_3.2H_2O$	36		X
Lead	Galena	PbS	86	X	
	Cerussite	$PbCO_3$	77		X
	Anglesite	$PbSO_4$	68		X
Zinc	Sphalerite	ZnS	67	X	
	Smithsonite	$ZnCO_3$	52		X
	Hemimorphite	H_2ZnSiO_5	54		X
	Zincite	ZnO	80	X	
Tin	Cassiterite	SnO_2	78	X	?
	Stannite	$Cu_2S.FeS.SnS_2$	27	X	?
Nickel	Pentlandite	$(Fe,Ni)S$	22	X	
	Garnierite	$H_2(Ni,MG)SiO_3.H_2O$	..		X
Chromium	Chromite	$FeO.Cr_2O_3$	68	X	
Manganese	Pyrolusite	MnO_2	63		X
	Psilomelane	$Mn_2O_3.xH_2O$	45		X
	Braunite	$3Mn_2O_3.MnSiO_3$	69	?	X
	Manganite	$Mn_2O_3.H_2O$	62		X
Aluminum	Bauxite	$Al_2O_3.2H_2O$	39		X
Antimony	Stibnite	Sb_2S_3	71	X	
Bismuth	Bismuthinite	Bi_2S_3	81	X	X
Cobat	Smaltite	$CoAs_2$	28	X	
	Cobaltite	$CoAsS$	35	X	
Mercury	Cinnabar	HgS	86	X	
Molybdenum	Molybdenite	MoS_2	60	X	
	Wulfenite	$PbMoO_4$	39		X
Tungsten	Wolframite	$(Fe,Mn)\ WO_4$	76	X	
	Huobnerite	$MnWO_4$	76	X	
	Scheelite	$CaWO_4$	80	X	

There may be only one ore mineral from one deposit. Chromite is ore mineral from chromium deposits, but there may be large number of ore minerals from same deposits. For example in deposits of copper ore minerals may be chalcopyrite, tetrahedrite, chalcocite and covellite and in some cases also secondary minerals such as malachite, azurite etc.

Ore deposits contain another component which is known as the gangue. The most common gangue minerals are silica, silicate, feldspar, mica, pyroxene, amphiboles, accessory silicates,

Table 5.2: Common Gangue Minerals

Class	*Name*	*Composition*	*Primary*	*Supergene*
Oxides	Quartz	SiO_2	X	X
	Other silica	SiO_2	X	X
	Bauxite, etc.	$Al_2O_3.2H_2O$		X
	Limonite	$Fe_2O_3.H_2O$		X
Carbonates	Calcite	$CaCO_3$	X	X
	Dolomite	$(Ca,Mg)CO_3$	X	X
	Siderite	$FeCO_3$	X	X
	Rhodochrosite	$MnCO_3$	X	
Sulphates	Barite	$BaSO_4$	X	
	Gypsum	$CaSO_4+2H_2O$		X
Silicates	Feldspar	. . . .	X	
	Garnet	. . . .	X	
	Rhodonite	$MnSiO_3$	X	
	Chlorite	. . . .	X	
	Clay minerals	. . . .	X	X
Miscellaneous	Rock matter		X	
	Fluorite	CaF_2	X	
	Apatite	$(CaF)C_{3\text{-}4}(PO_4)_3$	X	
	Pyrite	FeS_2	X	X
	Marcasite	FeS_2	X	X
	Pyrrhotite	$Fe_1{-}_1S$	X	
	Arsenopyrite	FeAsS	X	

tourmaline, garnet, apatite etc. The secondary minerals *e.g.* clay, talc, serpentine, zeolite may be as gangue minerals. There may be carbonate gangue minerals like calcite, dolomite, siderite, magnesite; sulphate gangue minerals like gypsum, anhydrite and barytes. Fluorite and chlorite may be considered as gangue minerals. Sometimes, pyrite and marcasite *i.e.* sulphides themselves can be gangue minerals.

Host Rock-Ore Relation

Host rocks or country rocks refer to the recepticle in which the ore deposit is found. The host rock may be sedimentary, igneous or metamorphic. Host rock and ore relation gives interesting clue to the implacement process of ore body. The ore may be syn-genetic *i.e.* the ore is formed at the same time as the host rocks or the ore may be introduced much later than the host rock; such ore host rock relation is called epi-genetic relation. Another type of relation is peri-genetic relation *i.e.* partly syn-genetic and partly epi-genetic. In many cases the host rock becomes gangue.

Intensity of Epigenetic Mineralisation: Following formula has been given by Russian geologist Belevstev for intensity of epi-genetic mineralisation:

$$Q \alpha c$$
$$\alpha p$$
$$\alpha v$$
$$\alpha t$$

Where

Q = intensity of mineralisation
c = porosity of rock
v = velocity of solution diffusion in the opening
t = temperature inside earth.
p = directed pressure (stress)

Thus, Q is the function of c, p, v and t. Other factors which control the intensity are:

(i) Chemical property of host rocks.
(ii) Chemical action and presence of certain elements in minor quantity also affects the intensity of mineralisation.
(iii) Continuous supply of ore ion in ore deposition.

Mineralisation

It is a broad term used for the formation of economic minerals. Mineralisation is not uniform in the history of more intense mineralisation. These are known as the epoch of mineralisation or metallization. Similarly, mineralisation is not uniformally distributed on earth. There are places which are more mineralised than others. These are known as mineralization provinces or metallogenic provinces. There are certain periods of mineralization known as metallogenic period.

Quality of mineralisation

It depends on various factors :

(i) Percentage of metal
(ii) Association with other metals
(iii) Types of ore
(iv) Nature and character of host rock and gangue.

Grade

It is a term roughly used to indicate in general the quality and character of deposits, both metallic and non-metallic. For example, Mn-deposites may contain metallurgical grade, chemical grade and battery grade. Similarly, limestone deposits may be of cement grade, chemical grade, metallurgical grade (flux grade limestone), lime grade, building grade etc.

Another use of the term grade is in run of mine (r.o.m.). The term cut off grade means the quality below which the ores cannot be mined. It is a dynamic concept. It changes with time, technology, improvement in method of mining, new use of metals etc. In India cut off grade for Fe is 50%, for Cu 0.8% and for Pb and Zn it is 3 to 4%.

Tenor

This term is used almost synonymously for quality. A low grade deposit may contain a good tenor because of particular association or because of particular usefulness of metals.

Protore

Protore is the term used for a low grade ore or country rock which contains ores in uneconomic quantity. For example, B.H.J. (Banded Hematite Jasper) contains only 15% FeO, and hence is a protore and not used as an ore. But this B.H.J. is, in many cases precursor for the formation of rich hematite ores because it has been formed ultimately from B.H.J.

Similarly manganese silicate rocks (Gondite rocks) are protore for Mn-ores of our country (with 10% MnO_2). Many rich Mn-deposits have formed by oxy-alteration of gondite protore. Sulphide schists are protore for many sulphide deposits.

Nature and Composition of Ore Fluid

In the formation of ore-deposits, metallic constituents in the form of liquid, or gas or a mixture of liquid and gas (*i.e.*, fluid) play a very important role. Beside these ore materials may in very special cases be present in native conditions. What the conditions was for the Ore bearing fraction of the magma that is primarily concerned with the formation of ore deposits.

From the point of view of genesis of ore fluids there are atleast three possibilities :

Ore fluids are related directly or indirectly with a magma which is a liquid rock melt capable of intrusion as well as extrusion. It is highly complex liquid and may be formed in the mantle below the Moho line or it can be generated in the crust and in some cases even near the surface. Magma is a liquid of variable composition, physical nature and pH *i.e.* H^+ concentration and Eh potential (redox potential). Compositionally magma consists of silica and oxides like Al_2O_3, CaO, MgO, alkali oxides K_2O and Na_2O.

Besides containing small amounts of such metals as Ni, Cr, Mn, Ti and Co, magma also contains large number of gaseous substance like H_2O vapour, Cl, B, F, P, As, Li, C and H. The refractory high temperature constituents include silicates of Mg, Fe, Ca and Zr. There are refractory oxides like Al_2O_3, ZrO and sometimes rare earth oxides. The non-refractory lower temperature constituants are quite different in nature and include alkali silicates, silicate-bearing lithium and a large number of phases containing water vapour and other volatiles. The metal components present in magma may exist in the form of simple electrically charged particles (ions) or more commonly they are found in the form of metal ion complexes *i.e.* hydrosulphides, polysulphides etc. Besides these, metals may be present in the form of chloride particularly in salts like NaCl, the metal ion bearing chloride solutions are supposed to be quite stable. But in the presence of metal sulphides metal chloride complexes become unstable.

Magmatic ore fluid may behave as ferous silicate melt or on the other extreme it may be similar in physical appearance to trully aqueous solution. Between these two extremes, magma may show all types of physical characters. From the temperature point of view magma may show a temperature much higher than the critical temperature of 400°C. At this high temperature magma behaves like a super ciritical fluid. It is well known that super critical fluids whether gaseous or liquid cannot be distinguished in terms of a truly liquid or a truly gaseous phase. For such fluids distinction of pneumatolytic phase from hydrothermal phase becomes meaningless.

Eh and pH variation of Magma

It is well known that magma can be acidic (acidic means a substance which in solution produces hydrogen ions and pH < 7) or basic alkaline (alkali being a substance which in solution releases hydroxyl ions and pH > 7) or the magma may be truly neutral. A magma originally acidic may become alkaline due to reaction with carbonate rocks. Conversely a magma which is orginally alkaline may become acidic in course of journey through particular types of host rocks.

Regarding Eh, magma generated deep within the crust of earth is generally oxygen deficient *i.e.* it has negative Eh. Magma generated in the upper levels may have neutral Eh or it may have positive Eh value. It is quite possible for the original reducing condition of magma to be changed to nutral or even oxidising condition. In some cases reverse may also take place.

Temperature and Pressure

Magmatic temperature like other properties is extremely variable. It may range from as high as 1200°C to as low as 100°C. At the high temperature magma is truly silicate (acidic) melt. At low temperature magma behaves more like a truly heated watery solution. Similarly the pressure at which magma solidifies may be few 10s of atmosphere at great depth or at other extreme, near the surface magma may be subjected to a pressure of little >1 atmosphere. Besides these overburden pressure, magma may be subjected to directed pressure or stress due to the earth movements accompanying magmatism. A third type of pressure plays a very important role in most magmas which are hydrous and rich in volatile constituents. This is the vapour pressure or the internal pressure or the partial pressure of magma. These volatile pressures are commonly met within silicic magma having a comparatively low temperature, but there are magmas which are almost dry or anhydrous and these are known as dry melts. Usually early fractions of magma cystallise under dry conditions while the later fractions of magma show a greater concentration of volatile constituents.

In extreme cases a fluid may be composed entirely of ore material. Such an ore enriched magma is known as ore magma. The concept of ore magma was first propounded by Spurr in 1923. He thought that by extreme differentiation a portion of magma may be composed entirely of ores either oxidic or sulphidic. This ore magma may intrude the country rocks in the form of ore dykes and ore sills. Fissher in 1950 has been able to experimentally produce the ore magma in laboratory. In the field dykes of chalcopyrite have been observed intruding the country rocks in Chile. The exact mechanism by which ore liquid gets separated from these silicate melts during the cooling of magma is not well defined. Alternatively the ore liquid being heavy may accumulate as a layer in lower portion of the magmatic chamber to be intruded later on by tectonic forces. Ores of magnomagnetite and magnesioferrite have been reported from pre-Cambrians of Ural mountains in USSR. These deposists are underlain by beds of rock salts or halite and the reaction between basaltic rocks and halite beds have given rise to iron-chlorides ($FeCl_3$) along with some magnesium. Oxidation of Mg-bearing Fe-Chlorides have given rise to deposits of magnomagnetite and magnesioferrite.

Ore fluids of Meteoric Origin

It includes ground water, connate water and also it may be mixed with juvenile water of magmatic origin. These three sources may contain in solution appreciable amounts of metals, cations and anions. Under favourable conditions this type of ore fluid may give rise to ore deposits including hydrothermal deposits. Hydrothermal deposits may be magmatic or meteoric. A variation of this type of ore fluid is examplified by the fluid of submarine volcanic or submarine exhalative origin. In this case ore fluid is derived by volcanism taking place in the bottom of ocean. Ore fluids react with the sea water and precipitate the metallic contents in the form of oxides and hydroxides. There are deposits of red hematite of this type of submarine exhalative origin. Similarly there are deposits of sulphides like copper, pyrite, sphalerite and other sulphides of submarine volcanic origin. These deposits are also known as sedimentary or hydrothermal origin. According to one view pre-Cambrian iron ore formation of some parts of world including India may be partly submarine exhalative orgin. Again, deposits of barytes in the form of layers associated with dolomite and limestones may be sedimentary/hydrothermal origin.

Ore Fluids of Metamorphic Origin

It is well known that during metamorphic deformations of sedimentary sequences important amount of water is sweated out of the parent metamorphic rocks. This water which is essentially of metamorphic origin may contain appreciable amount of ore metals during its passage through rocks. These types of rocks may contribute their metallic content to metamorphic ore fluid giving rise to ore enriched fluids. Cooling and crystallization of this type of ore fluids, may gives rise to a number of metallic and non-metallic deposits. For example-deposits of kyanite and quartz are found to occur as segregated veins in mica-schist or quartz-mica schist country rocks. Deposits of sulphide ores may occur in the forms of lensoid seggregations or veins in rocks known as sulphide schists. Deposit formed by this way is also known of metamorphic-hydrothermal origin.

There are a number of evidences both experimental as well as in field which give an insight in to the nature of ore fluids:

(1) ***Springs of Fumaroles and Geysers*:** These are common in areas of active or even extinct volcanoes.This gives an idea of the composition and nature of ore bearing fluids. These springs are not only hot but also contain a large number of metallic as well as non-metallic constituents. Deposits of powdery silica found around springs are known as kieselgur, opal; carbonate deposits are called as travertine; iron deposits like spathic iron ores, oxide deposits like limonites are found as bog. Oxides of Mn, sulphide deposits include AsS_2, Sb_2S_3, native sulpher, silver and other deposits like halides, chloride, iodide, borax. Such occurrence definitely shows the magma containing a host of metals and non-metals.

(2) ***Hot Brines***: Brines are highly saline bodies of H_2O. These have been found to contain a large number of heavy metals *i.e.* Pb, Zn, Cu, Fe, Cd, etc. They have been reported from the red sea where in deep bottom rich concentrations of metals are known to occur in brine portions. Another example is the sea in California which also contains high concentration. These occurrences show that ores may be carried in solution to an appreciable extent by brines.

(3) ***Fluid Inclusion***: Fluid inclusions in minerals and rocks represent minute quantities of the fluid from which minerals or rocks were deposited. Actually, during deposition of minerals and rocks these fluids get entrapped inside the minerals. These fluid inclusions may be gaseous or gas and liquid or entirely liquid. As they represent samples of the original ore fluid, their analysis gives important evidences regarding nature and composition of ore fluids. There are special techniques to isolate and collect fluid inclusions occurring in minerals like quartz, fluorite, feldspathoids, feldspar, micas prior to submit them to detailed chemical analysis. Such studies have shown that ore fluids may be initially acidic or basic or may be highly saline.

Transport and Migration of Ore Fluid

Ore fluid is generated in the depth of earth either in the level of crust or in the depth of mantle. In almost all cases the fluids once formed do not usually remain in the place of origin. It moves out either upward or sideways or even migrates below. Problems of transport and migration of ore fluid involve dynamics of fluid motion through solid particles. There is no direct evidence of the mechanism of the motion of ore fluid. Again, there is considerable controversy regarding the state in which the ore liquid migrates. The problem is complicated by the fact that the ore fluid if initially in a gaseous condition may be converted during its transport to a liquid condition. So deposition takes place from a liquid state. Again, there is controversy regarding the viscosity and fluidity of ore liquid. An ore liquid initially viscous might become highly fluid due to admixture of ground water during the journey. Evidences from the nature of fluid inclusions are rather inconclusive because fluid inclusions may be totally gaseous or partly gaseous and partly liquid or even completely liquid similar to watery solution. Evidences from solubility studies do not throw the final light on the nature of ore fluid because solubility of metallic sulphides, oxides or even metals is greatly different in the three states *i.e.*, gaseous, fluid and liquid states. It is known that certain substances are easily transported in the gaseous or vapour condition. Such is the case with compounds of B, As, Sb, S, Sn and few others. On the other hand there are substances which are highly refractory and can never be transported in a gaseous condition, such as compounds of Cr, Pt, Ca and Mg, some iron compounds and many base metal (Pb-Cu-Zn) sulphides. Again, there are certain metals and non-metals and their compounds which can be transported only in the solution conditions. Examples are various carbonates ($CaCO_3$, $MgCO_3$, etc.) and chlorides (NaCl, KCl, $MgCl_2$ etc.). The ore materials can be carried in solution which may be alkaline (ore in hydroxides), acid (ore carried in NaCl, complex sulpher ions), cyanide complexes etc. There are also neutral solution which may carry the ore material. Particularly difficult to explain are the metallic and non-metallic deposits of such highly insoluble metals as Hg, Ba, Sr and few others. For such deposits a huge amount of solution is needed and obviously the solution mechanism can not explain the formation of these deposits. In such cases the only way of transport is the form of ionic complexes (*i.e.*, HgS_2) which later on break down and form the simple sulphide deposits. Similarly, among the metals of Cu, Pb and Zn, Pb is carried in complex sulpho salts which however break down to give rise to ultimately simple sulphides.

Some metallic oxides and sulphides are carried in colloidal solution in which ultramicroscopic particles remain suspended in either colloidal gel or in the form of suspension i.e., solution. Deposition from the colloid takes place by floculation or coagulation when colloidal charge is mineralised. In the near surface portion of crust migration of ore fluid is easy to understand because the crustal rocks are permitted by fractures, joints and by shearing and other types of weak planes. These provide an easy channel way for the transport and migration of fluid. In the deeper parts of crust of earth such fractures are less common. Hence, the passage is not readymade. In such cases there are other mechanisms to force the ore liquid upwards and sideways.

(1) ***Gas action*:** It is well known that magma contains a considerable amount of gases and vapours which are kept under pressure at depth. As soon as there is release of pressure these vapours tend to move up and carry the ore material along with them. The metal particles get attached to the bubbles and they ride on the bubbles and are carried upwards along with vapours and gases. This mechanism of gaseous transport is one of the most important mechanisms of the migration of ore fluid.

(2) ***Tectonism*:** Tectonic pressures during a folding or orogenic movement cause the ore liquid to move up or sideways wherever space is available. This mechanism is known as filter pressing and many of the pegmatite ore deposits are known to have been emplaced by the mechanism of filter pressing action.

(3) ***Action of Capillary fissures*:** In this case migration over a short distance may be due to capillary action. Particularly aqueous ore solution may move a little upward by this mechanism.

(4) ***Diffusion transport*:** Diffusion may take place from one end of the rock body to the other and due to differences in the concentration of ions on the two sides of the column. This may also be induced by differences in temperature. However, magnitude of transport is only minor in nature but it is definitely a mechanism where transport is required in a colloidal medium particularly through impervious clays. Diffusion has been demonstrated in laboratory experiments to have been operative in sulphide bearing gels whereby Cu^{++} have diffused and produced Cu-Si gel mineral (Chrysocolla). Similarly, many clay complexes with adsorbed metals and metallic sulphides are known to have diffused through a colloidal medium.

Formation Precipitation of Minerals

Formation of minerals in nature happens to be a complicated process and often involves a number of steps like chemical, physical and biogenic. Various reactions involved in mineral formation have been duplicated in laboratory with partial success. But conditions in nature are too complicated to be reproducible under laboratory conditions. Reactions in nature leading to formation of minerals may involve liquid-liquids, gas-gases, liquids-gases, gases-solids and lastly gases-liquids. Mineral formation is controlled by a multitude of reactions and factors which include chemical, physical and combined physico-chemical, biological, bio-chemical and bio-physical processes.

Chemical Factors

Chemical factors important in precipitation of minerals are :

(i) Chemistry of mineral bearing solutions.
(ii) Chemistry of surrounding medium.
(iii) Oxidation potential (Eh) also known as oxidizing or reducing capacity of the medium.
(iv) Hydrogen ion concentration (pH of medium) also known as acidity/alkalinity.
(v) Trace quantity of element.
(vi) Salinity of medium.

Physical Factors

Physical factors controlling mineral precipitations are :

(i) Change in pressure.
(ii) Change in directed pressure or stress.
(iii) Temperature changes
(iv) Change in the nature of the electrical charge on colloidal particles in a colloid system
(v) Grain size.

Combined Physico-Chemical Factors

In many cases of mineral precipitation one or more of the above mentioned chemical factors may combine with one or more of physical factors to affect precipitation of minerals.

Biological Factors

Biological factors responsible for mineral precipitation may be the following:

(i) Plants including angiosperms and gymnosperms as well as some giant algal forms.

(ii) Animals including various groups of invertebrates and vertebrates including man.

(iii) Microscopic organisms both plants as well as animals such as bryophytes, fungi, mosses, thallophytes.

Bio-chemical Factors

These include one or more of the biological factors along with one or more of the chemical factors.

Bio-physical Factors

These include combined biological and physical factors.

Chemical Factors in Mineral Precipitation

A large number of chemical factors and reactions of diverse nature are involved in the precipitation of mineral phases. Most of the reactions have been reproduced under laboratory conditions in controlled environments.

Chemistry of Mineral bearing Solution

Mineral bearing solutions may be of diverse nature and diverse composition. They may be simple silicates or oxides or sulphides or sulphosalts etc. The solutions may be true solutions (salt + water) or may be a colloidal system or in some cases the components are present in ionic-complexes. In the case of true solutions either silicates or oxides or sulphides precipitation may take place when solutions of different composition meet and react with one another. In laboratory it is well known that the solutions of barium precipitates the insoluble sulphate minerals when comes in contact with dil. H_2SO_4. In nature precipitation of baryte is known to occur when barium containing solution encounters an acid medium. Similarly, deposits of gypsum near surface may raise out by the reactions of lime bearing solution with an acidic environment, the acid being produced by oxidation of iron sulphide present in the soil or in the rock. Examples of reactions involving a solid and a liquid in the laboratory are furnished by the well known precipitation of copper from a $CuSO_4$ solution when iron is introduced in the sulphate solution. In nature local precipitation of $CuSO_4$ takes place in the cavities and cracks of rocks wherever $CuSO_4$ solution encounters iron rich rocks. Another example commonly found in underground mines is the precipitation of grey or whitish iron sulphates on the mine wells wherever iron from limonite comes in contact with groundwater containing H_2SO_4.

Precipitation also takes place in laboratory conditions of a large number of metals when H_2S gas is allowed to pass in a solution containing the metals. In nature solutions of Cu, Pb and Zn precipitate the sulphide when sub solutions encounter H_2S gas produced either by petrifaction or decomposition of organic material or from pyrite.

Composition of Surrounding Medium

In nature it is well known that a number of carbonates of metals are precipitated when solutions containing these metals come in contact with carbonate rocks like dolomite, limestones, magnesite. Many occurrences of $CuCO_3$, $ZnCO_3$, $PbCO_3$ are found to be caused by precipitation in carbonate host rocks.

Oxidation Potential or Redox Potential or Eh

Changes in Eh of the medium are responsible for precipitation of a large number of minerals in nature. A good example is iron which is transported in solution in the form of iron bicarbonates-$Fe(HCO_3)_2$. This solution under reducing environment causes precipitation of large amount of iron in the form of trivalent insoluble iron hydroxide mineral known as limonite when the solution reaches in oxidising environment. Another example is uranium which is readily soluble in the hexagonal form under oxidising condition. Such hexauranyl solution is believed to precipitate uranium in tetravalent form as soon as the solutions come in contact with a reducing environment produced by organisms, logs of wood or by iron sulphide rich medium.

Change in Hydrogen ion Concentration (pH)

Changes in pH of mineral bearing solution may cause enough precipitation of minerals. Many naturally occurring mineral solution including ore fluids are slightly acidic in nature. As soon as there is change in pH from acid to alkaline, precipitation takes place. There are many examples of naturally occurring sulphides which are acidic in nature precipitating the sulphides as soon as acidity becomes neutral or it changes to slightly alkaline condition by reaction with the carbonate rocks. The opposite is also sometimes true. The ore fluids are sometimes alkaline in nature. As soon as this ore fluid becomes neutral or acidic, precipitation of minerals takes place. A very good example is afforded by various forms of silica. Quartz is carried away or transported in alkaline environment. As soon as silica bearing solutions become neutral or acidic, precipitation of silica takes place in the forms of jasper, chalcedony or opal.

Trace quantity of Elements

Solutions and precipitation of a major metal is very much influenced by the presence of trace quantities of such metals as Bi, As, Sb, Cd etc. In some cases incorporation of these trace elements increases the solution of the main sulphides. For example, presence of trace amount of V, Mo may enhance the solubility of sulphides. On the other hand presence of trace amount of Cd in ZnS, Ag in PbS, Ni in Cr ore, may cause an early precipitation of mineral phases.

Salinity of Medium

It is well known that many ore bearing fluids are saline. Salinity of ore bearing fluid may be due to its contact with such soluble rocks as NaCl or other evaporite. Such saline ore fluid contains a good amount of heavy metals like Pb, Zn, Mo, Se, Cd and precipitation of these heavy metals takes place in the bottom of the basin where salinity is less than the upper part. A very good example is Red sea where concentration of heavy metal is found in the bottom. In USA Satton sea precipitates a good amount of heavy metals in the bottom layers.

Physical Factors in Mineral Formation

Change in Pressure

Pressure change, in some cases directly affects the formation and precipitation of minerals while in other cases it indirectly influences precipitation of minerals by affecting chemical reactions. Pressure changes as cause of mineral formations are most effective in systems involving liquid and gaseous phases. A very good example is afforded by the solution of $CaCO_3$ under high CO_2 pressure. Precipitation of $CaCO_3$ takes place as soon as the pressure is reduced due to nearness to surface when CO_2 gas is released. Common examples are inorganic precipitations of many carbonate minerals in near surface environments and in shallow marine zone. Release of CO_2 affects mineral precipitation in another way by changing the pH from slightly acid condition of the medium to slightly alkaline, other examples of mineral precipitation due to change in pressure are afforded by such volatile bearing minerals as fluorite, tourmaline, apatite (chlor and flor apatite) and the hydrous group of minerals namely zeolites in geodes mostly in trap rocks.

Change in Directed Pressure or Stress

It is well known that in tectonic deformation on a geologic body the pressure may not be uniform. Such non-uniform pressure is known as directed pressure or stress.

Precipitation of many silicate minerals around original grains may take place by this type of process in metamorphic environments. Similarly, the fresh precipitation of sulphide phases may take place around initial grains of sulphide in many sulphide deposits subjected to metamorphism and deformation.

Change in Temperature

Changes of temperature play a dominant role in minerals solution and precipitation. Solubility of many compounds is increased with the rise of temperature except in those systems where a gaseous phase is involved. Temperature also indirectly influences mineral formations by its influence on the chemical properties of a chemical system i.e., pH, Eh, rate of chemical reactions and in many other ways. It is well known that a chemical reaction which takes place with the evolution of heat is called exothermic. In such reactions application of heat or rise of temperature leads to the formation of minerals with their minimum heat of formations. Again, in endothermic reactions application of temperature leads to the formation of minerals with their maximum heat of formation (Le, Chatelier principles). Many sulphide systems show a decreasing solubility with a decrease in temperature. Formation of successive zones of sulphide minerals around an igneous source is directly controlled by the decrease of temperature away from the source. Similarly, in many saline evaporite basins temperature controls whether gypsum or anhydrite will be precipitated. Many sulphide minerals have a number of phases depending on the temperature of formations. A high temperature forms silver sulphide (acanthite) and low temperature forms argentite. High temperature form of Cu_2S is isometric chalcocite and low temperature form is hexa, tetra chalcocite. Rise in temperature decreases insolubility in some cases involving gaseous phases. Example is precipitation of $CaCO_3$ in warm climates.

Electrical Charge

The colloid system on coagulation or floculations leads the fomation of mineral deposits. Colloids are suspensoids or suspension/emulsoids or sols or gels. It is also well known that colloids are electrically charged. Positively charged colloid systems are Al_2O_3, $Fe(OH)_2$, FeO (OH), KOH, NaOH. Negatively charged colloid systems are SiO_2 gels. When positively charged system comes in contact with a negatively charged colloid system, the system is neutralised and there is immediate formation of new minerals, for example formations of hydrated aluminum silicates (clay minerals, oxides-gibbsite, boehmite). Formation of iron-silicate deposits with some aluminum is due to colloid reactions. Precipitation of number of salts may take place whenever the river water comes in contact with the marine sea water containing NaCl as an electrolyte. The electrolyte NaCl floculate the colloidal water.

Grain Size

Grain size is an important physical factor of rocks. It is well known that the grain size affects mineral solubility to a great extent. Smaller grain of a particular minerals are usually more soluble than the bigger grains. Many cases of solution of fine grained silica and its re-precipitation in the form of granular forms are directly related to grain size.

Physico-Chemical Factors in Mineral Precipitation

In nature physical processes act closely in corporation with the chemical processes and in many cases it is not possible to isolate the physical factors from the chemical factors. Hence, most of the processes of mineral formation in nature are due to the combined physico-chemical factors.

Biological factors in Mineral Precipitation

Organisms play a dominant role in the domain of geological processes including weathering, sedimentation, formation of minerals, concentration of minerals, destruction of minerals. These organisms include mostly the microscopic forms of plant life, animal life and bacterial life and it is known from geological history that microorganisms existed on this earth almost right from its beginning to the solid existence of earth. Studies have revealed forms of bacteria, algae, conodonts, acritarchs, and trace fossils right from the upper part of pre-Cambrian rocks (Proterozoic rocks). Doubtful cases of early pre-Combrian life have been reported from some parts of the world. Hence, organisms have played role in the mineral formations right from upper Proterozoic times (Huronian) up to recent.

Plant and Mineral Formation

It is well known that many salts and crystals are synthesized in plant tissues by metabolic activity of various type of plants including modern plants like angiosperms, gymnosperms, pteridophytes and some giant forms of algae. The common minerals that are thus formed by the life activity include oxides (i.e. mostly silica, little iron oxides, rarely manganese oxides), carbonates (i.e. calcite including ferroan calcite, dolomite including ferroan dolomite) and occasionally sulphates (like gypsum, anhydrite etc.) and lastly salts of organic acids (oxalates). The net contribution of plant life in the formation of minerals is negligible as compared to the minerals formed by other process. The general name given to such crystals-minerals is bio-minerals or bio-crystals and the process is known as bio-mineralization or bio-crystallization. Such bio-mineralisation is mostly confined to the shallower surface parts of the earth where terrestrial plants grow. Unusual concentration of carbonate minerals takes place in the root zone of certain plants. These are purely fresh water carbonates. Other forms of mineral salts in microscopic dimensions developed in the roots, underground stems and leaves of certain grasses and leafy vegetables. Fine crystals of SiO_2 developed in grassy blades. A concentration of carbonates is affected by the thallus of giant alga Chara in shallow marine environment.

Animals and Mineral Formation

A variety of stones, minerals and salts are synthesized by various groups of invertebrate animals including modern bivalves, gasteropods, echinoderms and also by the vertebrate groups including man. In case of the invertebrates the mineral forms are mainly carbonates especially the fibrous aragonitic forms and also the finely micro-crystalline chalky form. The carbonates form an external shell and are extracted by the organism from sea water during its life process. In a few cases particularly in organisms like corals, foraminifers, sponges, the carbonates form mostly internal skeleton or frame work but in some cases an external test is developed. These may be carbonates or silica. Hard organic substances (chitin) are also formed as an exoskeleton by some animals of invertebrate groups. Human being also synthesize various salts and under certain abnormal conditions manufacture crystals/minerals inside the human systems. These stones found in human body may be carbonates or oxalates or phosphates. Organic phosphates as well as inorganic phosphates $Ca_3(PO_4)_2$ are synthesized in bones and teeth of vertebrates. These are also known as biominerals or biocrystals. The process of bio-mineralization and bio-crystallisation taking place in various animal groups is a subject of active research in many biological/palaeontological institutes.

Micro-Organisms and Mineral Formation

Microscopic groups of plants and animals are known as bacteria and algae. These micro-organisms have been operative almost since early Proterozoic times and doubtfully even earlier. They are active even today in various environments like lakes including saline lakes, springs, shallow coastal zone and inland seas. Although their activity is somewhat limited within a certain range of temperature and pH their total contribution to the mineral budget is significant.

For example minerals precipitated by bacteria and algae are ZnS (sphalerite), PbS (galena), sulphides of Ni, Co, V, Mo, Fe and Cu. Oxides being formed by the activity of micro organisms are silica (Chert), hematite, limonite, bog iron ore, MnO_2 (pyrolusite & psilomelane), uranivite, hydro-magnesite, dolomite, iron-rich carbonates (brunerite, ankerite, siderite), spathic iron and, sulphates formed by micro-organisms (gypsum, anhydrite, baryte, strontianite).

Bacteria and algae responsible for mineral formation are:

1. Iron bacteria (Chlamydo-bacteria), leptothrix.
2. Desulphurization bacteria-produces H_2O from sulphate minerals (desulphuricans) and sulphuretum, known as C-heterotrophic bacteria.
3. Thiorhodaceae (a) sulpher purple microbes-precipitates S and SO_4 from H_2S under negative Eh. (b) Chloromicrobes.
4. Beggiatoaceae - Thiobacteria thioparus precipitates S and SO_4 from H_2S under positive Eh known as C-autotrophic micro-organisms.

Fungi: It works under aerobic and anaerobic environments. By their activity iron which is mobilised and transported as salts of organic acid or complex is precipitated again as sulphides, $FeCo_3$ or as Fe $(OH)_3$ depending on pH or Eh after breaking up organic iron compound by their life activity.

Algae: Blue-green algae (Cyanophyceae) :

These groups of algae are photolithotrophic and grow abundantly in shallow water:

Phytoflagellates: They precipitate Fe $(OH)_3$ as limonites, goethites etc.

Chemo-lithotrophic: These are aerobic in nature having pH range of 25 to 6. They oxidise and precipitate $Fe_2(SO_4)_3$ and $Fe(OH)_3$ from solutions containing iron sulphides.

Mixo-trophic: (Gallionella (Leptothrix), siderocapsa, metallogenium). They are aerobic to micro-aerophilic, with pH range 6 to 8. They oxidise iron brought in from iron bicarbonate solutions and precipitate it in the form of Fe $(OH)_3$.

Following are some examples of deposits in which role of micro-organisms is evident.

Sulphide Deposits

Role of sulpher micro-organisms in the formation of sulphide deposits have been reported from different parts of world. Jejunow, W.M. (1897) described bacterial colonies and microbes particularly from modern environment in the form of suspended bacterial plates (Chemo-lithotrophic type) of colourless sulpher microbes. These bacteria developed in fluctuating transition zone between the oxygen containing surface water and H_2S containing low level water. He also observed the pattern and continuity of the process of precipitation of iron and copper sulphide.

In geomicro-biological studies of the genesis of Mansfield copper shale (Permian) in Germany heavy metal concentrations in the oceanic sediment of reservoirs have been observed. Also there are examples of biological sulphide precipitation in an environment of hot brines in Red sea. Similar precipitation has been observed in hot sulpher lakes. Micro-organisms thus can contribute to the formation of sedimentary sulpher and sulphide deposits under anaerobic or aerobic conditions. There are evidences of pyrite microspheres from Mississippi valley Pb-Zn deposits similar to the bacterial colony. In India definite evidences of micro-organism activity are present in various pre-Cambrian sulphide deposits such as Khetri, Rajpur-Dariba, Chitradurg etc., and isotopic studies have also shown definite contribution of bio-genic sulpher in the formations of some of these sulphide deposits.

Oxide Deposits

Classical chemo-lithotrohpic to mixotrophic $Fe(OH)_3$ precipitating iron bacteria are well known since the work of Winogradsky (1880) on iron bacteria. Many other micro-organisms contribute to the precipitation of Fe and MnO_2. Relation between these micro-organisms and mineral formation is still a subject of great controversy. Banded iron stones and ferruginous chert of pre-Cambrians of S-Africa, USA, Brazil, Australia and India give doubtful activity of bacteria in the precipitation of both, chert and iron oxides. Oolithic SiO_2 rich and sometime carbonate rich deposits of Mesozoic (Jurassic) have been found to contain evidences of bacterial activity particularly in the form of oxides probably of organic origin. Again, iron silicate of early Palaeozoic (ordovician/silurian) also have yielded remains of bacteria and algae. Carbon containing flints of pre-Cambrian deposits in banded hematite have been found to contain rich residue of cyanophyceae and bacteria. Cloud (1968) has described the possible contribution of microorganisms in the formation of pre-Cambrian iron oxide deposits. In India, role of cyanophyceae in the formation of rich phosphorite deposits has been found in pre-Cambrian Aravalli Matoon formation of Jhamarkotra and other places in Udaipur district. Association of algal structures in the form of stromatolytes with the phosphorite deposits is too consistence to be dismissed as a chance coincidence. Similarly, association of $MgCO_3$ and phosphatic $MgCO_3$ in pre-Cambrian calc-zone of Almora and Chamoli district with stromatolytes is a pointer to the role of algae in the precipitation of rich carbonates.

Micro-organisms and Evaporite Deposits

Occurrence of anaerobic desulphurization process is reckoned to the date of pre-Cambrian. So far oldest indirect evidence is the occurrence of black carminative with pigmented colours with FeS and sheath of iron bacteria transformed into FeS in Permian soil deposits. Recent occurrence in sediments and brines may correspond to the occurrence in salts, brines and sediments of palaeozoic deposits. In the sediments of the highly concentrated continental salt lakes these bacteria yield H_2S to permit development of population of sulphurettum which can tolerate high salt content, high temperature and high pH values in association with the transition from sulphate lakes to carbonate lakes.

6

Classification of Mineral Deposits

Many schemes of classification of mineral deposits have been made by various workers based mainly on size, shape, form, mode of origin, value etc. Many of them are not much useful now. Around the beginning of 20th century much attention was directed to the schemes of classification mainly based on origin. Examples of them are as shown in Table 6.1.

Table 6.1: Early Classification Schemes

Beck, 1904	*Bergeat-Stelzner, 1904*	*Irving, 1908*
1. Primary	Protogene	Bedrock deposits
A. Syngenetic	Syngenetic	Syngenetic
(i) Magmatic Segregations	With eruptive rocks	Igneous
(ii) Sedimentary ores	With sedimentary rocks	Sedimentary
B. Epigenetic	Epigenetic	Epigenetic
(i) Veins	Cavity fillings	Cavity fillings
(ii) Epigenetic deposits not veins	Replacements	Replacements
	points not veins	Contact-metamorphic deposits
2. Secondary deposits	Secondary	Disintegration deposits
A. Residual	Residual	Mechanical
B. Placers	Placers	Chemical

A. *Syngenetic : Those formed at the same time as the enclosing rock.*
B. *Epigenetic : Those formed later than the enclosing rock.*

Beck in 1909 changed his classifications as follows:

1. Magmatic segregations
2. Contact metamorphic deposits
3. Fissure veins
4. Bedded deposits
5. Stocks

6. Secondary alterations
7. Sedimentary ore deposits
8. Detrital deposits.

Lindgren's Classification (1911)

This is a genetic scheme of classification of ore deposits (Table 6.2) which proposes two broad types of deposits :

1. Deposits formed by Mechanical Processes.
2. Deposits formed by Chemical Processes.

Table 6.2: Genetic Classification

Process	*Temp °C*	*Pressure*
1. In surface waters		
(i) By reactions	0-70	Medium to high
(ii) Evaporation		
2. In bodies of rocks		
(i) Concentrations of substances contained within rocks		
a. By weathering	0-100	Medium
b. By groundwater	0-100	Medium
c. By metamorphism	0-400	High
(ii) By introduced substances		
a. Without igneous activity	0-100	Medium
b. Related to igneous activity		
(a) By ascending waters		
(i) Epithermal deposits	50-200	Medium
(ii) Mesothermal deposits	200-300	High
(iii) Hypothermal deposits	300-500	High +
(b) By direct igneous emanation		
(i) Pyrometasomatic deposits	500-800	High +
(ii) Sublimation	100-600	Low to Medium
c. In magmas by differentiation		
(i) Magmatic deposits	700-1500	High +
(ii) Pegmatites	575 ±	High +

Lindgren in 1922 proposed an additional terminology as follows:

3. Deposits dependent upon the eruption of igneous rocks:

Hydrothermal deposits

(a) Epithermal
(b) Mesothermal
(c) Hypothermal

Emanation deposits

(a) Sublimates
(b) Exudation veins, surface type
(c) Pyrometasomatic deposit
(d) Exudation veins, deep seated type

Magmatic deposits

(a) Orthotectic
  (i) Differentiation *in situ*
  (ii) Injected
(b) Pneumotectic
  (i) Differentiation *in situ*
  (ii) Injected

Beck and Berg Classification (1922)

They presented a classification which is as follows :

1. Magmatic ore segregations
2. Contact deposits
3. Epigenetic sulfide deposits
4. Ore veins
5. Epigenetic ore stock works
6. Gossan formations.

Niggli Classification (1925)

He classified on the basis of "plutonic" and "volcanic" similar to that made for igneous rocks which is as follows :

1. Plutonic	A. Hydrothermal B. Pegmatitic-pneumatolytic C. Ortho-magmatic
2. Volcanic	A. Exhalative to hydrothermal B. Pneumatolytic C. Ortho-magmatic

Schneiderhohn Classification (1932)

Later on more extended genetic classification was introduced by Schneiderhohn as below:

(a) *Magmatic rocks and ore deposits*

(a) *Intrusive magmatic*
  (i) Intrusive rocks and liquid magmatic deposits
  (ii) Liquid magmatic – pneumatolytic

(iii) Pneumatolytic
1. Pegmatite veins
2. Pneumatolytic veins & impregnations
3. Contact pneumatolytic

(iv) Pneumatolytic hydrothermal

(v) Hydrothermal

(b) *Extrusive magmatic*

(i) Extrusive-hydrothermal

(ii) Exhalation

(b) *Sedimentary deposits*
1. Weathered zone (oxidation, enrichment)
2. Placers
3. Residual
4. Biochemical-inorganic
5. Salts
6. Fuels
7. Descending groundwater deposits

(c) *Metamorphic deposits*
1. Thermal contact metamorphism
2. Metamorphic rocks
3. Metamorphosed ore deposits
4. Rarely formed metamorphic deposits

Bateman Classification (1942)

He gave a classification which is based on genesis as under:
1. Early magmatic deposits
2. Late magmatic deposits
3. Pegmatitic deposits
4. Contact metasomatic
5. Hydrothermal deposits
6. Submarine exhalative
7. Chemical sedimentation deposits
8. Mechanical concentration deposits
9. Residual deposits
10. Sublimation
11. Bacteriogenic
12. Supergene enrichment deposits
13. Deposits formed by metamorphic process.

Mac Diarmid Classification (1964)

More recent and accepted classification of ore deposits is proposed by Mac Diarmid.

Igneous deposits:

1. Magmatic segregation deposits
2. Pegmatites
3. Igneous metamorphic deposits
4. Hydrothermal :
    - a. Hypothermal
    - b. Mesothermal
    - c. Epithermal
    - d. Telethermal
    - e. Xenothermal
5. Sedimentary deposits
    - (i) Mechanical accumulation
    - (ii) Chemical precipitates
6. Deposits due to weathering
7. Supergene sulphide enrichment
8. Metamorphic deposits

Mineral deposits, on the basis of their origin are classified into two types: **1. Primary or Hypogene deposits:** These are formed at depth by magmatic action, ascending solution, heat and pressure.

2. Secondary or Supergene deposits: These are formed at or near surface by the action of weathering, descending solutions and surface water.

Table 6.3: Genetic Classification (after Sharma and Ram)

Type of Deposit	*Temperature (°C)*	*Pressure*	*Example*
1. Magmatic	High (700-1500)	High	Platinum, Chromite, Magnetite, Ilmenite, Ni-Cu Sulphide
2. Pegmatite or Pneumatolytic	High (575)	High	Mica, Feldspar, Beryl, Lithium minerals and Cassiterite
3. Pyrometasomatic or Contact Metasomatic or Replacement	Elevated (500-800)	High	Cassiterite, Wolframite, Magnetite, Graphite and Sulphide of Iron-pyrite (FeS_2), Pyrrohtite (FeS); Cu; Pb-Galena (PbS) and Zn-Sphalerite (ZnS)

(*Contd...*)

4. Hydrothermal: (a) Cavity filling (b) Replacement	(i) Hypothermal: High (300-500)	High	Cassiterite (Sn deposit), Molybdenite (Mo deposit), Gold (Audeposit)
	(ii) Mesothermal : Intermediate (100-300)	High	Cu-Pb-Zn deposit
	(iii) Epithermal : Low (50-200)	Medium	Native Cu, Native Ag, Sulphides of Hg, As, Sb
	(iv) Telethermal :	Low	U-V and Pb-Zn deposits
5. Volcanic and Fumarolic	Low	Low	Sulpher
6. Meteoric or Circulating water	Moderate to Shallow depth: Low (0-100)	Moderate	Phosphate, Baryte, Pyrite, Celestite, Flint
7. Residual	Low (0-100)	Lower Moderate	Residual Clays, Bauxite and some Fe and Mn Ore deposit
8. Sedimentary			
(a) Evaporation			Salt, Gypsum, Anhydrite, Alkalis, Nitrates and Borates
(b) Chemical Reactions			Sulphides, Fe and Mn Ores, Limestone
(c) Accumulation of Material Inorganic or Organic			Clays, Limestone, Sandstone, Coal, Oil Shale, Diatomite and Chalk
(d) Detrital or Placer			Au, Pt, Sn Ore (Cassiterite), Ilmenite ($FeTiO_3$), Monazite, Gemstone
9. Oxidation and Secondary Enrichment			CO_3^{-2} and SO_4^{-3} of Cu and Pb, Sulphides of Cu-Borates, Covellite, Chalcocite

Modern Classification

Following is a simple genetic classification that is usable for the field or laboratory (Table 6.4):

Table 6.4: Genetic Classification

Process	*Deposits*
1. Magmatic concentration (high T and P)	1. Early magmatic A. Disseminated crystallization B. Segregation C. Injection 2. Late magmatic A. Residual liquid segregation B. Residual liquid injection C. Immiscible liquid segregation D. Immiscible liquid injection
2. Sublimation (low T and P)	Sublimates
3. Contact metamorphism (low-high, T&P)	Contact metasomatic iron, copper, gold etc.
4. Hydrothermal processes A. Temp and Pressure (low to high) 1. Telethermal 2. Epithermal 3. Leptothermal 4. Mesothermal 5. Hypothermal 6. Xenothermal	
B. Cavity filling	Open space deposits A. Fissure veins B. Shear zone deposits C. Stock work D. Ladder vein E. Saddle reef F. Tension-Crack filling G. Breccia filling (a) Volcanic (b) Tectonic (c) Collapse
C. Replacement (low to high T&P)	Replacement deposits A. Massive B. Load fissure C. Disseminated
5. Sedimentation exclusive of evaporation, (low T&P)	Sedimentary : Iron, manganese, Phosphate, etc.
6. Bacteriogenic	Bacterial product or reduction
7. Submarine exhalative Volcanism, (low T&P)	Submarine volcanic
8. Evaporation (low T&P)	Evaporites A. Marine B. Lake C. Ground water
9. Residual and mechanical concentration, (low T&P)	
A. Residual concentration	Residual deposits
B. Mechanical concentration	Iron, manganese, bauxite, placers A. Stream B. Beach C. Eluvial D. Eolian
10. Surficial oxidation and supergene enrichment Oxidized, supergene sulphide. (low T&P)	
11. Metamorphism (low to high T&P)	A. Metamorphosed deposits B. Metamorphic deposits.

Genetic Classification according to mode of Origin

Hydrothermal Deposits

These deposits have been classified into 3 categories :

1. Epithermal Deposits: Epithermal deposits include those deposits of hydrothermal origin which are formed more or less in the shallower portion of Earth crust within a depth of 2000 ft from the surface. Because of the shallow depth of formation, these deposits are preserved mostly in the younger rocks of the geological period *i.e.*, tertiary. The connection with the parent source igneous rocks is found in many cases while in only a few examples the direct igneous relationship can not be established. In their mode of occurrence these deposits mostly occur as vein fillings with all possible types of open space structures such as vugs and crustified veins, both symmetrical as well as asymmetrical and the breccia filling. Replacement deposits are also present but to a minor extent. Mineralogically these deposits mostly consist of self arsenites, self-antimonites, tellurites, selemites of silver, gold and rarely base metals. The others include antimony, stibnite, and Hg Cinnabar, Native Hg, in the case of gold and silver Electrum (Ag + Au). The gangue minerals are commonly carbonate, dolomite, rodochrosite and silica (low temp). High temp minerals like pyroxenes, tourmaline garnets are usually absent. Wall rock alteration is typical for these deposits and includes chloritization, pyritization, sericitization and propylitization (propylite = mixture of sericite, albite, chlorite silica, pyrite + epidote). Texturally these ores are often fine grained unless they recrystallize by later metamorphism. Structurally the deposits are emplaced along faults, fissures and brecciated zones. Colloidal structure is commonly found and given evidence of low temperature deposits of 50 to 100°.

2. Mesothermal Deposits: Mesothermal deposits include, as the name indicates, those deposits of hydrothermal origin which are formed at intermediate depths and at moderately high temperature. In many instances, they exhibit intermediate characteristics between deep seated hypothermal and relatively shallow epithermal deposits. These deposits in general show some characteristics more or less similar to mesothermal deposits particularly in their lower part and some character similar to epithermal deposits in their higher parts.

The depth of their original formation ranges from 300 ft below the surface to 8000 ft, and the temperature range varies from 350° to about 200°C. In their mode of occurrence they characteristically show typical vein and fracture fillings particularly in brittle host rock like quartzites. They also occur as replacement deposits in carbonate host rocks. In their structural setting these deposits are usually found in highly folded and faulted pre-Cambrian rocks affected by Laramide orogeny. Besides sedimentary rocks, igneous rocks like monzonites, diorites and monzodiorites may also serve as the host rocks while the metamorphic host rocks include quartzites and rarely schists. In their mineralogy these deposits principally contain silver, lead, zinc, copper and some uranium, bismuth, arsenic etc. Common ore minerals include pyrites, chalcopyrite, arsenopyrite, galena, sphalerite, uraninite, gold and silver. The commonly found gangue minerals are quartz particularly ribbon quartz formed by the alteration of laminated and foliated host rock silica, carbonates, serricite, chlorite, orthoclase and rarely baryte. High temperature minerals such as tourmaline, topaz, pyroxene, and garnet are usually absent while low temperature minerals such as zeolites are not found.

Most of these deposits show a zoned structure both in time as well as space on a regional scale as district scale as local scale. This zoning is caused by the thermal gradient or is controlled by the wall rock chemistry.

Wall rock alteration: The typical wall rock alteration shown by these deposits includes sericitization, which produces a bleached zone. Chloritization, pyritization, feldspathization (orthoclase), silicification are also common.

Ore host rock relation: Ore may be emplaced prior to the main orogenic deformation of the host rock (pretectonic ore) or it may be more or less simultaneous with the tectonic event (syntectonic relation) or the ore may be developed much later than the organic deformation (post-tectonic ore).

Origin: Geothermometric and Geobarometric studies show clearly that mesothermal ores are formed at moderately high temperature (200 to 350°C) and moderately high pressure (3 to 7 Kb). In many cases a connection of ore deposits with a parent intrusive rock can be established. While in some there may not be any apparent connection between the deposits and intrusive body. Paragenetic studies have shown that in most cases pyrite and arsenopyrite are the earliest sulphides to be crystallized followed by sphalerite and galena. While copper sulphides are the last minerals to crystallize. The gangue minerals may be deposited in all the three stages *i.e.*, pre ore gangue, syn ore gangue and post ore gangue. Wall rock alteration studies have shown that alteration may take place prior to the period of metalization or it may be synchronous with the main period of mineralization or some time wall-rock alteration may post date the main phase of mineralization. Studies of texture and structure have shown that these deposits may show both cavity filling structure as well as replacement structure. Study of ore intrusive body relationship shows that while in many cases the basic intrusives are mineralized as they are pre ore, in other cases the intrusive like lamprophyres are post ore and hence not mineralized. Regarding metamorphic effects on these ores, they are not prominent because the ore and the gangue are already formed at high temperature and pressure.

3. Telethermal Deposits: Telethermal literally means away from thermal heat source. These are the deposits which are formed far away from the igneous intrusive source. In fact in many cases, connection between igneous activity and the ore deposition can not be established in these deposits. There are surface-shallow depth deposits having many features related to ground water circulation.

Mode of occurrence. Most of these deposits occur as vein fillings with the typical structures like open veins, comb structure, crustified veins etc. These deposits are also found to occur in the pore space of sandstones and conglomerates and some times as a cementing material of the rock also. In the case of limestone host rock replacement type of deposits are also found.

Texture and Structure. Texturally these deposits may vary in size from extremely fine grained to very coarse grained, indicating these by the influence of crystallization from ground water. Open work structures such as cavity, breccia fillings, veins with comb structure, vugs are commonly encountered. Disseminated replacement structures are also some time met with.

Mineralogy. Mineralogically these deposits consist mostly of shallow oxides and hydroxides of metals like uranium, vanadium, molybdenum, and some time also sulphides of various metals.

U- Minerals (secondary)

Carnolite (K_2 $(UO_2)_2$ $(VO_4)_2$ nH_2O)

Tyuyaumenite ($Ca(UO_2)_2$ $(VOC_1)_2$. nH_2O)

Autunite (Ca (UO_2) $(PO_4)_2$ 10-12 H_2O)

Torbanite (Cu (UO_2) $(PO_4)_2$ 8-12 H_2O)

V- Minerals	Hewettite (CaV_2O_{16} $9H_2O$)
Mo- Minerals	Ilsemannite (Mo_3O_8. nH_2O)
Cu- Minerals	Tenonite CuO
Cuprite Cu_2O	
Fe- Minerals	Limonite FeO (OH)

Gangue minerals are commonly calcite, carbo dolerite, fluorite, in these deposits. In some telethermal deposits we get lead and zinc.

Wall rock alteration: Typical wall rock alteration found in the case of epithermal deposits is absent here. The only local alteration includes pyritization, silicification, carbonatization. In the case of porous sediments quartization is found. Bleached zones so commonly found in high temperature hydrothermal deposits are not present here.

Origin. Telethermal deposits represent the most controversial deposits in economic geology because of following reasons :

1. Relationship with any igneous intrusive rock is rarely found. However, in certain deposits there is presence of typical hydrothermal minerals as molybdenite and metal sulphides like galena, sphalerite, chalcopyrite, pitchblende. All these indicate some hydrothermal activity particularly at deposits far away from any igneous intrusive bodies.
2. In most of the cases evidence of slightly heated meteoric or ground water evidences of the activity are present. These evidences include association of these ores with zones of leaching from such metal bearing bodies such as metal bearing sandstone or granite.
3. In certain cases there are evidences of syngenetic or diagenetic precipitation of the metal along with the enclosing sediments. Some field geologists say that it is a typical type of syngenetic deposits.
4. These deposits are formed by the percolation from the overlying or inter layer tuffs or volcanic rocks.

The problem of origin can not be solved unless age determinations are made of the ore deposits as well as of the sediments. If the ore deposits are found much younger than the enclosing host rock then there is possibility of their being epigenetic in origin.

7

Forms and Modes of Occurrence of Ore Deposits

Ore deposits of diverse origin occur in different types of host rocks and in different types of structural environments. Like any other geologic body they show a multitude of forms, some simple, some complex and some which can not be put in any other category. Forms can be considered from the point of view of the structural background of the host rock in which the deposits occur:

(a) Forms in structurally simple host rock *i.e.*, horizontal beds, simple folded beds, simple faulted beds.
(b) Forms in structurally complicated host rocks *i.e.*, host rocks with tight oppressed folds, host rocks with refolding and cross folding, host rocks with a complicated systems of faults, fractures and joints.
(c) Forms which cannot be put in the above categories.

Forms can also be classified in relation to their attitude with respect to the host rock. There are again three categories:

(1) Forms which are dominantly concordant with the attitude of the planer-elements like bedding planes, foliation planes etc.
(2) Forms which are transgressive or discordant with the attitude of the S-planes.
(3) Forms which are partly concordant and discordant.

Forms can also be considered in terms of the genesis of the ore deposits:

(a) Forms in syngenetic ore deposits.
(b) Forms in epigenetic ore deposits.
(c) Forms in lateral seggregated types of ore deposits.

(d) Forms in the metamorphosed and mobilised ore deposits.
(e) Forms in other genetic types.

Some other types of form :

(1) Veins, Veinlets, Stringers, Lodes, Streaks
(2) Crustified veins (symmetrical and asymmetrical) with vugs
(3) Lensoid types
(4) Stockwork, saddle reefs, ladder veins
(5) Disseminated forms, impregnations
(6) Encrustations
(7) Pipes, necks, plugs
(8) Seggregations
(9) Bedded, laminated
(10) Pitches and Flats
(11) Replacement
(12) Blanket or sheet
(13) Shoe Strings
(14) Pockets, sacs
(15) Beaded
(16) Dykes and sills, lacolith, Phacolith, lopolith, bysmoliths
(17) Irregular forms.

Some of the important forms have been discussed here.

Crustified Veins

Crustified veins are supposed to be typical form of a few epigenetic hydrothermal deposits where deposition takes place in more or less regular fashion in open channels and in open cavities. Crustification refers to the development of typical toothed appearance in the quartz veins and associated ores. The internal arrangement of the ore mineral and the accompanying quartz may be symmetrical with respect to the walls of the cavities or their arrangement may be highly asymmetrical. This is open space available in the upper surface of the earth's crust. This is empty vertical cavity and known as drusy cavity or vug. There may be two possibilities, an ore vein (marcasite or pyrite) may run continuously or may dilate due to tectonism. This is first generation quartz.

The crystals are punched on the pyramid of second generation. It may be possible that there may be second stage old dilation or there may be deposits of magnetite vein between the second and third generation of quartz. Excellent examples of simple crustified quartz veins are commonly found in the sulphide region of Jharkhand (Mosabani, Siddheswar block).

Stockwork, Saddle reef, Ladder veins

In epigenetic deposits the mode of occurrence of the mineralisation is controlled among other factors by the tectonic deformation of the host rock.

Saddle reef: It is a very characteristic occurrence of veins particularly quartz veins (reef) carrying particles of gold and occurring in acutely folded schistose rocks. The kolar gold field is the typical example of similar to saddle reefs.

Ladder veins: It is a mode of occurrence of epigenetic hydrothermal deposits. This form is controlled by the attitude of the fracture planes in the host rocks. Fractures are developed in rocks by tensional forces and they may develop in a different direction. In a vertical body of igneous rocks there may be transverse fractures. If these fractures are mineralised in host rocks then mode of occurrence will look like a ladder. Crystal asbestos vein filling in the form of ladder veins in the serpentinite ultrabasics of Singhbhum, Jharkhand and Kuliphum, A.P. are good examples.

Stock Work: Again in the domain of the epigenetic mineralisation crusting and granulation may play an important role in controlling the form of mineral deposits. In some cases joint systems develop in the host rocks. The joint system may sometimes show a dendritic or reticulated system. These joints and fractures are mineralised. This type of mode of occurrence of mineral deposits is known as stockworks. The veins are thin and paper like. Such type of deposits is only economical for Ag, Pd and some Pb etc. for example copper deposits of Malanjkhand in Bilaspur district. This is sulphide ore filling in the quartz, silver, cobalt, bismuth, copper epigenetic deposits in Germany.

Pipes, Necks and Plugs

This is usual mode of occurrence of plutonic igneous rocks which have been tectonically forced to be pushed up slightly. It also refers to actual sulphide ore deposit which is not in the form of veins but in the form of chimney or cigar shape. Such ore body is known as pipes or plugs. Neck is not common.

Pipes may be vertical, inclined and sometimes may be horizontal which is rare. Pipe shape of ore body may be due to its limitations in the occurrence of the epigenetic ore material along the intersection of the fractures or along the intersection of prominent fault and their fractures or even such pipe shape may result due to the intersection of bedded deposit and a fault. In the case of country rocks pipes may be due to circular fractures. These pipes or plugs are mostly deep seated and their importance is due to the fact that diamond is entirely confined in the pipe rocks and occurs as disseminated crystals, (example-kimberlite). Similar rocks are found in the form of pipes around Malanjkhand. Some sulphide deposits occur in the form of pipes.

Pitches and flats: They are the mode of occurrence of the epigenetic hydrothermal deposits in limestone country rocks. Carbonate rocks develop in solution cavities, example : Mississippian limestone in U.S.A.

Blanket or Sheet: This is mostly the mode of occurrence of sedimentary or residual types of deposits. Sand may occur as sheet called sheet sand. It is rarely epigenetic, hydrothermal deposits shows this types of deposit. Sheet is layer or strata of ore body. It is used for magmatic deposits. Sheet is used in the hydrothermal deposits called sheet fractures or parallel sheet planes. An enriched surface of Nickel deposit is called 'Mantas.'

Shoe-strings: The sedimentary deposits like sand occur in the form of lace like strings of the shoe.

Pegmatitic ore Deposits

Pegmatitic ore deposits are those formed during the pegmatitic stage of crystallisation of magma. The magma consists of volatiles and non-volatile constituents in the ortho-magmatic and late magmatic deposits. In some cases the volatiles are also present but they are not concentrated and hence they do not exert a critical influence in the formation of these deposits. However, in the non-metallic pegmatite phase of crystallisation, the volatiles become enriched sufficiently to play a dominant role in their formation. Ortho and late magmatic deposits are usually formed at great depths and at low Eh (redox potential) and slightly acidic pH. The temperature range is between 1200°C & 800°C. The contact metasomatic or pneumatolytic deposits are usually formed at moderate depths in a slightly reducing to slightly oxidising environment and with temperature range between 800°C to 400°C. The ore liquid is slightly acidic to alkaline particularly in contact with carbonate rocks. The pegmatite deposits are formed at comparable depths under mostly oxidising condition except those pegmatites containing important sulphide minerals. The pegmatitic fluid is much more liquid and much less viscous than the other types of ore magma and is usually slightly acidic to slightly alkaline. The temperature range of formation is between 700°C and 250°C. It is thus clear that the three stages of magmatic crystallisation overlap with each other in respect of depth of formation, temperature, pressure and in their Eh and pH condition. Some petrologists doubt independent existence of a pneumatolytic stage. Because at temperature of 400°C and above there is no real distinction between a gaseous and a liquid phase. At such high temperature the properties of liquid and vapour phase can not be distinguished and hence a separate high temperature gaseous phase can not be distinguished from the high temperature liquid. In ore geology the pneumatolytic ore deposits are referred to more appropriately as contact metasomatic or igneous metamorphic deposits. For the same reason there is actually no difference between a high temperature pneumatolytic and a high temperature pegmatite deposits.

Pegmatites and their associated ore deposits deserve a special category in ore classification because of their peculiarities of texture, structure and mineral composition. Because of wide range of temperature of formation of pegmatites they have been classified into :

(i) high temperature magmatic pegmatite (500°C), and

(ii) low temperature hydrothermal or replacement pegmatites.

Pegmatite ore deposits have also been classified according to the mineral paragenesis into :

(a) Simple pegmatites

(b) Complex pegmatites

Simple pegmatites are those of one or two minerals. On the other hand complex pegmatites are comprised of a great variety of mineral assemblage sharing complex replacement between earlier and later formed minerals. Pegmatites have also been classified on the basis of their parentage such as granite pegmatite, granodiorite pegmatite, syenite pegmatite etc. Pegmatites of ultrabasic parentage are almost unknown because of the paucity of volatiles in the ultrabasic magma. Similarly volcanic association of pegmatites are unknown. Pegmatite deposits can also be classified according to genesis into three main types:

(i) Magmatic pegmatite

(ii) Hydrothermal replacement pegmatite

(iii) Metamorphic or secretionary pegmatites formed by metamorphic differentiation process during the metamorphism of basic as well as of acid rocks.

These pegmatites are small in size and regular in shape, are mostly uneconomic and are not genetically connected with any intrusive igneous rocks.

Mode of Occurrence

Pegmatites in their mode of occurrence are very similar to the igneous rocks and this is controlled in case of igneous rocks by the

(i) composition of pegmatitic fluid,

(ii) structure of host rocks, and

(iii) inherent force of intrusion.

Pegmatites may occur as dykes, sills, sheets, small lacolith and other irregular bodies. However, the most common mode of occurrence is a vein, similar to the veins of hydrothermal deposits. In fact sometimes it is very difficult to distinguish between a deep seated catathermal vein deposits from a pegmatite deposit. Like the hydrothermal veins the pegmatitic veins show a most complex variation in the vein geometry and in the pattern of the veins, like :

(i) Veins showing pinch and swell structure.
(ii) 'En echelon' veins.
(iii) Veins faulted and foldeld.
(iv) Veins in the forms of detached rocks.
(v) Veins showing parallel pattern.
(vi) Veins showing reticulate pattern.
(vii) Intersecting veins.
(viii) Veins displaced or dislocated due to movement or orogeny.

Texture of the Pegmatitic Ores

Like the pegmatitic rocks, the pegmatitic ore whether metallic or non-metallic, oxides or silicates are invariably coarse and the ore crystals are well formed known as euhedral forms. Fluid inclusions are very commonly found in the pegmatitic ores. Because of large size and beautiful shape of these minerals or pegmatitic crystals, these pegmatites are called as paradise of hunting ground.

Structure of the Pegmatitic Ores

Pegmatites represent the filling of material in open spaces that is in joints, cracks and fissures in the rock. These open spaces may be connected with the surrounding medium or they may represent a close system. In both the cases the pegmatites may show an internal structure known as zoning of pegmatites.

On the other side, the chilled zone is present and below it a continuous zone is present called outer zone. It consists of usual pegmatitic minerals of very coarse grains but is not rich in economic minerals. The innermost zone mostly consists of dark colour quartz known as core zone of pegmatite. Between the central core and outer zone lies the most important zone of pegmatites known as intermediate zone. This may be divided into three or more sub-zones-outer, middle and inner. Economically the contact of intermediate zone with outer zone is more important for the formation of mica books. The entire intermediate zone contains variety of economic minerals, rare earth minerals, precious and semiprecious minerals. So, it is studied in great detail. It is mostly found that zones replace each other. The core zone is replaced by intermediate zone and it is replaced by border zone. Origin of the zone structure may be explained by one or more of the following theories.

Theory of Fractional Crystallisation

In this case the original magma is supposed to crystallise and precipitate minerals according to well known reaction series of Bowen. The quartz represents the last product of crystallistation of pegmatitic magma and the gradual change in the composition from core to the border is correlated with the earlier stages of magmatic crystallisation. Salvage zone represents earliest product of crystallisation and is same as the magma composition. The inner most and intermediate zones show progressively differentiated magma. Pegmatites associated with basic rocks contain basic plagioclase (labroadorite), amphibole and intermediate plagioclase (andesine). The outer most zone consists of only mica, biotite etc.

Contact Metasomatism or Pyrometasomatism

In magmatic cycle an important process by which economic deposits are formed includes contact metasomation, also known as pyrometasomatism, or igneous metamorphism or even pneumatolysis. Contact metamorphism refers to the high temperature matasomatic action in the contact zone of igneous rocks. The term pyrometasomatism indicate metasomation at high temperature due to the heat of intrusive magma. The term igneous metamorphism indicates metamorphic action in the vicinity of intrusive igneous rocks while the classical pneumatolysis stands for pneumatolytic action that is action of various vapours and gases derived from the magma in the country rock. Contact metasomatism is often distinguished from contact metamorphism which represents contact metamorphic action of heated magma on the country rock.

It is similar in many respects to contact metasomatism with the important difference that practically no fluids are introduced in the country rock. The main effect is of heat. Under the influence of heat recrystallisation takes place without the addition or subtraction of materials. A contact 'aureole' is also developed in contact metamorphism but no important economic deposit results from contact metamorphism except the formation of some deposits of andalusite, graphite and garnet. Contact metasomatism differs from the metasomatism in the fact that the metasomatic replacement takes place at high temperature. Metasomatism is essentially a process of simultaneous solution of one component and precipitation of other component in capillary pores so that the structure of the host rock is often well preserved. This metasomatic replacement may take place even at room temperature. We can discuss contact metasomatism under following heads:

Mode of occurrence

Contact metasomatic deposits mostly occur either at the margin of silicic igneous intrusions like granodiorite, tonalite, monzonite or in the immediate contact lying near host rocks preferably of reactive type like limestones, dolomites and impure carbonates. The deposits are often small but highly concentrated and occur in the form of tabular bodies, lenses and pods or irregular masses.

Texture and structure

In texture, deposits are granular (usually coarse grained) and consist of aggregates of anhedral to subhedral crystal. A country rock is also recrystallised to crystalline limestone. The important structure is the development of zones of silicate in vicinity of the igneous body. This is known as 'Skarn Zone' or 'Zone of Tektite'. Many contact metasomatic deposits are located within the skarn or a little away from it. However, the skarn zone is sometimes not developed. Replacement structure is often present. The ore minerals either replace the skarn rock or even the carbonate country rock itself.

Mineralogy

The mineralogy of contact metasomatic deposit is rather simple. The deposits may consist of oxides like magnetite, hematite, ilmenite, rutile. Oxide of tin and wolframite, hubnerites, ferberite, sulphide minerals (pyrite, chalcopyrite, galena, sphalerite), sulphosalts (sulpher), arsenite etc. are rare, so also native metals are. However, the composition of associated skarn rock may be extremely interesting because of the lime with silica giving rise to various calc-silicate minerals like wollastonite which forms at a temperature of above 550°C. Lower temperature lime silicates form when impure carbonates are present including tremolite, actinolite, epidote, zoisite, forsterite. High temperature silicates include anorthosite, larvinite and the few other rare minerals like larnite, monticellite besides silicates typical of skarn zone. There may be the development of special minerals due to the activity of vapours and gases *e.g.* tourmaline, flxinite (boron metasomatism) topaz, fluorapatite (fluorine metasomatism), scapolite family (chlorine metasomatism); Micas (zinnwaldite), iron rich garnet (almandine), lime rich garnet (spessartite), andalusite, grossularite, pyrope etc.

Origin of contact metasomatism

Contact metasomatic deposits are genetically related to igneous intrusive rocks mostly of silicic composition like granodiorite, quartz-diorite, monzonite etc. The reason why they are not associated with basic and ultrabasic rocks is that the latter are relatively poor in volatile content and hence cannot be effective to produce metasomatic reactions. These deposits are also very rarely associated with granitic rocks. The reason is not very clear. Normally the granites contain good amount of volatiles. Their temperature of formation varies from 500°C to 750°C. The absence of contact metasomatic deposits in the contact aureoles of granitic rocks of granite needs further research for classification.

Role of vapour and gases in formation of pegmatite deposits

In the formation of these deposits water-vapour, CO_2 and other gases play a dominant role. Similarly, in case of contact metamorphic deposits vapour and gases are all important. These volatile constituents work in different ways:

(a) They help in the transport of ore materials because of their high mobility. Particles of ores may be carried a long distance riding on the bubbles of vapour and gases.

(b) They facilitate reactions between the ore materials and the country rock, between ore and skarn and sometimes between one ore and another. Reaction and recrystallisation are greatly facilitated in presence of H_2O vapour, carbon dioxide, halogen and other gases.

(c) Vapour and gases combined with metallic ions to form volatile metallic component. *e.g.*

(a) $Fe + Cl_2$ → $FeCl_2$ (in absence of O_2)
→ $FeCl_2$ (in absence of O_2)

These compounds are gaseous at high temperature and carried or transported in vapour form.

(b) $Sn + Cl_2$ → $SnCl_2$
→ $SnCl_4$

Both are in gaseous form and carried to a distance. Similarly compounds of phosphorus and boron are in the gaseous forms.

Role of invaded rocks (host rock)

Although contact metasomatic deposits may sometimes occur in marginal zone of igneous intrusive rocks of appropriate composition and appropriate forms as stocks, batholiths and boss. More commonly they are found in the host rock of appropriate composition (i.e. limestones, dolomites, and limestones + pure dolomites) either in immediate vicinity or they are found in 'skarn' rocks (garnet rocks) developed by the contact action of the intrusive on the impure carbonate rocks. The reason why these deposits are located in carbonate rock may be as follows:

(a) Carbonate rocks are more soluble than rocks like sandstones, shales and other igneous and metamorphic rocks. The solubility causes the formation of pore cavity and open spaces in the country rocks which are simultaneously filled by the ore materials derived from the magma.

(b) Carbonate rocks indicate an alkaline environment of the deposition. So, when low pH ore solutions come in contact with carbonate rocks, the ore solution is neutralised and there is immediate precipitation of ores.

(c) In some contact metasomatic deposits the carbonate host rocks may supply some ingredient of the economic deposits:

(a) $CaCO_3 \rightleftharpoons Ca^{+2} + CO2\uparrow$

$Ca + F \text{ (magma)} \rightleftharpoons CaF_2$

(b) $FeCl_3 + CaCO_3 \rightleftharpoons FeO + Ca + Cl$

$SnF + H_2O \longrightarrow SnO_2$ (Cassiterite)

Pressure and temperature condition

Contact metamorphic deposits are usually formed at shallow to high depth at a pressure of a few kilometers. Near surface contact metamorphic deposits are not known probably because gases and vapours escape. Similarly at very great depth (5 km) contact metamorphic deposits do not form because the gases and vapours are held in solution at these depths. Hence, these contact metamorphic deposits are mostly associated with Mesozoic and Tertiary igneous intrusives and not with Pre-cambrian. Temperature range for these deposits is 400-800°C and from this point of view contact metamorphic deposits may very well resemble the katathermal (deep seated hydrothermal) deposits. The higher temperature of formation of 800°C makes them at times resemble the true magmatic deposits or liquid magmatic deposits and ortho-metasomatic deposits. Vapour pressure is always high during this formation. In this respect they very much resemble the high temperature magmatic pegmatites which are formed at more than 400°C. Examples are hematites and magnetite of Fierra, New Mexico; Sulphide deposits of Bisbee (U.S.A) and tin deposits of Burma.

Ore solution, be sideway or downward or even upward, important mineralization may take place at a junction of porous or pervious beds with impervious beds *e.g.* junction of sand with shale. Mineralization is confined in many regions towards the bottom as junctions. Mineralization may be localized along that part of the bed which is less porous than the adjoining portion.

Chemical Control

Chemistry of host rock along with its mineralogy controls localization of many ore deposits both of syngenetic as well as epigenetic types. In many instances it has been found that a mineralization which passes through igneous rocks like monzonite through granite and gabbro is maximum enriched through monzonites. A mineralization passing tuffs, basalts and rhyolites, the maximum concentration is found only in tuff portions, the reason is probably chemical.

Dolomites contain Pb-Zn sulphides. It is found that ZnS is mostly confined to siliceous portion of dolomite. In other portions (non-silicified) Pb is more common. Silica grains in dolomites have caused precipitation of sphalerite and silica rich portion.

8

Structures, Tectonics and Mineralization

In the case of epigenetic deposits whether late magmatic or hydrothermal or igneous or metamorphic, structure and tectonics play a dominant role in localisation and emplacement of these desposits. Tectonics include such broad regional structures as geanticlines, geosynclines, regional faults and thrusts, regional folds and junction of plates in the modern concept of plate tectonics. Structure includes local field features and deformations as small scale folds, faults, fractures and joints, plunge of folds and foliation planes including bedding planes.

Aspects of Mineralisation

(a) A look at the metallogenic map of various continents in which distribution of various metallic and non-metallic deposits in space and time is shown clearly reveals that the major distribution of the deposits is controlled by such large scale tectonic features as orogenic belts (belts of folded mountains), geosynclinal basins, regional faults and thrust, junction of plates and other tectonic features. Similarly, a study of the distribution of mineralisation in the mineralised districts of any country clearly demonstrates that epigenetic deposits are located either along the bedding planes or other types of S-planes of rocks or around the crest of folds or along local fault planes or along fractures and joints or even along the hinges or terminations of plunging folds. There is another aspect of the structural studies of mineralisation.

(b) Ore deposits whether syngenetic or epigenetic can be affected by folds, faults and thrusts. Effect of these structures on ore deposits is many fold. Their studies are of great significance not only from theoretical point of view but of great practical significance in mining geology where ore veins or beds may be abruptly terminated against a fault and location of the missing portions poses an accute problem for mining geologists. The effects may be beneficial or disastrous. In the case of supergene sulphide enrichment process and superficial oxidation process structures of the host rocks along with other factors have a dominant role in the process of enrichment.

(c) A third aspect of the study of structures relates to the investigation of depositional characteristics and metamorphism of ore deposits whether syngenetic or epigenetic. Such deformation and metamorphism induce the development of new textures, new structures and new mineral assemblages in the deposits. Thus, deformed just as the case with the deformation of other geologic body, this aspect of study assumes special importance in the case of pre-Cambrian deposits during which period majority of mineralisation has taken place throughout the world. Such metamorphosed, deformed and mobilised ore deposits form a class by themselves and need special methods of investigations.

Thus, in relation to the kinematic relationship of the ore deposits with tectonism there may be three types *i.e.* (i) deposits may be pre-kinematic *i.e.* the mineralisation has taken place more or less before the tectonic events, or (ii) deposits may be syn-kinematic with respect to deformation, (iii) the deposits may be post kinematic in which the deformation episodes have preceded the mineralisation.

There may be gradational relationship between mineralisation period and period of tectonism. In a mining area there may be more than one period of deformation and folding.

Structural and Other Controls of Ore Localisation

Structural controls of ore localisation discuss various structures of the host rock in relation to localisation of epigenetic ore deposits. These controls play a relatively minor role in the localisation of syngenetic magmatic or syngenetic sedimentary deposits. Syngenetic magmatic includes the early magmatic deposits which have crystallised along with the enclosing intrusive magmatic rocks like bedded chromite deposits, bedded Cr, Pt and Ti deposits, disseminated deposits of diamond. Syngenetic sedimentary includes typical chemical precipitate, residual deposits, placer deposits, evaporite and deposits formed by superficial oxidation and enrichment processes.

Structural Control of Epigenetic Deposits

The controls of structures in the localisation of the deposits can be studied from two point of view:

(a) Regional structural control

(b) Local or detailed structural control

Regional or broad structural controls

This type of controls determine why some terrains of earth contain epigenetic mineral deposits while other regions are free or devoid of mineralisation. The controls by the regional or broad structure is necessarily ill defined that means the control is less significant. A look at the metallogenic map of any part of the world showing the relation of mineralisation in time and space is convincing enough to demonstrate that broad zones of mineralisation coincide with the broad or regional structures like orogenic belts, regional faults, regional thrusts and regional folds including igneous intrusion.

Orogenic belts or belts of fold mountains: Some of the prominent orogenic belts of pre-Cambrian include Turonian (Proterozoic). Appalachian orogenic belts of palaeozoic include caledonian, variscan, hercynian. Mesozoic and Cenozoic orogenic belts include Alpine, Andes, Rockies and Himalayan orogenic belts. In India there are atleast three pre-Cambrian orogenic belts *i.e.* Dharwar orogeny, E-ghat-Satpura orogenic cycle in between Singhbhum orogeny correlated with Aravalli orogenic cycle, (Delhi). Pre Aravalli is correlated with lower Dharwar. Representative of the equivalent of variscan, caledonian, hercynian are not well developed. The representative of Alpine is Himalayan orogenic belt. In our country mineralisation is found broadly controlled by pre-Cambrian orogenic cycle and again there is mineralisation in Himalayan orogenic belts. For example Kolar Gold Field (KGF), chromite belt, nickeliferous chromite are related to E-ghat-Satpura orogenic cycle and Cu-Pb-Zn of Rajasthan with 'U' and 'Fe' in Singhbhum orogenic belt. Pre-Cambrian Fe and Mn belts of India are also distinctly related with the doubtful iron-ore geosyncline or iron-belt geosynclines. Himalayan sulphide deposits are related to Himalayan orogeny.

Regional faults and ore localization: Regional faults are the large scale faults running for 100s of Kms and usually of deep seated origin. Such regional faults are loci of igneous intrusion as well as mineralization. For example, there is a major fault in Michigan state (USA) along which mineralization is localised. In the peninsular shield a large number of deep faults have been located by aerial mapping. These faults have been found to be coinciding with the location of carbonatite intrusive and associated mineralisation (fluorite, columbite etc.)

Regional Thrust: Thrust planes are usually low angle reverse faults and they mark the junction of two or more orogenic terrains. According to plate-tectonics hypothesis, such thrust planes represent lines of suture indicating the junction of plates where one plate has gone down below the other (subduction) and in tectonics it is known as underthrusting. Examples of a prominent thrust plane controlling the localisation of sulphide minerals as well as apatite, magnetite is the nearly 800 Km long Singhbhum thrust plane where deposits of Cu, U and apatite are located.

Regional folding and intrusion

Folding on a regional scale is usually associated with the intrusion of Batholiths which are the source of the multimetals. These batholiths are usually of granite, monzonitic or granodioritic composition and occur in highly folded belts. Examples are Sierra Nevada batholith of USA, Singhbhum granodiorite batholith of India, Erinpura granitic batholith and Tertiary batholith of Karakoram mountains.

Local or detailed structural controls

Detailed or local structural controls explain how ore deposits are localised only in certain parts of regional structures. Detailed structural control includes the local structures like local folds, faults and thrusting, local fractures and joints and locally developed other types of cavities and pores.

Local folds: In many mineralised orogenic belts, concentration of ore deposits is found in areas showing intense folding and deformation including cross-folding, superposed-folding, plunging folds and fold terminations. It is well known that deposits are localised in the fracture planes of folded (strata) structures.

Deposits are localised along axial plane cleavages, herring bone cleavage or strain slip cleavage or crenulation cleavage. Deposits may be located where there is sudden change in the plunge directions or at the nose of plunging folds. For example in Mosabani mines rich Cu-deposits are located at the depth where plunge suddenly changes. In Ketri Cu-belt fracture plane mineralisation is often encountered. In Zawar mines emplacement of sulphide ores is controlled by intersection of fracture plane and cleavage plane.

Local Faults: Local faults exert a strong control on localisation of ore deposits. Many rich loads are found to occur along minor faults traversing the host rocks. The intersection of fault with fracture is responsible for the formation of ore suites (enriched portion of the ore). Examples of some of the mineralised sulphide deposits in Rajasthan copper belts particularly in pyrite deposits of Saladipura and some portions of the deep gold mines of K.G.F. (Karnataka) are significant.

Local Fractures: Fractures and joints are places of ore-localization. These joints may be of tensional origin or by other mechanisms. Fracture filling is a common example of ore control particularly in the case of sulphides. Intersection of fracture may be the place for the formation of pegmatite bodies containing gem varieties of beryl *i.e.* emarald in Rajasthan. The fracture planes are responsible for mica books in many pegmatite bodies of India.

Local Controls by Thrust Sheets

A number of parallel thrust planes may control localisation and concentration of sulphide deposits as is the case with Zawar Pb-Zn deposits, but is controversial.

Controls by cavities, pores either primary (inherited) or secondary (induced)

Blow holes or vesicles of volcanic lava, porosity of host rocks, cavities by leaching or cavity due to crystal defects, capillary cavity around grain boundary or tensional cracks, also control mineralisation. All these are the places of epigenetic mineralization e.g. Deccan traps. A good example of pore filling sulphide deposits is Himalayan Pb-Zn deposits.

Other Controls of Ore localization

Control of localization of ore deposits in case of most of the epigenetic deposits is mainly due to structural factors but in case of many syngenetic deposits either of magmatic origin or sedimentary origin, the localization of ores is controlled by other factors particularly stratigraphic assemblages (stratigraphic control, chemistry and mineralogy of host rocks).

Stratigraphic Control of Ore-localisation

It is well known that sedimentary rocks contain siliceous assemblages (orthoquartzite, chert, sandstone etc.), carbonate assemblages (limestone, dolomite) and pelitic or argillaceous assemblages (shales, mudstones, sandy shales, calcareous shale etc.) In pre-Cambrian terrain these lithological assemblages are present as their metamorphic equivalents *i.e.* quartzites, marbles, calc-silicates, paraschists, paragneisses, slates, hornfelses, phyllites. Besides these they also occur as products of volcanism deposited in the manner of sediments in the ocean basin. These are known as pyroclastic sediments and examples are various types of tuff rocks, welded tuffs (ignimbrites) and some porcellanites (volcanic chert but of other origin). Sedimentary syngenetic ore deposits particularly of sulphides which occur as stratiform deposits are mostly confined to carbonaceous or bituminous organic shales or their metamorphic equivalents (carbon phyllites, carbon schists often with graphite and fine grained pyrite). For example in lower Vindhyans in Dehri-on-sone area sedimentary pyrite is present at the base of Khenjua stage. The black pyrite is confined to shales. Another example is the restriction of concentrate association of Zn-Pb-Cu deposits within the horizon of graphitic mica schists or within calc-granulites in the polymetallic sulphide deposits of Rajpura-Dariba. A very good example of stratigraphic control is the copper shales of Mansfield of East Germany where CuS are entirely confined within the black shales in the form of thin parallel beds. Pyrite deposit of Saladipura, Rajasthan, is also confined to black shales to schists.

Oxide Deposits

Deposits of Uranium oxide in the tetravalent forms are confined to either sandy layers or also to black pyritic shales. The best example is sandstone type Uranium deposits found in USA. Indian Uranium deposits do not belong to this category but some of Uranium in Jadugoda mines of Jharkhand is confined to dark colour schist rocks. Deposits of Mn are associated with clays and sandy clays which on metamorphism gives rise to Mn bearing schists and gneisses known as gondites. Similarly, chemical sediments of iron are confined within fine grained siliceous rocks known as chert in which the iron oxide bands alternate with chert bands, the rock known as BHJ (Banded Hematite Jasper). In some other cases sandy layers alternates with layers of iron oxides known as BHC (Banded Hematite Chert).

Regional Stratigraphic Control

Regionally such broad stratigraphic units as basins (geosynclines) and broad stratigraphic breaks as unconformity are usually places of ore deposition, *e.g.* Himalayan basin, Dharwar basin, Aravalli basin, Rocky basin etc. Examples of unconformities are the well known unconformities between the stages of upper Vindhyans where conglomerates contain diamonds. Mineralisation is confined to alluvial sandy channels. Although subsequently the host rock and mineralisation is serially altered by deformation and metamorphism. Blind river U-deposits (USA) are confined to unconformities. In Lake Superior district as many as 4-5 unconformities have been identified in pre-Cambrian sedimentary sequence which contain the largest deposits of rich iron ore in the world.

Another factor of stratigraphic importance in the localisation of mineralisation is the changing porosity and permeability of the host rocks. Porosity and permeability control the movement of ore solutions of whatever origin. During the passage of ore solution sideways or downwards or even upwards important mineralisation may take place at a junction of porous and pervious beds with impervious beds e.g. junction of sand with shale. Mineralisation is confined in many regions towards the bottom at such junctions. Mineralisation may be localised along that part of a bed which is less porous than the adjoining portion.

Chemical Control

Chemistry of the host rock along with its mineralogy controls the localisation of many ore deposits, both of the types—syngenetic as well as epigenetic. In many instances it is found that a mineralisation which passes through igneous rocks like monzonite through granites and gabbro is maximum enriched through monzonites. A mineralisation if passing through tuffs, basalts and rhyolite, the maximum concentration is found only in the tuffs portion, the reason is probably chemical.

Dolomites are the host rocks of Pb-Zn sulphides. It is found that ZnS is mostly confined to the siliceous portion of dolomites. In the other portion (nonsilicified) Pb is more. Silica grains in dolomite have caused the precipitation of sphalerite in silica rich portion.

9

Supergene Sulphide Enrichment

When ore deposits become exposed by erosion, they are weathered along with the enclosing rocks. The surface waters oxidize many ore minerals and yield solvents that dissolve other minerals. An ore deposit thus becomes oxidized and generally leached of many of its valuable materials down to ground water table, or to a depth where oxidation cannot take place. The oxidized part is called zone of oxidation. The effects of oxidation, however, may extend far below the zone of oxidation. As the cold, dilute, leaching solutions trickle downward they may lose a part or all of their metallic content within the zone of oxidation and give rise to oxidized ore deposits. If down-trickling solutions penetrate WT their metallic content may be precipitated in the form of secondary sulphides to give rise to a zone of secondary or supergene sulphide enrichment. The lower, unaffected part of the deposit is called primary or hypogene zone (Table 9.1). This zonal arrangement is characteristic of many mineral deposits that have undergone long-continued weathering. In places, supergene sulphide zone may be absent, and in rare cases, oxidized zone is shallow or lacking as in glaciated areas or regions undergoing rapid erosion. Special conditions of time, climate, physiographic development, and amenable ores are necessary to yield the results as shown in Fig. 9.1, but they are sufficiently common for oxidized and enriched supergene ores to occur in most of the nonglaciated land areas of the world.

Effect of oxidation has been either to render barren the upper parts of many deposits or to change ore minerals into more usable or less usable forms, or even to make rich bonanzas.

The effect of supergene enrichment has been more far reaching, for it has added much where there was previously little; leaner parts of veins have been made rich and unworkable protore has been enriched to ore grade. Indeed, many of the great Cu-areas would not have come into existence except for the process of supergene sulphide enrichment.

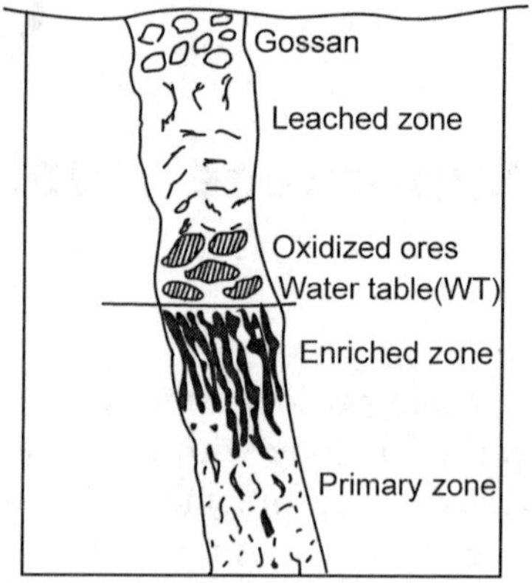

Fig. 9.1: Diagram of Zones of a Weathered Vein, with Oxidized, Supergene Enrichment, and Primary Zones

Table 9.1: Common Hypogene, Supergene and Oxidized Ore Minerals

Metal	*Minerals Generally Hypogene*	*Minerals Generally Supergene*	*Minerals Generally Originating in Oxidized Zone*
Copper	Chalcopyrite Bornite Enargite* Tetrahedrite* Tennantite*	Chalcocite Sooty chalcocite* Covellite	Native copper Malachite* Brochantite* Antlerite* Atacamite* Azurite* Chrysocolla* Cuprite* Tenorite*
Silver	Tetrahedrite* Tennantite*	Native Silver Argentite Pyrargyrite Proustite Stephanite Polybasite Pearcite	Cerargyrite* Embolite* Bromyrite*
Gold	Native gold Gold tellurides*	Native gold	Native gold?
Zinc	Sphalerite Willemite*	Wurtzite	Smithsonite* Hemimorphite* Hydrozincite*

(*Contd...*)

Lead	Galena*		Cerussite* Anglesite* Pyromorphite* Leadhillite
Iron	Pyrite* Marcasite Pyrrhotite* Arsenopyrite Magnetite* Hematite Specularite* Siderite	Marcasite	Goethite* Iron sulphates* Hematite
Manganese	Rhodochrosite* Rhodonite* Manganite (?) Alabandite*		Psilomelane Pyrolusite Braunite
Nickel	Millerite Pentlandite* Niccolite*	Bravoite (?)	Garnierite*

*Always hypogene or supergene or oxidize according to the column in which they are placed.

Supergene oxidation and enrichment go hand in hand. Without oxidation there can be no supply of solvents from which minerals may later be precipitated in the zones of oxidation or of supergene sulphides. The process resolves itself, therefore, into three stages :

(1) Oxidation and solution in the zone of oxidation
(2) Deposition in the zone of oxidation
(3) Supergene sulphide deposition.

Oxidation and Solution in Zone of Oxidation

Effects of oxidation on mineral deposits are profound. Minerals are altered, and structure is obliterated. Metallic substances are leached, or are altered to new compounds that require metallurgical treatment for their extraction quite unlike that employed for unoxidized materials. The texture and type of deposits are obscured. Compact ores are made cavernous. Ubiquitous limonite obscures everything and imparts to the gossan that familiar rusty color which from earliest times has attracted the curiosity of miners. One can only infer what lies beneath, but in this inference one can be guided by the character of oxidation products themselves.

Water with dissolved and entangled O_2 is the most powerful oxidizing reagent, but CO_2 also plays an important role though locally chlorides, iodides, and bromides play a part. These substances react with certain minerals to yield strong solvents, such as $Fe_2(SO_4)_3$ and H_2SO_4. H_2SO_4, in turn, reacting with NaCl yields HCl, which with iron yields the strongly oxidizing $FeCl_3$.

Chemical Changes Involved

There are two main chemical changes within the zone of oxidation :

(1) Oxidation, solution, and removal of valuable minerals

(2) Transformation *in situ* of metallic minerals into oxidized compounds.

Most metallic mineral deposits contain pyrite. This mineral under attack readily yields sulphur to form iron sulphate and sulphuric acid; pyrrhotite does same. The following reactions indicate, without intermediate steps, their general trend :

(1) $FeS_2 + 7O + H_2O = FeSO_4 + H_2SO_4$

(2) $2FeSO_4 + H_2SO_4 + O = Fe_2(SO_4)_3 + H_2O$

Reaction (2) passes through intermediate stages during which S, SO_2, and $FeSO_4$ may form. Sulphur may oxidize to sulphuric acid. The ferrous sulphate readily oxidizes to ferric sulphate and ferric hydroxide :

(3) $6FeSO_4 + 3O + 3H_2O = 2Fe_2(SO_4)_3 + 2Fe(OH)_3$

Ferric sulphate hydrolizes to ferric hydroxide and sulphuric acid:

(4) $Fe_2(SO_4)_3 + 6H_2O = 2Fe(OH)_3 + 3H_2SO_4$

Ferric sulphate is also a strong oxidizing agent and attacks pyrite and other sulphides to yield more ferrous sulphate.

(5) $Fe_2(SO_4)_3 + FeS_2 = 3FeSO_4 + 2S$

Ferric sulphate, in addition, changes to various basic sulphates.

The above reactions indicate the importance of pyrite, which yields the chief solvents, ferric sulphate and sulphuric acid, and also ferric hydroxide and basic ferric sulphates. Moreover, ferric sulphate is continuously being regenerated not only from pyrite but also from chalcopyrite and other sulphides. Ferric hydroxide changes over to hematite and goethite and forms the ever-present limonite that characterizes all oxidized zones. Basic ferric sulphates may be deposited as such, but generally limonite is the end product.

The part played by ferric sulphate as a solvent is evident in following equations. Although the end products are obtained, it is not established in all cases that the reactions given below are those that actually take place.

(6) *Pyrite* — $FeS_2 + Fe_2(SO_4)_3 = 3FeSO_4 + 2S$

(7) *Chalcopyrite* — $CuFeS_2 + 2Fe_2(SO_4)_3 = CuSO_4 + 5FeSO_4 + 2S$

(8) *Chalcocite* — $Cu_2S + Fe_2(SO_4)_3 = CuSO_4 + 2FeSO_4 + CuS$

(9) *Covellite* — $CuS + Fe_2(SO_4)_3 = CuSO_4 + 2FeSO_4 + S$

(10) *Sphalerite* — $ZnS + 4Fe_2(SO_4)_3 + 4H_2O = ZnSO_4 + 8FeSO_4 + 4H_2SO_4$

(11) *Galena* — $PbS + Fe_2(SO_4)_3 + H_2O + 3O = PbSO_4 + 2FeSO_4 + H_2SO_4$

(12) *Silver* — $2Ag + Fe_2(SO_4)_3 = Ag_2SO_4 + 2FeSO_4$

Similarly other minerals are dissolved yielding, except for lead, soluble sulphates of the metals. Sulphuric acid also attacks various sulphides, yielding sulphates of their metals. Chlorides, bromides and iodides, chiefly of silver are also formed.

Economic minerals redeposited within the zone of oxidation are chiefly native metals, carbonates, silicates, and oxides. The metals of greatest commercial importance are Cu, Zn, Pb, Ag, V and U of which large and valuable deposits are known. Less valuable deposits of Co, Mn, Fe and other substances are similarly formed. For the most part, the ores are deposited in the lower part of the zone of oxidation.

Methods of Precipitation

Various processes, acting single or together, operate to bring about precipitation of the metals in zone of oxidation.

Evaporation and Saturation: By simple saturation attendant upon evaporation, particularly in arid climates, metallic compounds are deposited as veinlets, blebs, and paint. Such compounds are chiefly sulphates. This process has operated at Chuquicamata, Chile, the largest copper deposit in world, where all the production has come from the oxide zone. Chief ore minerals are various copper sulphates. Antlerite is the dominant ore mineral, but brochantite, chalcanthite, and krohnkite are common. Other sulphates of Fe, Cu, Mg, Na and K also occur.

Oxidation and Hydration: Oxidation and hydration yield oxides and hydrous oxides of Fe, Mn, Cu, and other materials. The formation of ferric hydroxide and finally goethite is an example.

Reactions between Solutions: Sulphate solutions formed in zone of oxidation may encounter various other solutions that bring about precipitation within oxidized zone.

Chloride Solution: Sodium chloride, a widespread constituent of soil and ground water of arid regions, reacts with sulphates and other compounds to form chlorides of Ag, Cu, Pb and other metals. Silver chloride, or cerargyrite, the ore of most commercial importance, is supposedly formed as follows :

$$Ag_2SO_4 + 2NaCl = 2AgCl + Na_2SO_4$$

Similarly embolite, bromyrite, and iodyrite, bromides and iodide of silver, are formed in less abundance.

Carbonated Solutions: Bicarbonate solutions react with sulphates and sulphate solutions to form carbonates of metals. Thus, reaction with copper sulphate yields the common basic carbonates of copper, azurite and malachite; with zinc sulphate, zinc carbonate, smithsonite; with lead sulphate, lead carbonate, cerussite, is formed, an ore mineral of great commercial importance. Also native copper and cuprite are attacked to yield copper carbonate.

Other Solutions: Cuprite (Cu_2O) and tenorite (CuO), common ore minerals of oxidized zone, are possible formed by reactions between copper sulphate and ferrous and ferric sulphate. Native copper is also formed, among other ways, by reaction between cupric and ferric sulphate solutions. Likewise, native silver is formed by reaction between silver sulphate and ferrous sulphate.

Reaction with Gangue or Wall Rocks: Solutions of copper or zinc sulphate react with carbonate minerals to form malachite and azurite-carbonates of copper or smithsonite-carbonate of zinc. These solutions may also react with available silica to form chrysocolla-copper silicate, or hemimorphite-common zinc silicate. Hydrozincite and monheimite may also be formed. Iron carbonate is commonly precipitated along with zinc carbonate and generally becomes oxidized, adding iron impurity and limonite color to zinc ore.

The formation of lead sulphate, which is practically insoluble, results, not in a solution, but in the deposition of indigenous anglesite, and carbonate, cerussite, is formed from the sulphate. No silicate of lead is known, but oxides of lead form in zone of oxidation.

Silver sulphate solution does not give rise to natural carbonates or silicates of silver by reactions with wall rocks. Either the silver solutions formed in zone of oxidation form silver salts or pass through oxidized zone into supergene sulphide zone.

The availability of carbonate minerals, such as common calcite or limestone, to form copper or zinc carbonates offers no problems; but the availability of silica to form copper or zinc silicates is less simple. Quartz is inert to solutions of oxidized zone, and answer is probably colloidal silica set free during weathering.

Supergene Sulphide Enrichment

Changes in ore deposits produced by oxidation are so clear and prominent that the major features have long been recognized, although cropping interpretation is relatively recent. Secondary sulphide enrichment, however, although mentioned by DeLaunay, was not clearly understood until clarified by Emmons *et al.* in 1901 and Ransome in 1910. The next 15 years saw the greatest advance, led by Graton, Emmons, and others, in the knowledge of supergene sulphide deposition.

Their study, greatly aided by development of ore microscope and use of reflected polarized light, is very significant.

The metals in solution that escape capture in oxidized zone trickle down to where there is no available oxygen, generally water table, and there undergo deposition as secondary sulphides. The metals removed from above are thus added to those existing below, thereby enriching the upper part of sulphide zone. This forms zone of secondary enrichment or, as it is now more generally referred to, supergene sulphide zone. It in turn is underlain by primary or hypogene zone. Progressive erosion permits deeper oxidation, and after a time, supergene sulphides themselves become oxidized and their metal content is then transferred to downward progressing enrichment zone. Primary ore may, thus be enriched to as much as ten times its original metal content. Rich ores are made richer, lean ores are made valuable, and noncommercial primary material or protore is built up to commercial grade. The process is, therefore, not only of great scientific interest but also of far-reaching economic importance to mineral industry.

Favourable conditions must exist for supergene sulphide enrichment to take place, but they are sufficiently common that sulphide-enriched deposits are widely scattered over the nonglaciated areas of Earth. The process is of greatest importance with Cu and Ag deposits.

Requirements for Supergene Enrichment

Oxidation. Oxidation of ore deposits may occur without attendant sulphide enrichment, but enrichment cannot take place without accompanying oxidation. Therefore, oxidation is the primary requisite, and the factors that have been shown to favour oxidation must be present.

Suitable Hypogene Minerals: The ore deposit must contain primary minerals which upon oxidation yield the necessary solvents. Iron sulphides are essential and deposits lacking them rarely contain supergene sulphide zones. Chalcopyrite alone does not yield sufficient solvents and deposits where this mineral predominates in the primary ore, have negligible

Hg, Ag, Cu, Bi, Pb, Zn, Ni, Co, Fe and Mn.

He showed that any one of these metals in solution is precipitated as a sulphide by any one lower in the series. He attributed it to their relative order of affinity for sulphur. Later, Wells showed that Schurmann's series corresponded essentially with Weigel's order of solubilities of sulphides and that deposition is determined by their relative solubilities.

This list is arranged with the least soluble first and the most soluble last. This simple law clarifies the cause and order of deposition of supergene sulphides and why one supergene sulphide replaces another. For example, if silver sulphate solution encounters in supergene sulphide zone any sulphide lower in the series than itself, silver sulphide (argentite) will be deposited at its expense. Similarly, copper sulphide from copper sulphate solution will replace galena, zincblende and pyrite. The farther apart in the series they are, the more complete will be the replacement. Manganese sulphide, being the most soluble, could not be precipitated by any sulphides of Schurmann's series; this may be the explanation of its rarity in nature. Lead sulphate should yield lead sulphide, but it is insoluble. Soluble zinc sulphate, however, should yield zinc sulphide in contact with nickel, cobalt or iron sulphides. Its general absence as a supergene sulphide is, therefore, striking. It is probable that relative concentrations of ferrous and ferric ions in the solutions also play a part in the sequence of deposition.

In copper deposits, both chalcocite and covellite are found to replace bornite, chalcopyrite, enargite, galena, sphalerite, pyrrhotite, pyrite, tetrahedrite, and tennantite and other more complex minerals. Chalcocite is the most common supergene sulphide, but covellite is locally dominant. Cupric sulphide (covellite) tends to be precipitated where the ratio of ferric to ferrous iron is relatively high and cuprous sulphide (chalcocite) where the reverse ratio exists. Chalcocite is replaced by supergene covellite and hypogene covellite is also replaced by supergene chalcocite.

Supergene sulphide replacement is invariably volume for volume and is not, therefore, a molecular interchange between supergene enrichment. Primary ore also must contain met that can undergo supergene enrichment. Pb or Zn, for examp do not ordinarily yield supergene sulphides; Cu and Ag d Sn apparently does not, and little is known of supergen enrichment of Ni and the less common metals; much researc still remains to be done in this field.

Permeability: Permeability of ore deposit is essential to enable trickling solutions to penetrate beneath zone of oxidation. If the walls are quite pervious and the deposit only slightly so, the solutions may enter the walls and in the absence of suitable precipitants become dissipated. Impervious inclined faults may bar the entry of downward-moving solutions into the underlying deposit.

Absence of Precipitants in Oxidized Zone: Oxidized zone must be free from precipitants, such as carbonate rocks, that fix the metal content of enriching solutions in the form of oxidized compounds.

Zone of No Available Oxygen: Supergene sulphides can be deposited only where oxygen is excluded. Generally this is below the water level. Under certain circumstances, oxygen may also be lacking within oxidized zone and sulphide deposition may then take place there.

Precipitants: It is widely and mistakenly thought that supergene copper sulphides are deposited beneath WT upon quartz, gangue or any other minerals. They are deposited from sulphate solutions only at the expense of other sulphides, arsenides or similar minerals and by replacing them. Hydrogen sulphide also effects precipitation. Thus, underlying ore minerals, generally hypogene ores, are essential for deposition of supergene sulphides. If the down-trickling sulphate solutions do not encounter sulphides in zones lacking available oxygen, the sulphates become dissipated, and no supergene sulphide enrichment occurs.

Cause and Method of Deposition

Schurmann determined that certain metals in solution are deposited as sulphides in the presence of other sulphides and in order as follows :

replacing and replaced substances. If replacement of pyrite by chalcocite, for example, were a molecular interchange, a grain of denser chalcocite should occupy only a portion of the volume of a grain of pyrite. This is not the case in nature. Hence, customary balanced chemical equations depicting such changes do not correctly represent what has taken place. As used here, they are given to indicate the trend of change rather than the exact chemical interchange in all instances.

Chemistry of Supergene Sulphide Deposition. Since the chemistry of copper enrichment has been more thoroughly investigated than that of other metals, it can be used mainly to exemplify the changes that take place. Under proper conditions, copper sulphate will react with other ore minerals to form copper sulphides, as is indicated in the following equations, some of which have been established in the laboratory.

(1) $PbS + CuSO_4 = CuS + PbSO_4$ (*Covellite*)

(2) $ZnS + CuSO_4 = CuS + ZnSO_4$ (*Covellite*)

(3) $5FeS_2 + 14CuSO_4 + 12H_2O = 7Cu_2S + 5FeSO_4 + 12H_2SO_4$ (*Chalcocite*)

(4) $4FeS_2 + 7CuSO_4 + 4H_2O = 7CuS + 4FeSO_4 + 4H_2SO_4$ (*Covellite*)

(5) $CuFeS_2 + CuSO_4 = 2CuS + FeSO_4$ (*Covellite*)

(6) $5CuFeS_2 + 11CuSO_4 + 8H_2O = 8Cu_2S + 5FeSO_4 + 8H_2SO_4$ (*Chalcocite*)

(7) $Cu_5FeS_4 + CuSO_4 = 2Cu_2S + 2CuS + FeSO_4$ (*Chalcocite, covellite*)

(8) $5Cu_5FeS_4 + 11CuSO_4 + 8H_2O = 18Cu_2S + 5FeSO_4 + 8H_2SO_4$ (*Chalcocite*)

(9) $5CuS + 3CuSO_4 + 4H_2O = 4Cu_2S + 4H_2SO_4$ (*Chalcocite*)

Copper sulphate is also known to react with marcasite, pyrrhotite, enargite, tennantite, tetrahedrite and other minerals. Supergene chalcopyrite and bornite are also formed as transition products in the alteration of pyrite or chalcopyrite to chalcocite or covellite. They are formed only in microscopic amounts, and, except for one instance, he has never observed either bornite or chalcopyrite as supergene minerals in amounts large enough to be seen by naked eye. Certain reactions also yield native copper. In general, among copper minerals, those highest in copper content are more readily replaced by supergene sulphides than those of low copper content. No established supergene sulphides of Pb, Bi, Mo, W, U or V have been recorded.

Degree of Enrichment

It is the extent to which hypogene minerals are replaced by supergene sulphides. There are three stages of enrichment from an initial stage to almost complete replacement. These stages are insipient, partial and complete.

(1) Insipient stage indicates a beginning stage and represents either weak enrichment or the bottom of zone of enrichment. It is characterized by a thin coating of supergene sulphides on the outside of hypogene grains.

(2) Partial enrichment represents about 1/4 - 2/3 replacement of hypogene ore minerals. The coatings are thicker.

(3) Complete enrichment represents maximum replacement of hypogene minerals by supergene sulphides.

Sometimes enrichment is selective and may take place in two ways. Either (a) only veins of hypogene minerals are affected and disseminated grains are unaffected, or (b) only certain minerals are replaced. While at other times in pervasive enrichment supergene sulphide pervades the entire ore body. It is a vigorous and strong enrichment.

Relation of Enrichment of Water Level

Sulphide enrichment starts at or below WT. Thus the position of enrichment zone is controlled by WT. Arising WT causes cessation of sulphide enrichment.

Factors Affecting Sulphide Enrichment

There are various factors which affect sulphide enrichment.

(a) **Water Level:** High WT lessens sulphide enrichment while a deep WT develops a well enriched and thicker supergene sulphide.

(b) **Primary ores:** Enrichment depends upon the type of hypogene/primary minerals. In general, the higher the grade of primary ore, the greater the degree of enrichment.

(c) **Wall Rocks:** Wall rocks also affect the process of enrichment. Carbonate rocks generally inhibits enrichment by precipitating the metals in oxidized zone.

Structure : Faults, specially the larger ones generate deep roots of enrichment.

Topography: Depth of enrichment zone depends upon the topography of the area. Although the top of enrichment zone conforms with topography, its conformation is with the topography that existed at the time of enrichment.

Climate: The climatic factors that favours or retard oxidation also apply to sulphide enrichment.

Erosion: The optimum condition of erosion for thick enrichment is a rate just about equal to the rate of oxidation and solution. This provides a continuous supply of enriching solution.

Time: Considerable geological time, say millions of years, is necessary for extensive sulphide enrichment.

SECTION-C
ECONOMIC MINERALS

10

Minerals and Metals

There is an uneven distribution of minerals around the world. Countries, which are fortunate enough to have supplies of more than one mineral can develop an advantageous economic position. This is certainly true where the mineral is regarded as 'strategic', with an assured market and limited producers, such as chromium. A country dependent for foreign earnings on one mineral can be vulnerable to sudden movements of world price. If a country relies heavily on mineral imports, stockpiles are often built up to guard against possible interruptions in supply. One effect of this has been moderate to sharp fluctuations in world prices.

Processing of a mineral can be far more profitable than its extraction. Many developing countries do not have plants to process their minerals. International companies are an important factor, as they are often more financially powerful than the country in which they operate.

Mineral resources are basically classified into two major types (Fig. 10.1):

(i) Fuel Minerals

(ii) Non Fuel Minerals

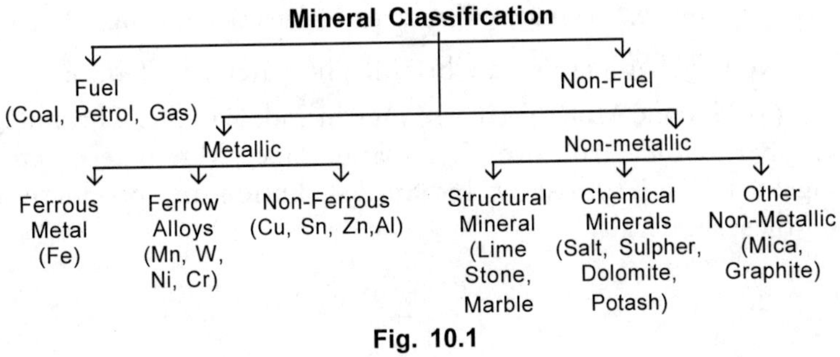

Fig. 10.1

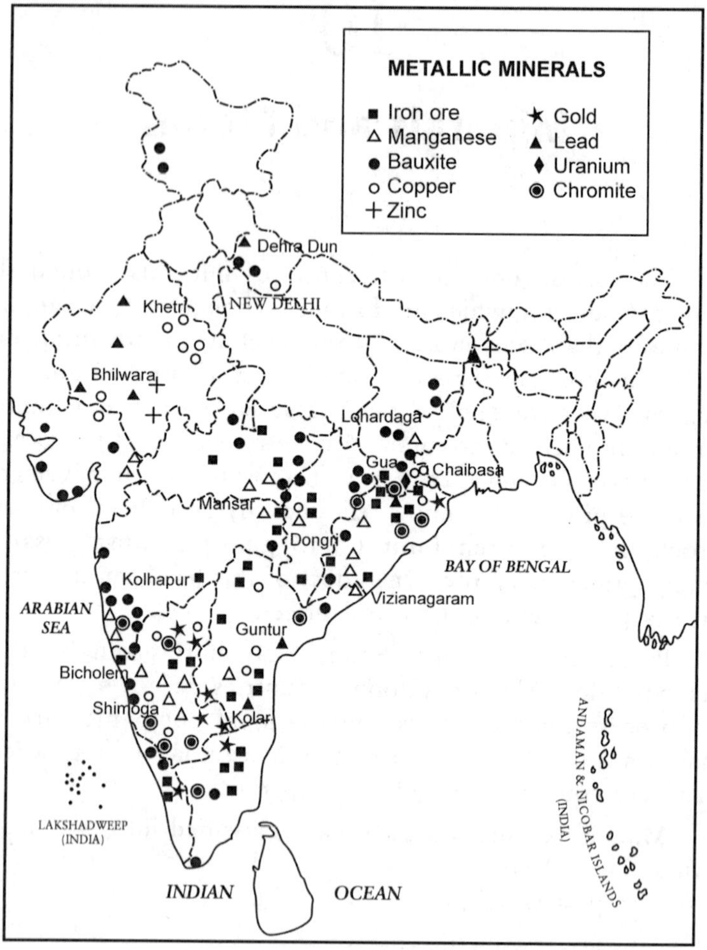

Fig. 10.2: Distribution of Mineral Resources in India

Non fuel minerals can broadly be categorised into:

(a) Metallic Minerals: These may include Bauxite, Chromite, Copper, Gold, Iron-ore, Lead and Zinc, Manganese, etc. Figs. 10.2 and 10.3 show Indian distribution and production of these minerals.

(b) **Non-metallic minerals:** These may include Asbestos, Barytes, Clay, Diamond, Feldspar, Fireclay, Granite, Graphite, Gypsum, Limestone, Magnesite, Mica, Phosphate, Pyrite, Quartz, Salt, Tungsten, Uranium etc. Fig 10.4 and 10.5 show Indian distribution and production of these minerals.

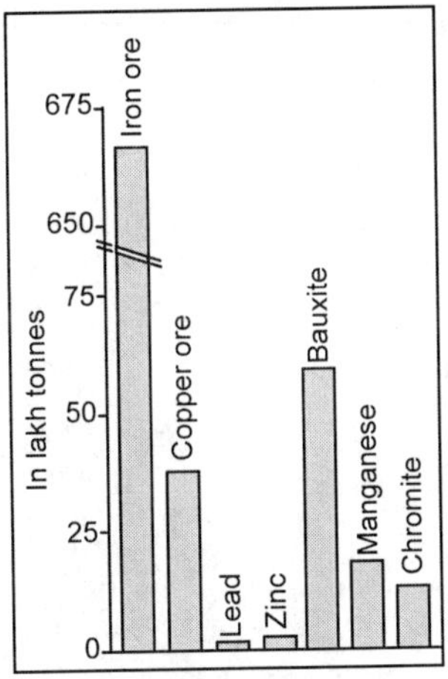

Fig. 10.3: Pattern in Production of Metallic Minerals

Minerals can also be classified into following categories:

(a) **Precious metals and gems:** These include Diamond, Gold, Platinum, etc.

(b) **Iron and ferro-alloys:** These include Iron, Chromium, Cobalt, Manganese, Molybdenom, Nickel, Tungsten, Uranium etc.

(c) **Base metals:** These include Copper, Lead, Zinc, Bauxite, Mercury etc.

World distribution has been given in Fig. 10.6 and production of these minerals in Latin American countries has been shown in Fig. 10.7.

Economy of a nation largely depends upon the mineral resources it has been gifted with as these minerals are the backbones of various industries like fertilizer, iron and steel, cement, asbestos, ceramics, glass, petroleum, aluminium, brass, copper, refractories etc. Major mineral based industries of India are shown in Fig 10.8. In this section important precious metals and gems, iron and ferro-alloys, base metals and some other important economic deposits like mica, limestone, phosphates etc. have been discussed. A list of important ore forming group of minerals is given in Table 10.1.

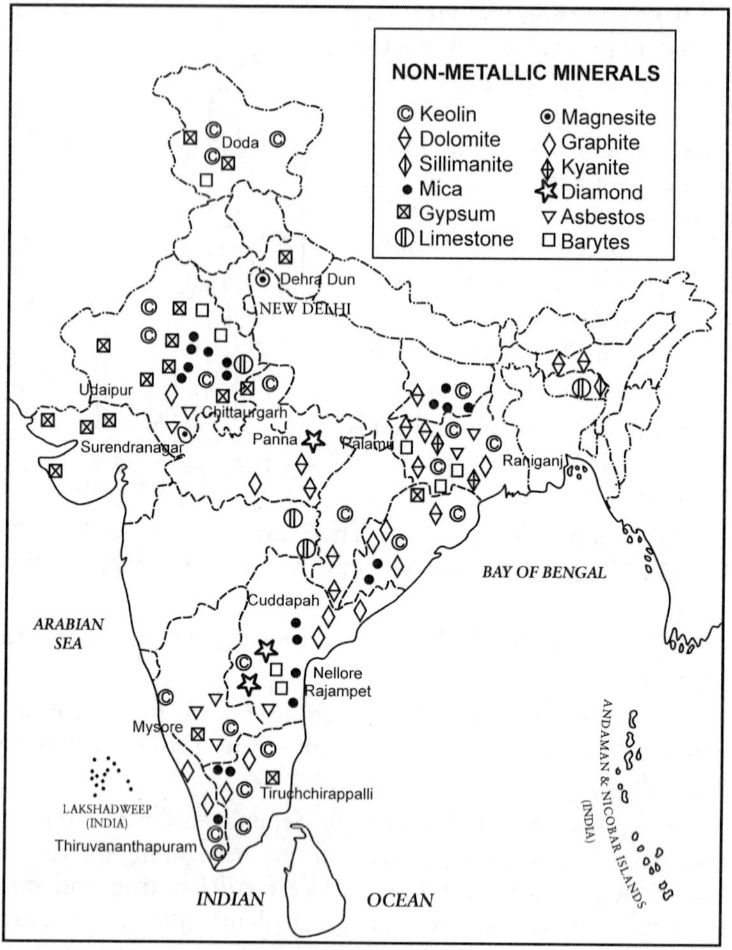

Fig. 10.4: Distribution of Non-metallic Minerals in India

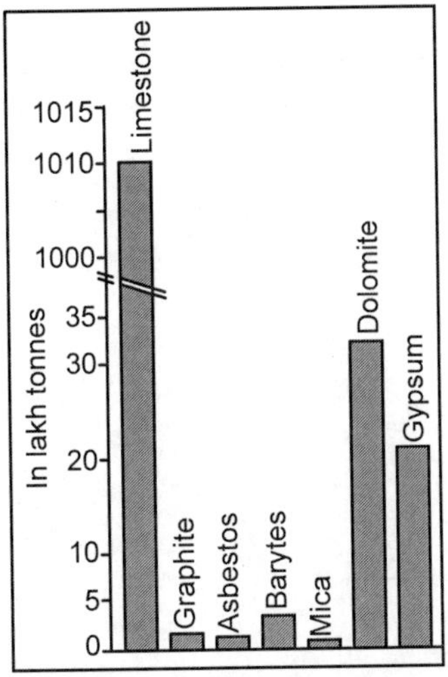

Fig. 10.5: Pattern in Production of Non-Metallic Minerals

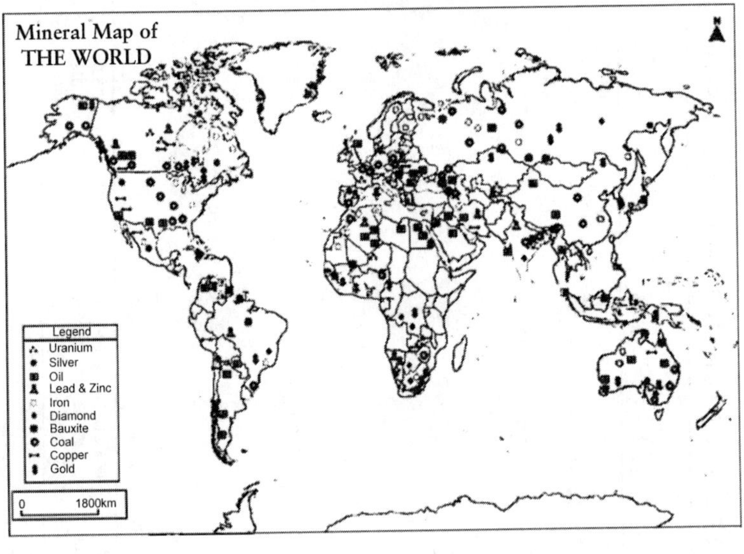

Fig. 10.6: World Distribution of Minerals

Fig. 10.7: Metals & Minerals Production in Latin America (*in WO metric tons, unless otherwise specified)*

Country	*Iron Ore*	*Bauxite*	*Zinc (Zn Content)*	*Copper (Cu Content)*	*Lead (Pb Content)*	*Nickel (Ni Content)*	*Tin (Sn Content)*	*Silver (Ag Content) Metric Tons*	*Gold (Au Content) Kilograms*
Argentina		—	27.2	180.1	12.8			248.2	44,131.0
Bolivia	-	-	172.7	-	12.0	-	177	472.2	9.628.0
Brazil	317,800.0	22.700.0	185.2	147.8	260	82-5	12 9	38.2	45,000.0
Chile	8.629 0	-	36.2	5.361.0	0.7	-	-	1.607.0	42.100.0
Colombia	644.2	-	-	2.9	-	94.1	-	8.4	15,700.0
Costa Rica	-	-	-	-	-	-	-	-	1.210.0
Cuba	7.8	—	-	-	-	75.0	-	-	-
Dominican Republic	-	NA	-	-	-	46.5	-	-	-
Ecuador	-	-	-	-	-	-	-	-	5,500.0
El Salvador	-	-	-	-	-	-	-	-	-
French Guiana	-	-	-	-	-	-	-	-	2.000.0
Guatemala	-	-	-	-	-	-	-	49.7	5,036.0
Guyana	-	1.558.0	-	-	-	-	-	-	6.406.0
Honduras	-	-	37.6	-	11.8	-	-	55.0	4.100.0
Jamaica	-	14.865.0	-	-	-	-	-	-	-
Mexico	10,983.0	-	453-9	334.1	135.0	-	N.A.	2,569-5	38.961.0
Nicaragua	-	-	-	-	-	-	-	2.9	3.395.0
Panama	-	-	-	-	-	-	-	-	-
Paraguay	-	—	-	-	-	-	-	-	-
Peru	7.138.0	-	1,201.8	1,049-9	313.3	-	385	3,471.0	202.834.0
Suriname	-	4.924.0	-	-	-	-	-	-	9,362.0
Trinidad and Tobago	-	-	-	-	-	-	-	-	-
Uruguay	15.5	-	-	-	-	-	-	-	3.200.0
Venezuela	23.000.0	5,928.0	-	-	-	20 0	-	-	12,400 0
Latin America Total	368,217.5	49,975.0	2,114.7	7,075.9	511.5	318.1	69.6	8,522.2	450,963.0
Share of World Total	20.1%	26 3%	21.1%	46.6%	14.0%	14.9%	23.6%	42.6%	184%
World Total	1,830.000.0	190,000.0	16,666.6	15,200.6	3,660.6	2,130.6	293.6	20,000.6	2,456,666.6

Note: Annual production volume as of 2006; world total rounded
Source: United States Geological Survey

Fig. 10.8: Major Minerals in India

Table 10.1: Ore-forming Group of Minerals

Elements	*Mineral*	*System of Crystallisation*	*Chemical Composition*
Al	Bauxite		$Al_2O_3.2H_2O$
Sb	Stibnite	Orthorhombic	Sb_2S_3
As	Realgar	Monoclinic	AsS
	Orpiment	,,	As_2S_3
	Arsenopyrite (Mispickel)	orthorhombic ,,	FeAsS
Ba	Witherite	,,	$BaCO_3$
	Baryte	,,	$BaSO_4$
Bi	Bismuthinite	,,	Bi_2S_3
Cr	Chromite	Isometric	FeO Cr_2O_3
CO	Cobaltite	,,	CO AsS
Cu	Chalcopyrite	Tetragonal	$CuFeS_2$
	Malachite	Monoclinic	$CuCO_3$ $Cu(OH)_2$
	Azurite	,,	$2CuCO_3$ $Cu(OH)_2$
Fe	Pyrite (fools gold)	Isometric ,,	FeS_2
	Hematite	Hexagonal	Fe_2O_3
	Magnetite	Isometric	Fe_3O_4
	Limonite	Amorphous	$2Fe_2O_3$ $3H_2O$
	Siderite, Chalybite, Spathic Fe	Hexagonal	$FeCO_3$
Pb	Galena	Isometric	PbS
	Anglesite	Orthorhombic	$PbSO_4$
	Cerussite	,,	,,
Mg	Magnesite	Hexagonal	$MgCO_3$
Mn	Pyrolusite	Orthorhombic	Mn_2O_3 H_2O
	Psilomelane	Amorphous	Hydrated Oxide of Mn (+oxide of Ba, Na, K)
Hg	Cinnabar	Orthorhombic	HgS
Mo	Molybdenite	Hexagonal	MOS_2
Ni	Millerite	,,	NiS
	Niccolite (Kupfer nickel)	,,	NiAs
Ag	Argentite	Isometric	Ag_2S
Na	Halite	,,	NaCl
Sn	Cassiterite (Tin Stone)	Tetragonal	SnO_2
Ti	Ilmenite	Hexagonal	FeO. TiO_2
	Rutile	Tetragonal	TiO_2
W	Wolfram	Monoclinic	(Fe, Mn) WO_4
Zn	Sphalerite (Zn-Blende)	Tetragonal	ZnS
S	Sulhur	Orthorhombic	S
C	Graphite (Plumbago)	Hexagonal	(C + Fe^{+3}Oxide + clay as impurity
	Diamond	Isometric	C
Au	Gold	,,	(Au+Ag).

Resources of Important Minerals and Metals

So far 47 ores and minerals have been covered under the national mineral inventory in India. The known world reserves of some of these minerals and metals are also shown in Table 10.2 to facilitate comparison. But, it needs to be emphasised again that these figures without any reference to their geologic assurance or techno-economic feasibility do not signify the real picture of affluence or deficiency. These figures may at best serve as a starting point for carrying out further research and analysis. Moreover, inventory of mineral resources being of dynamic nature, the reserves estimated at a given point of time may soon become outdated.

Table 10.2: World and Indian Reserves of Some Selected Minerals and Metals

Mineral/metal	*World reserve (MT)*	*Indian reserve (MT)*
Asbestos	110	1.50
Barytes	166	71.26
Bauxite	21,800	2,330
Chromite	1,030	51.55
Coal	13,779,000	1,58,464
Copper metal	352	3.31
Diamond (Industrial)	980 M carats	1 M carats
Fluorspar	220	1.87
Gold metal	0.42	0.55
Graphite	21.03	4.65
Iron ore	147000	12000
Lead metal	70	2.33
Manganese ore	816	154
Phosphate rock	13,855	115
Zinc metal	147	7.5

Table 10.3 shows production of selected minerals and metals during 1989 in India and World. Production of minerals and metals in India during 1997-98 to 2001-02 is shown in Table 10.4 and 10.5.

The per capita consumption of important minerals are very low in India as compared to developed countries. To cope up with the population pressure India needs a technological revolution resulting in substantial growth of industrial production in 21st century. A comparative picture of production during past 60 years for important minerals is shown in

Table 10.6 which indicates substantial increase in production for various minerals as compared to that of 1951.

Table 10.3: Production of Minerals and Metals

Mineral	*Production (t/MT)* *India*	*World*
Asbestos	36,502 t	4.37
Barytes	0.55	5.30
Bauxite	4.34	91.63
Chromite	1.00	11.96
Coal	199	4,882
Copper metal	41,032 t (cathode)	8.83
Diamond (industrial)	15, 451 carats	52.70 M carats
Fluorspar	5,176 t (graded)	5.43
Gold metal	1.872 t	1,910 t
Graphite	47,731 t	653,000 t
Gypsum	1.54	95.44
Iron Ore (haematite & magnetite)	51.43	943.10
Lead metal	34,729 t (primary & secondary)	3.45
Limestone	63.24	118.84 (Lime)
Manganese ore	1.33	24.04
Phosphate rock (apatite & rock phosphate)	0.70	170
Zinc metal	71, 572 t (ingots)	7.04

Table 10.4: Mineral Production in India (1997-2002)

	Minerals	*Unit*	*1997-98*	*1998-99*	*2000-01*	*2001-02*
1.	Coal	MT	296	293	314	328
2.	Lignite	MT	23	23	23	24
3.	Natural Gas	Mm^3	19,703	19,522	27,860	27,863
4.	Petroleum (Crude)	MT	33	32	32	32
5.	Bauxite	000't	6,112	6,451	7,993	8,585
6.	Cu-Mineral	000't	4,499	4,253	164	164
7.	Gold	Kg	2,636	2,463		
8.	Gold (Primary)	,,	1,952	1,958	7,712	10,251
9.	Gold (Secondary)	,,	684	505		
10.	Fe-Mineral	000't	32,753	29,810		
11.	Fe-Mineral	,,	35,789	34,797	80,762	83479
12.	Refined Fe-Mineral	,,	7,181	6,076		
13.	Lead (Refined)	000't	60	63	54	52
14.	Manganese Mineral	000't	1,641	1,525	1,595	1,553
15.	Tin Mineral (Refined)	Kg	35,143	35,851	—	—
16.	Zinc (Refined)	000't	292	349	366	399
17.	Diamond	Carrat	30,944	34,579	57,407	81,448
18.	Limestone	MT	110	109	127	130

19. Mica (Raw)	t	1,697	1,492	1084	1266
20. Mica	t	907	889		
21. Apatite	000't	7	14	11	12
22. Phosphorite	000't	1,219	1,609	1,351	1,057
23. Asbestos	000't	—	—	15	11
24. Barytes	000't	—	—	845	916
25. Dolomite	000't	—	—	3,078	3,088
26. Fireclay	000't	—	—	441	368
27. Fluorite (conc)	000't	—	—	3	7
28. Fluorite (graded)	000't	—	—	44	48
29. Garnet (abrasive)	000't	—	—	232	282
30. Gypsum	000't	—	—	2,644	2,888
31. Kaoline	000't	—	—	871	808
32. Laterite	000't	—	—	606	572
33. Magnesite	000't	—	—	318	280
34. Shale	000't	—	—	828	272
35. Silica Sand	000't	—	—	2,358	1,428
36. Sillimanite	000't	—	—	15	15
37. Steatite	000't	—	—	596	549
38. Wollastonite	000't	—	—	122	135

Table 10.5: Production of Non-Ferrous Metals in India (000't) 1992-2002

Metals	*1992-93*	*1994-95*	*1996-97*	*1998-99*	*2001-02*
Aluminium	483	476	5,260	5,42	8,585
Copper	48	50	41	37	164
Copper (Refined)	42	46	38	35	—
Gold (Kg)	1,850	2,373	2,710	2,463	10,251
Lead (Primary)	38	34	38	50	52
Lead (Secondary)	10	14	4	—	—
Zinc	126	149	140	172	399

Table 10.6: Production/Quantum Jump of Important Minerals (000't) 1951-2002

Minerals	*Production*				
	(1951)	*(1989)*	*Increase*	*(2002)*	*Increase*
1. Iron ore	4,150	51,400	12 fold	83,479	20 fold
2. Bauxite (Al-ore)	68	4,345	64 ,,	8,585	126 ,,
3. Copper Ore	374	5,126	14 ,,	—	—
4. Lead (Conc)	2	46	23 ,,	52	26 ,,
5. Zinc (Conc)	2	126	63 ,,	399	200 ,,
6. Chromite	17	1,003	59 ,,	1,811	107 ,,
7. Manganese Ore	1,398	1,334	Static	1553	Static
8. Limestone	29,60,000	63	212 fold	1,30,000,000	44 ,,
9. Dolomite	14	2,353	168 ,,	3088	211 ,,
10. Kyanite	43	39	Decline	—	—
11. Gold (Kg)	4,995	1,829	,,	10251	2 ,,

India is producing nearly 60 minerals excluding coal and fuel minerals. We have surplus in 10 minerals, self sufficient in 16 minerals, deficient in 9 minerals and we have poor inventory for about 15 important minerals. The life index is more than 100 years, computed as mineral production at 2000 AD, in case of bauxite, bentonite, corundum, dolomite (non-refractory), fire-clay, garnet, iron ore (hematite iron), kaolin, limestone, magnesite, sillimanite (beach sand), and zircon. In chromite (all grade), pyrophyllite, rock phosphate (usable including beneficiable grades), wollastonite, zinc (all economic and marginal deposits), life index is about 50 years. As against this asbestos, baryte, calcite, chalk, copper metal, feldspar, fluorspar, fullers earth, gold, gypsum, kyanite, manganese ore, steatite, vermiculite have life index less than 50 years.

Estimation of Life Index

Life index of mineral resource, rather than the mineral resource as such, is the real and reliable yardstick of affluence or deficiency. In India, Planning Commission has introduced this approach in formulation of five year plans. In 1984-85, "Expert Group on Inventory of Mineral Resources" was constituted which ushered in a realistic approach in estimation of life indices of mineral resources of India.

Essentially the method consisted in finding the quotient between the resource and the annual production. However, in actual methodology there are certain problems which need to be addressed to. The Expert Group adopted the following approach:

(i) Instead of *in situ* geological resource, the mineable/recoverable reserve, not only at present, but also in foreseeable future, was considered.

(ii) Instead of considering total recoverable reserve and total production, different grades of both reserve and production were considered separately for estimating separate life indices of different grades of mineral resources.

(iii) Not only the current annual production but also future projections of production were considered.

(iv) Total recoverable reserves under all the categories, *i.e.*, proved, probable and possible were considered. There appears to be some argument in favour of this, because, though the reserves under possible category have a low degree of assurance at present, these may take care of unforeseen discoveries that may be made in near future.

(v) Reserves of uncertain grade were excluded for the purpose of life index estimation.

Life index of mineral resources is actually a derivative of mineral inventory and is a valuable and reliable tool in policy planning involving minerals. Like mineral inventory itself, the life indices of mineral resources are also not static. With every unforeseen change in production pattern, with every new discovery of mineral deposits and with every change in technology of mining and recovery, estimation of life index will undergo change. Life indices of some selected minerals as estimated by Expert Group in 1985 are shown in Table 10.7.

Table 10.7: Life Indices after 1995 of known Reserves in India for Some Selected Minerals
(Estimation based on the data available in 1985)

Sr. No.	*Mineral*	*Balance life after 1995 (Years)*
1.	Bauxite	301
2.	Chromite	
	(i) Total reserves	139
	(ii) Mineable reserves	49
3.	Coal	
	(i) Coking	93
	(ii) Noncoking	115
4.	Copper	
	(i) Only producing mines and projects	18
	(ii) All economic and marginally economic deposits	36
5.	Dolomite (Usable grade)	496
6.	Gold	10
7.	Iron ore	
	(a) Haematite	
	(i) High grade (+65 % Fe)	32
	(ii) Medium grade (62-65 % Fe)	182
	(iii) Low grade (–62% Fe)	98
	(b) Magnetite (Usable grade)	29

8.	Kyanite		14
9.	Lead		
	(i)	Producing/developing mines	20
	(ii)	All deposits	38
10.	Limestone (S.M.S. grade)		
	(i)	Overall reserve	446
	(ii)	Reserve in present supply base	90
11.	Manganese ore		
	(i)	High grade	15
	(ii)	Low grade	31
12.	Pyrites		149
13.	Rock phosphate and apatite		
	(i)	Usable grade	58
	(ii)	Only high grade	7
14.	Sillimanite		
	(i)	Massive	7
	(ii)	Beach sand	300
15	Zinc		
	(i)	Producing/developing mines	36
	(ii)	All deposits	50

Cut-off-grade: A more important economic decision in assessment of mineral resource concerns the choice of cut-off grade. This is the minimum grade that should be included in resource under assessment. The value chosen is usually somewhat below the actual minimum economic grade as it is generally found that the higher grades in the deposit will raise the overall average to the economic grade required. Small changes in cut-off grade used might make big differences in overall tonnage, and this could affect the economic viability of the deposit. In addition to the current economic grade, it is important to visualise future economic grade also; because, with improvement in technology and market conditions, lower grades of a mineral deposit may become economically viable. As investigation in USA has shown that while the average grade in porphyry copper ores, that were being produced in early 1900s was about 3%, the same in 1950 was only 0.8 or 0.9%; now it is even less than 0.4%. In India, currently the economic cut-off of Cu ore being mined by OC method is 0.45% Cu; however, it is foreseen in some quarters, that in near future a cut-off of 0.2% Cu may become economic. Until recently, 30% Cr_2O_3 used to be considered as economic cut-off grade for estimation of chromite resources in India.

Now, with the advent of beneficiation/briquetting technology, it has been possible to up-grade lower grade ore, and consequently there is now a tendency to take 10% Cr_2O_3 as economic cut-off grade. However, decisions regarding cut-off grade are linked with other techno-economic considerations like average mill-feed grade, depth of mining, nature of mineral, method of processing, etc. Table 10.8 shows economic cut-off grade of some important minerals along with other associated conditions as presently in vogue in India.

Table 10.8: Economic cut-off Grades of Some Minerals in India and the Associated Conditions

S.No.	*Ore/Mineral*	*Economic cut-off grade*	*Associated conditions*
1.	Apatite	18% P_2O_5	Average grade should be above 20% P_2O_5 for direct use in soil reclamations.
2.	Bauxite	40% Al_2O_3	SiO_2 should be < 5%
3.	Chromite	10% Cr_2O_3	Average grade should be about 20% Cr_2O_3 which is currently beneficiable.
4.	Copper ore	0.45% Cu	Average grade should be above 1% Cu applicable to large resources mineable by OC method.
		0.5% Cu	Average grade should be above 1% Cu; applicable to resources mineable by UG methods.
5.	Fluorite	10% CaF_2 5% CaF_2	Mineral meant for beneficiation Mineral processed by hand sorting.
6.	Gold	3 gm/t	Average grade should be 3.5-4.0 gm/t in case of shallow mines.
7.	Graphite	12% Fixed carbon	This is the minimum mill-feed grade for beneficiation plants in India, and the mineral is used after beneficiation.
8.	Kyanite and Sillimanite	40% Al_2O_3	Lumpy mineral; average grade should be at least 54% Al_2O_3.
9.	Lead-zinc ore	3% Pb+Zn	Ore should be simple and not ploymetallic; average grade should be above 5% Pb+Zn.
10.	Manganese ore		25% Mn
11.	Rock Phosphate	11-16% P_2O_5	Cut-off varies depending upon the design of beneficiation plants.
12.	Vanadium ore	0.8-1.0% V_2O_5 0.28-0.035% V	Based on limited mining experience (in USA cut off grade is 0.7% V_2O_5 or 0.25% V)

Concept of Small Mines

Mining, Geological & Metallurgical Institute of India (MGMI) has proposed a definition based on production as the criterion. As in USA, the ceiling of annual production, *i.e.* 50000 t has been accepted in India. However, different limits have been proposed for different minerals, depending upon their price & overall level of production in individual Indian mines. For this purpose, minerals have been grouped under 7 broad groups with an upper limit of annual production for small mines of each group of minerals. Again, MGMI has suggested that the annual production limit should be multiplied by a factor of 60 to arrive at what should be the maximum geological reserve in small deposits (Table 10.9).

Table 10.9: Maximum Limit of Geological Reserves and Production in Small Deposits of Different Groups of Minerals in India

Group	*Minerals and grades*	*Total geological reserve in a small deposit (t)*	*Max. limit of production (t/yr)*
Group-I	(i) Asbestos (A & B qualities)	15,000	250
Group-II	(i) Mica ore (3% mica basis)	60,000	1,000
	(ii) Fluorite (85% CaF_2 basis)		
Group-III	(i) Gold ore (5 ppm basis)	150,000	2,500
	(ii) Sillimanite		
Group-IV	(i) Kyanite	300,000	5,000
	(ii) Barytes		
	(iii) Graphite (60% C basis)		
Group-V	(i) Phosphorite/apatite	600,000	10,000
	(ii) Manganese ore (25% Mn)		
	(iii) Chromite		
	(iv) Magnesite		
	(v) Talc/soapstone (Powdery)		
	(vi) Tungston ore (0.2% WO_3)		
	(vii) Calcite		
	(viii) Chinaclay		
	(ix) Fireclay		
Group-VI	(i) Lead-zinc ore	1,500,000	25,000
	(ii) Copper ore		
	(iii) Gypsum		
	(iv) Feldspar		
	(v) Quartz		
	(vi) Silica sand		

Group-VII	(i) Bauxite (ii) Iron ore (iii) Limestone (iv) Dolomite (v) Stowing sand (vi) Stone aggregate (vii) Brick earth (viii) Ochre (ix) Other minor minerals (x) Coal (xi) Lignite	3,000,000	50,000

Knowledge of mineral resource changes with changes in technology as well as in changes in economic conditions that affect shifting of resource from one class to another (Table 10.10).

Table 10.10: Change in Quantitative Knowledge of Mineral Resources of India

Minerals		*Approximate known resources*				
	Unit	*1967*	*1971*	*1975*	*1980*	*2002*
Bauxite	(MT)	160	249	550	2500	2953
Chromite	(MT)	8	14	17	111	90
Copper ore	(MT)	—	280	243	456	538
Gold metal	(t)	48	47	42	65	68
Gypsum	(MT)	1098	1216	1205	1249	238 (Recoverable)
Iron ore	(BT)	10	10	13	18	13
Lead-Zinc ore	(MT)	148	107	212	351	176
Limestone	(BT)	18	50	63	73	76

11

Diamond Deposits

India has been the centre of diamond mining and cutting industry. The diamonds have been known to and used by ancient Indians since the earliest times. They have been mentioned in the Vedas, which dates back 10,000 B.C., and other Hindu texts of great antiquity in considerable detail. Indian folk lore and mythology is replete with stories about diamonds and the famous Koh-I-Noor was in possession of King Karna, who ruled Anga empire in Mahabharat times, about 5000 years ago. There is no doubt that India was the original supplier of diamonds to the kings of ancient and medieval world.

With the discovery of diamonds in Brazil in 1725, pre-eminence of India was lost and production declined sharply and practically ceased by the end of 19th century, by which time the diamond production in S-Africa was in full swing. The industry was revived to some extent after First World War, but serious attention was paid to it only after Independence.

The most famous diamond deposits in India are popularly known by the name of 'Golconda mines'. This is a misnomer since no mines ever existed near Golconda, and it was only a mart, being the capital of the kingdom of this name. Their preeminence attracted the attention of many people and several descriptions were published. Regular mining of these deposits ceased long ago. The only area where there is a steady production of diamonds at present time is Panna diamond field in central India, for which authentic and geological records are available.

There are old narratives of diamond workings in Sambalpur (Orissa) and adjoining territory, near Wairagarh in Chandrapur (Maharashtra), in some areas of Palamu and Ranchi (Jharkhand), but hardly any sign of them remain now. GSI has been carrying out investigations in all the areas where diamonds were reported to occur, and has met with some success like the discovery of new diatremes in Panna field and in Anantpur (AP). Its greatest achievement, has been the find of kimberlite bodies in the Jungel area of Mirzapur (UP), which was a virgin territory with no past lore of diamonds at all.

Prospecting, mining and beneficiation for diamonds are entirely different from those for other minerals. The diamond content in the host rock is extremely low, 0.2-1ppm, and to extract this minuscule constituent vast amounts of parent material needs to be treated. Because of this and other factors the prospecting methods are different. There are no chemical or physical indicators for direct detection of diamonds, and the only way to confirm whether a rock contains diamond is to actually extract it. There are, however, several indirect methods of prospecting. Various types of deposits require different approaches in prospecting and mining.

Diamond Reserves in India. A reserve of 0.98 M carrat has been estimated. In addition 26,50,000 carrats of restricted reserves have also been estimated. Conditional reserves are estimated 1.9 M carrat. (Table 11.1)

Table 11.1: Reserve and Production of Diamond in India

Year	*Reserve (in M Carat)*	*Production*	
	Recoverable	*Conditional*	*(in Carat)*
1999-2000	0.98	2.6	—
2000-2001	—	—	57407
2001-2002	0.98	1.9	81448
2004-2005	-	-	78316
2005-2006	-	-	44170
2006-2007	-	-	2769

Occurrence. Diamonds commonly occur in sands and gravels or cemented conglomerates, ranging in age from present river alluvium or marine beach deposits to pre-Cambrian rocks (sandstones and conglomerates). The primary home of diamond is, however, in ultrabasic igneous intrusive rocks like peridotite and kimberlite (altered peridotite, termed 'blue ground'), in which diamonds occur as crystals very sparsely disseminated. Diamond-bearing ultrabasic igneous rocks (or igneous breccias) usually occur as pipes, vertical or steep sided funnel shaped masses or as dykes. Diamond is usually regarded as a magmatic mineral crystallised *in situ* or its crystals were formed earlier at depth and carried upwards to the present site of cooling of magma. In some deposits diamonds are regarded to have been formed at great depth from inclusions of carbon-bearing phyllites or schists in original magma.

Distribution. Diamond fields of India are grouped into three regions; (i) *S-Indian tract* of A.P., comprising parts of Anantapur, Cuddapah, Guntur, Krishna, Mahbubnagar and Kurnool districts (ii) *Central Indian tract* of M.P., comprising former Indian States of Panna, Charkhari, Bijawar, Ajaigarh, Kothi, Pathar Kachhar, Baraunda, Chobepur and (iii) *E-Indian tract* of mostly Orissa, lying between Mahanadi and Godavari valleys.

(i) Diamond gems in *S-Indian tract* are associated with a bed of conglomerate at the base of Kurnool series (Vindhyan S.G.). This bed is found west of Banganapalle (Kurnool district) exposed over 25 Km in length and small crystals of diamond have been won in past, but most of the stones have been obtained from river alluvium and surface soil, derived from conglomerate or possibly from necks, pipes, dykes and sills of decomposed basic rocks some of which resemble agglomeratic tuffs. The best known locality is Wajra Karur in Anantapur district where good stones have been found in the fields from time to time after rains. A tuff agglomerate 'neck' with epidote rock was thought to be the source of these gems, but no diamond has been recovered from this rock. It is also suggested that epidote-bearing pegmatites, intrusive into hornblende and epidote granite-geneiss, which had been worked in this region, may be the source of diamonds.

(ii) Diamonds in *Central Indian tract*, especially in Panna occur in four distinct geological formations : (i) agglomeratic tuff of Majhgawan pipe, (ii) conglomerate deposits, (iii) deep alluvial deposits, and (iv) shallow alluvial deposits. It is now believed that the pipe of volcanic agglomerate of Majhgawan (which is possibly post-Kaimur in age) and such other pipes hidden under younger rocks and alluvium may be the original source of diamonds which are found in two horizons of diamondiferous conglomerate in Upper Vindhyan formations of this region the one occurring between Kaimur series and overlying Rewa series and the other between Rewa series and overlying Bhander series. These conglomerates are characterised by containing pebbles of red jasper; the lower conglomerate horizon being more productive than the upper. Panna and the neighbouring areas are promising sources and producing centres of diamonds in India.

Average quantity of diamond found in the pipe rock and conglomerate bed is 10 to 12 carats/100t of rock mined, and in gravel it is 5 carat/100t. Panna diamond area produces both industrial and gem varieties and their ratio may vary from 1:4 to 1:1.

(iii) In *E-Indian tract*, diamonds have been found in Mahandi alluvium (at Hirakud near Sambalpur, Orissa), in the grits at Wairagarh (Chanda district, M.P. and in the sands of Koel river, a tributary of Sone river (Palamu district, Jharkhand). The source of diamonds may be the extensions of Vindhyan formations in this region.

Panna Diamond Field

Total reserves of diamonds in Panna area have been estimated at 55.8 MT of Majhgawan pipe rock (with diamond content of about 3 M carats, 3 MT of Shahidan conglomerate bed (with diamond content of about 750,000 carats) and 3.35 MT of deep alluvial gravel beds in Bhagen river area (with diamond content of 870,000 carats). At present the entire production of diamonds in India is from M.P. and in 2000-01 the output of diamonds from Panna fields was about 157,000 carats. Biggest diamond found here was in 1961, and was named 'Vijay' and auctioned for over Rs. 5 lakhs. Diamond fields of M.P. are now being worked by NMDC.

This field extends in a belt trending NE-SW from Majhgawan railway station (24°55' : 80°49') in east to Keraia (24°38' : 80°00') in west. Northern boundary can be taken along the line formed by the edge of Kaimur scarp, and southern limit is indicated by Karauli Quartzite outcrop where it ends along the alluvial Haveli plain. These are the limits of known diamond workings, about 100 Km long and 25 Km wide. Although the same formation extends eastwards in Banda (U.P.) and westwards in Chhatarpur (M.P.), no evidence of ancient workings could be seen. In this area also the old workings are found along two linear zones, one on the table land formed by Dhandraul Quartzite, and the other on Karauli outcrop. Stray diamonds have also been found from localities farther than these limits, but they were believed by previous workers to have been washed down there.

Mining of diamonds is carried out by NMDC at Majhgawan and Ramkheria since 1961 when diamond mining was nationalised. Some private companies had diamond mining leases earlier. In addition, small plots were leased by M.P. Government on a small annual fee to individuals, who were allowed to dig shallow pits in certain designated lands; these working were not covered by Mineral Concession Rules or Mines Act. They have produced some of the largest diamonds in this field (Mathur, 1976).

Regional Geology

Diamond occurrences in Panna belt are associated with some formations of Vindhyans. Original diamond-bearing horizons in S-India and Mahanadi valley as well are considered as of the same Vindhyan age and affinity (Table 11.2).

Kimberlite diatremes are emplaced through Bundelkhand Granite (Angor) and through Dhandraul quartzite (Majhgawan and Hinota).

Classification of Diamond Deposits

In past, diamond deposits in Panna area were classified by some workers on the basis of the nature of workings, like deep, shallow and alluvial. This classification was neither scientific nor practical. A classification based on the genesis and mode of occurrence of different deposits has been proposed as in Table 11.3.

Table 11.2: Geological Succession in Panna Area (Mathur *et al*, 1970)

Alluvium Laterite and lithomarge Basic intrusives	Rewa Group	Karauli Quartzite Jhiri Shale Itwa Quartzite Panna Shale Paisuni Formation
Vindhyan Supergroup	Kaimur Group	Dhandraul Quartzite Pipartola Conglomerate
	Semri Group	Tirohan Limestone Palkawan Shale Ken Limestone Pandawa Fall Quartzite
	Unconformity Unconformity	Bundelkhand Granite

Table 11.3: Classification of Diamond Deposits (Mathur, 1981)

Genesis	*Mode of occurrence*
Primary:	Kimberlitic bodies
Secondary:	Conglomerates
	(i) Near the base of Jhiri Shale
	---- Unconformity ----
	(ii) Associated with Itwa Quartzite
	(iii) Near the top of Karauli Quartzite
Detrital	(i) Older alluvial gravels
	(ii) Superficial gravels, Alluvium on
	(a) Kaimur Plateau
	(b) Rewah Plateau
	(iii) Detritus in stream gorges, cutting through
	(a) Kaimur scrap
	(b) Rewa scrap
	(iv) Recent detritus

Conglomerates must have acquired their diamonds from some primary sources like Majhgawan kimberlite. Since the age of Majhgawan pipe is certainly pre-Rewa, and all the conglomerates belong to Rewa Fm, it will not be far wrong to presume that there must have existed in that period several kimberlite diatremes which furnished the stones, but they are not known now.

Kimberlites

Only one kimberlite was known from this area in the beginning. It was referred to as agglomeratic tuff by Sinor (1930) who recognised its affinity with the 'blue ground' of Kimberley area, S-Africa. It occurs as a pipe intrusive through Dhandraul Quartzite, and has been a steady producer of diamonds. Field studies indicate that there must be several more such pipes in the area, since kimberlitic bodies have a tendency to occur in clusters. As a result of magnetic and electrical resistivity surveys a new pipe was proved by trenching and pitting at Hinota (Kailasam, 1970). It occurs under a thin mantle of soil and is exposed in a small nala that cuts across it.

During the course of mapping, Jhingran and Puri (1956) located a rock very similar to that at Majhgawan in Bundelkhand Granite terrain at Angor (24°44' : 79° 25') in Chhatarpur (M.P.). Its geologic setting is entirely different from that of the other two. It is situated almost 100 Km west of Panna occurrences.

Majhgawan

This diatreme is situated about 19 Km SW of Panna town. It has a pear-shaped outline on the surface; the maximum length is about 490 m and at its widest it is about 305 m. It has been mined by both OC and UG methods, shafts have been sunk and gallery and drives have been made at 91.5 m level for sampling and mining purposes.

Majhgawan rock is a kimberlite breccia, composed mainly of serpentine after olivine, with subordinate phlogopite, chlorite, carbonate minerals, magnetite, ilmenite, hematite, leucoxene, iddingsite, and perovskite (Mathur and Singh, 1971). Garnet is rare though it is an abundant constituent of kimberlites from other parts of world. The top section of pipe rock is light greenish in colour and is traversed by subhorizontally disposed calcite veinlets and stringers in great profusion; they diminish in abudnance with increase in depth when the rock becomes hard and very dark coloured, greenish black.

Thermal and X-ray studies (Kresten and Paul, 1976) show that most abundant minerals are vermiculite, serpentine and chlorite in varying amounts; in some samples phlogopite and smectite are fairly abundant. There is also minor amounts of palygorskite. It is a micaceous kimberlite.

Earlier a tenor of 11.95 carats/100 t was estimated (Mathur and Singh, 1963). During actual mining the tenor has improved to 12.5 carats/100 t, but at depth it falls down to 9.6 carats/100 t in UG workings.

Hinota

This diatreme is situated about 3 Km NW of Majhgawan. On the surface it has a rough oval shape, measuring 215 × 180 m, elongated along a general NW-SE direction. The pipe body at depth is inclined eastwards.

Top portion of Hinota diatreme is composed of highly weathered clayey and calcareous material of yellowish green and grey colour; with increasing depth the rock becomes harder and darker. Kimberlite is mainly basaltic with small portions of micaceous variety and kimberlite breccia. Accidental xenoliths of different kinds of shales, greyish quartzite, occasional jasper, argillaceous limestone and vein quartz are sometimes distributed in the main mass. These are obviously derived from the formations which the kimberlite diatreme has passed through. No cognate xenolith has been noted. Thermal and X-Ray studies of some borehole samples of this kimberlite showed the presence of phlogopite and Mg- vermiculite in all the samples, and serpentine, chlorite and smectite in all but the deepest samples (Kresten and Paul, 1976).

The pipe rock was also treated, and a gem diamond weighing 0.97 carat was recovered (Mathur, 1961). Subsequently, a more detailed investigation was carried out by GSI during 1970-72. Treatment of kimberlite and OB gave a good crop of diamonds, but the exploration showed that diamond potential was poor, with a tenor of only 0.72 carat/100 t of kimberlite. Diamonds were also extracted from the surface soil, gravel and highly weathered clayey pipe rock near the surface.

Angor

This diatreme consists of several relict bodies within Bundelkhand Granite, the largest of which is 450 × 200 m in size. The pipe rock has a varied composition, peridotite at lower levels passing upwards through olivine pyroxenite to pyroxene rich gabbro at higher levels. These rocks cannot be considered of kimberlitic affinity, although some diamonds have been recovered from gabbroidal rock. It is concluded that though diamond-bearing but it does not appear to be of any economic significance.

Chemical Composition

Chemical analyses of Majhgawan kimberlite show that they are similar to other kimberlites. Silica is akin to S-African kimberlites, but is higher than in Siberian kimberlites. Magnesia increases with depth, and on the whole is lower than in typical kimberlites from S-Africa and Siberia, but titania is significantly higher than in other kimberlites. Lime percentage in upper part of the pipe is higher due to the profusion of calcite veins, but decreases with depth as calcite veins also diminish and then practically disappear.

Paul *et al.* (1975) after extensive analysis of this and other Indian kimberlites observed that in the composition from the same pipe a good deal of similarity is seen, and kimberlites from the same locality have distinct chemical characteristics as is noted from the analyses of Majhgawan and Hinota kimberlites. In the norms calculated from these analyses olivine is as high as 44% in Majhgawan and 42% in Hinota.

Trace Elements

Paul *et al.* (1975) in their study of micaceous kimberlites observed that rare-earth distribution patterns in all the kimberlites are linear including those from Panna field. Total REE abundances are higher in these kimberlites than in the basaltic Anantpur occurrences, though the average La/Yb ratios are similar in both types of kimberlites. Compared with mantle abundances, La has been enriched 30-95 times but Yb enrichment is only 2-5 times in kimberlites. Similarity of REE distribution pattern in kimberlites from different geological provinces implies that enrichment in light REE is inherited from their source region in the mantle, or that magmatic processes have been sufficiently similar in all the provinces modifying REE patterns in a similar way.

Distribution of noble metals in kimberlites provides some insight into the behaviour of these metals during melting in a mantle environment. Minute traces of palladium, irridium and gold have been detected in samples from Majhgawan (Paul *et al.* 1979).

The F and Cl of Indian kimberlites are related to their geographical distribution. In Panna kimberlites they are higher than those of Anantpur kimberlites (Paul *et al.*, 1976).

According to Mathur and Alexander (1981) some metal trace elements in Hinota kimberlite show narrow ranges and uniform distribution except for Ni. The dispersion of these elements in the soil over kimberlite body is several times higher than over the surrounding kimberlite. Trace elements in the ash of some selected trees do not show any anomalies.

An analysis of Majhgawan kimberlite by Morgan and Lovering (1971) yielded values of 2.36 ppm of U and 12.0 ppm of Th. This gave a Th/U ratio of 5.1, which is low enough for Majhgawan diatreme to be reasonably productive on the basis of parameters determined by Akimov and Semenov (1970). Th and U abundance studies by Paul *et al* (1977) on Majhgawan samples show a very wide range in Th/U ratios; some samples showed an average of 6.74, which is close to the limits of diamondiferous kimberlites, and agrees with the result of Morgan and Lovering (1971). Some samples that gave ratios of 13.29 and 16.63 could well have come from the barren portions of the diatreme. Since there is no control on diamond content of the samples it is really premature to draw this conclusion, but the temptation is strong.

Ages of Panna Kimberlites

Majhgawan diatreme is intrusive into Dhandraul Quartzite and is therefore younger than it. Post-Kaimur age of Majhgawan rock is also supported by radiometric dates. Kaimur sandstone on the basis of glauconite ages determined by K/Ar method is considered 940 ± 30 M.y. to 910 ± 30 M.y. (Vinogradov *et al.*, 1964). Rb/Sr analysis of phlogopite separated from Majhgawan kimberlite gave an age of 1130 ± 20 M.y. (Mathur, 1964) and K/Ar analysis showed an age of 1056 M.y. (Crawford and Compston, 1969).

Paul *et al.* (1975) have given somewhat younger ages to Panna Kimberlites. According to them Majhgawan surface samples gave K/Ar ages of 974 ± 30 M.y. and 1004 ± 40 and samples from UG workings 980 ± 39 M.y. and 1120 ± 45 M.y., while Hinota sample showed 1170 ± 46 M.y. They suggest that their determinations should be taken as the minimum ages, but their figures fit in the overall pattern of dates.

Conglomerate

In past conglomerates in Panna area were important sources of diamonds, but the production from them has now declined greatly. The earlier workers recognised two horizons only of diamondiferous conglomerates (Sinor, 1930), but Mathur (1962) has found that there are, in fact, three distinct stratigraphic horizons in which they occur. He has shown that these conglomerate horizons form integral parts of Rewa-Fm and occur as intraformational conglomerates in them, and are incapable of being mapped even as separate members much less as formations. They have not been noted beyond the confines of Panna diamond field.

There is another horizon of conglomerate—Pipartola conglomerate which occurs towards the western end of Panna diamond field; and forms the lowest formation of Kaimur Group in Bundelkhand. It is of restricted outcrop, but attains a thickness of about 15m within a short distance. Its pebble lithology is essentially similar to that of the thin diamondiferous conglomerate layers in Rewa Group, and one would have thought that it would be diamond bearing as well. But no diamonds are known to have been recovered from it.

Conglomerates in Jhiri Shale

One of the most productive conglomerate layer occurs towards the base of Jhiri Shale. A few cm or meter of shale generally separates the conglomerate from the base of the formation, but in some cases it is found to occur directly over the top of the eroded Dhandraul Quartzite. Usual thickness of this conglomerate is about a few cm to 50cm; in a rare case a layer of 1.5 m thickness was observed in a very small exposure. Sometimes another horizon of diamondiferous conglomerate is present a few cm above the main layer, it is much finer and muddy in appearance, only 10-15 cm thick, and is known locally as Ranjika.

The main conglomerate layer consists of pebbles of jasper, quartz, quartzite (white and blackish), hematite, limonite, and some clay pellets, generally 10-15 cm. In size though larger and smaller sized pebbles are not infrequent. Pebbles are set in a hard, siliceous matrix. The finer type consists of pebbles or granules of < 8-10 mm size embedded in a shaly, muddy matrix.

Conglomerate horizons occur mostly under layers of shale sometimes as much as 20 m thick; its exposures are rare since being of very small thickness it gets easily eroded in the absence of protective shale cover. Although originally spread of conglomerate must have been much larger, its present extent is fairly restricted to a narrow band along the strike.

Conglomerates in Itwa Quartzite

Diamondiferous conglomerate occurs in two or more layers in this formation as seen in the abandoned pits; some outcrops are also seen when the overlying strata have been removed by erosion. Exact number of conglomerate layers are not known but it is believed that there may be several in the nature of shoestrings occurring at different levels in the formation. Since this formation is believed to extend southwards beneath Jhiri Shale in Rewa scarp, it is possible that virgin layers of this conglomerate may be encountered there, increasing the diamond potential of this horizon.

Conglomerate layers are usually 30-150 cm thick. In some pits a sandstone parting may be seen dividing the conglomerate. It is very similar to the coarse one in Jhiri Shale, both in its compaction and pebble lithology, except that green, hard quartzite pebbles also occur occasionally and are considered a good indication to diamonds by local miners.

Conglomerate in Karauli Quartzite

On the top of Karauli Quartzite numerous old workings for diamonds are scattered as discontinuous, widely separated clusters in a narrow, linear band between Khanger (24° 46' : 80° 37') in East and Sakeria (24° 39' : 80° 17') in West. The narrow band of old workings indicates that conglomerate was confined to a zone of small width along the regional strike. Conglomerate occurs beneath a cover of 1-3 m of sandstone as seen in the old workings. Its thickness is not known with certainty, since it is not seen *in situ* anywhere in old pits. However, at Khangar a small outcrop is seen whose thickness is about 60 cm; it is possible that original thickness here was greater and some of it has been removed by erosion since it is exposed on the surface.

This conglomerate in appearance and pebble lithology is different from the other two associated with Jhiri Shale and Itwa Quartzite. This is a kind of boulder bed consisting of rounded subrounded boulders of quartz arenite 10-30 cm or even larger in size enclosed in a ground of a conglomerate consisting of about 5 cm size red iron-ore, jasper, chert, vein quartz and quartz arenite pebbles set in a ferruginous matrix.

Detrital Deposits

Diamonds have been obtained from almost the entire Kaimur plateau in Panna area. They occur in lateritic soil and alluvial cover, which may vary from a few cm to as much as 15 m. The gems in the placers appear to be derived from disintegration of the conglomerates, and to a small extent from the weathering of kimberlites. Quite often pay gravels are embedded in the joints and fissures in sandstone. A 0.5-1.5 m thick diamond bearing gravel also occurs in a layer under cover of 8-10 m alluvium deposited along the channel of Baghen river. This river and other streams washing through diamondiferous areas, carry some diamonds along with other particles during floods and deposit them in sand and gravel in their courses, particularly in narrow gorges after they descend into the plains from Vindhyan scarps. Assessment of this gravel showed that its diamond conent ranged from 23 to 30 carats/100 t of gravel (Mathur, 1956,) or 0.41-0.54 carat/m^3, but a more extended sampling by NMDC gave an average of 0.235 carat.

Superficial gravels attract the local miners who rarely go deeper than 1 or 2 m, and are called 'chilla' locally. They are generally associated with the soil cap occurring over Dhandraul Quartzite, and in quartzite fissures and joints. One notable feature of many groups of chilla workings is that they are situated along the sloping edges of ridges of Dhandraul rock.

Since the richest gravels occur on Kaimur tableland it is not unexpected that the streams flowing through it carry diamondiferous detritus. This tableland overlooks Bundelkhand plain at a height of 80 m or so, and the streams debouching from the escarpment have carved out narrow gorges with steep and spectacular falls at their heads. All along the gorges and in the valleys where they open out there are thick accumulation of alluvium and sand. All of these have supported diamond diggings in past, but there is no working now. The most well known of these are Salikpur Seha, north of Paharikhera (24° 51' : 80° 31') through which flows Baghain river, and Lakhanpur Seha through which pours Ranj river north of Panna. There is a deep pool of water at the foot of Baghain fall called Brahspatkund. The sediment accumulated at the bottom of this pool might pay profitable dividends by bringing up diamonds which must have had accumulated there over the centuries brought down by Baghain river. Other gorges along Kaimur scarp also deserve to be prospected for diamonds.

Gorges similar to those indenting Kaimur plateau cut through Rewa tableland, but they are much shorter and not as deep.

Physics of Panna Diamond

It is a well known fact among Panna miners and diamond merchants that the physical characters of diamonds occurring in Majhgawan kimberlite is distinctly different from those found in conglomerates and gravels. A local expert can tell the source merely by looking at a stone. These differences also have a far reaching impact on the question whether Majhgawan pipe was the original source which provided the conglomerates with their diamonds. If the diamonds in conglomerates are physically different from those in Majhgawan pipe, we must look for other pipes, not yet known as their birth place.

International grading of diamonds is into two qualities, namely, gem and industrial. But in Panna another class is added, that of off-colour; the stones which have a dark hue and are of a shade between the clear of the gem variety and dark of the industrial are put in this class. These are also cut for jewellery of cheaper variety.

After a study of the proportion of two main varieties of diamonds found in different types of deposits in Panna field, Mathur (1957) concluded that the percentage of industrials from Majhgawan ranged from 53 to 76% (average 65.5%), conglomerates of Shahidan ranged between 7 and 17% (average 14%), and from Ramkheria gravel from 9 to 22% (average 18.5%). This shows clearly that conglomerates and gravels yield better quality stones of gem variety. In computing these data the off colour stones have been included in gem category.

Analysing the size and value data of diamonds from shallow mines, Mathur (1976) observed that the average comes to 1 carat/1 diamond. This may be compared against the average of 0.443 carat/stone from Majhgawan kimberlite, and 0.528 carat/stone from Ramkheria gravel.

Crystal Form

Sinor (1930) carried out for the first time a crystallographic study of diamonds from Panna field. Most of the diamonds weighing >1 ratti (about 1 carat) belong to the hexakis octahedron or to tetrakis hexahedron. Most of the smaller diamonds occur in the hexakis-tetrahedron form. Also the forms such as simple octahedron, cube or dodecahedron are not found among Panna diamonds. Crystals with distorted forms with marked surface corrugations, flat crystals, barrel shaped etc., also occur. Some of hexakis-octahedra of this area appear almost spherical owing to their strongly curved contours, while in others, octahedral outline shows clearly owing to the six sided pyramids being somewhat shallow.

A detailed study of crystal forms of Panna diamonds was made by Ramasesan (1944), who recognised hexakis-octahedral and tetrakis-hexahedral as the main forms; he also found several twinned crystals. All crystals have surface markings of striations and pits.

Rounded forms of Panna diamonds have also been observed by Raman (1951) who says that their faces are smooth and lustrous, characteristic of the finest diamonds from that locality. They also show a pattern of sharp ridges which divide the curved faces into 24 triangular areas. The curved surface of these diamonds is not continuous but consists of distinct sections (Raman and Ramasesan, 1946).

The 52 uncut large diamonds strung in a necklace owned by king of Panna showed intense to very weak (and in two specimens unobservable) luminescence (Raman, 1944). In these and other Panna diamonds blue luminescence is characteristic of colourless and perfect specimens; imperfect stones sometimes show blue, greenish-yellow or mixed type of luminescence. Strongly luminescent diamonds emit a yellow phosphorescence when the exciting radiation is cut off. Raman and Randall (1944) noted that the perfect diamond which are blue luminescent and UV-opaque type are optically isotropic and their birefringence, if any, is <1 part in 1000,000. No systematic change in magnetic susceptibility with the variation in colour, absorption, luminescence and photoconductivity was found in Panna diamonds (Sigamony, 1944).

Most Panna diamond crystals contain inclusions; the solid ones occur mainly as discrete mineral crystals and few as aggregates, while the gaseous inclusions are of CO_2 (Phukan, 1971).

Andhra Pradesh

The richest diamond fields in the country, which gave some of the most famous diamonds to the world in past, occurred in A.P. The stones were obtained from gravels along Krishna and Penner rivers and conglomerates in Kurnool, A.P. Some diamonds were also picked up from Wajrakarur area, but they were from the soil in the surrounding area. True identity of Wajrakarur as a pipe was not recognised in the older times.

The S-Indian deposits are also classified into (i) primary kimberlitic (ii) conglomerates and (iii) placer and detrital, on the basis of their genesis. Each type is restricted to separate regions, unlike in Panna field in which they occur together in one belt.

A classification of diamond deposits in A.P. is given in Table 11.4.

Table 11.4: Genetic Classification of Diamond Deposits of A.P.

Mode of Occurrence		*Types and deposits*
Kimberlite Bodies	(a)	Diatremes (Anantpur district)
		(i) Wajrakarur -3 (ii) Lattavaram -2
		(iii) Muligiripalle-1
	(b)	Dykes (Kurnool District)
		(i) Chelima
Conglomerates	(a)	In Banganapalli Quartzite (Kurnool district)
	(b)	In Gollapalli Sandstone (Gondwana/Rajahmundri)
Sandstones		Tertiary Sandstone (Krishna district)
Quaternary Gravels	(a)	Deep seated along river valleys
		(i) Krishna river (ii) Penner river
		(iii) Tungbhadra river
	(b)	Superficial Gravels
		Jonnagiri-Kalyandurga area (Anantpur district)

Kimberlites

A single diatreme of kimberlite has been known from S-India at Wajrakarur (15° 01' : 77° 23') since past 100 years, almost from the time when H. Carvill Lewis in 1877 coined the term 'kimberlite' for the rock in whose matrix occurred the diamonds in S-Africa. Geological and geophysical studies have delineated several more ultramafic bodies in Wajrakarur area and in an entirely new tract around Lattavaram (14° 55' : 77° 16'), about 19 Km S-W of Wajrakarur; two more pipes were noted during 1974-75, one near Wajrakarur and another at Muligiripalle (14° 51' : 77° 19'), all in Uravakanda area of Anantpur. They are intrusive into Peninsular Archaean complex comprising hornblende gneisses, composite gneisses and granulites with dykes of microgranite, dolerites, and pegmatites, etc. Three of these pipes are diamond-bearing, but the incidence is so low that they can be considered as only of academic interest, prospecting is continued in an effort to establish a pipe with a diamond content high enough to make it a commercial success.

Pipe No. 1 (Wajrakarur)

Diamond mining in Wajrakarur area has been going on since times immemorial, not in pipe rock but in alluvial tracts. Though pipe rock itself has given disappointing results, the soil around the village throws up diamonds regularly after heavy rains. Some of the stones found are of considerable size but their origin has never been discovered.

Wajrakarur pipe has been investigated by geophysical surveys, trenching, drilling and beneficiation and it has been proved that the pipe is diamond bearing, but its economic viability is not established.

The pipe occurs close to the village. Four large oval patches of kimberlitic rock were demarcated after a study of few surface outcrops, exposures in old shafts and prospecting, pits. Geophysical work showed that all these patches are interconnected to form a diatreme over 1100 m long and 60 to 180 m wide on the surface aligned in a N 60°E direction (Rao and Phadtare, 1966). Matrix of this rock is highly altered and difficult to recognise; it is made up mainly of serpentine, bowlingite or xylotile, secondary amphibole, chlorite and zeolite-like prehnite; primary minerals like olivine are rarely seen.

An ilmenite megacryst from this pipe contains inclusions of diopside and Fe rich garnet. Giekilite content of ilmenite is 39% and is within the range found for other kimberlitic ilmenites (Akella *et al.* 1979).

Pipe No. 2 (Wajrakarur)

This body was discovered by geophysical survey carried out by GSI. It is situated 3.2 Km East of the main pipe, and is crescent shaped with convexity towards North; its maximum length is 300 m and it is 70 m at its widest. The pipe rock is very hard and compact, melanocratic and microcrystalline. It is made up mainly of olivine and iron oxides embedded in a cryptocrystalline groundmass. Olivine is highly altered to serpentine. Other minerals are garnet, bowlingite, iddingsite, ilmenite and carbonates (Rao and Phadtare, 1966). No diamond was recovered from the rock itself.

Surface samples of this rock contain olivine, phlogopite, serpentime, chlorite and a little calcite and apatite. Kresten and Paul (1976) suggest that chlorite exceeds serpentine. Chlorite, probably after olivine, constitutes the bulk of iddingsite reported.

Pipe No. 3 (Lattavaram)

This body, roughly bean shaped, 75 m at its longest and 25 m at its widest, is situated 1 km east of lattavaram. No rock outcrop is seen since the pipe is covered by a mantle of kunkery soil at least 50 cm thick. Under this lies a light bluish rock, soft and friable, in which are embedded very hard dark blue boulders and pebbles of fresh rock. The softer material has yielded diamonds.

Akella *et al.* (1979) found that this kimberlite consists of phenocrysts of olivine in a very dark, fine grained matrix of serpentine, calcite, phologopite and magnetite rich spinel; minor monticellite and pectolite are also present. Serpentinised olivine phenocrysts are interesting in that the veins of serpentine contain numerous minute needles of nickel sulphide crystals about 2 mm long with triangular cross-sections. Electron microprobe analysis of this phase showed a maximum content of 72% Ni by weight, strongly suggesting that it is hazelwoodite (Ni_3 S_2). Analysis of minerals from this pipe indicates that xenoliths from this pipe are eclogite, made up of Fe-rich garnet and diopside beween whose grains can be seen irregular patches of white alteration material (possibly clay).

Kiberlite samples examined by Kresten and paul (1976) contains large amounts of serpentine (>2/3 by weight), some calcite and little vermiculite; in contrast to other S-Indian kimberlites this does not contain detectable olivine.

Pipe No. 4 (Lattavaram)

This pipe, roughly egg shaped on the ground is situated 1.6 Km east of Lattavaram; its maximum length is 265 m and maximum width 130 m. Fresh, unweathered rock is dark blue, very hard and compact. It is coarse grained and porphyritic, and in composition very similar to rock from pipe No. 3; hornblende occurs in addition (Rao and Phadtare, 1966).

Pipe rock contains phenocrysts mostly of olivine (mostly serpentinised) and occasionally of garnet (Akella *et al.* 1979). Other observed phases are ilmenite, sphene and perovskite. Groundmass is made up predominantly of serpentine, calcite, minor phlogopite, spinel and ilmenite.

DTA and X-Ray studies of this pipe rock (Kresten and Paul, 1976) reveals the presence of olivine, calcite, phlogopite, serpentine, chlorite, and massive variety of palygorskite. This kimberlite does not contain detectable vermiculite, unlike those from Panna field or Pipe 3 of Lattavaram.

Pipe No. 5 (Muligiripalle)

A dyke-like body, tredning E-W about 250 m long and about 4 m at its widest was noted by GSI. It is a micaceous kimberlite composed essentially of olivine and phlogopite with perovskite, micas and magnetite as accessories, embedded in a groundmass of serpentine and carbonates. No diamond has been found in it.

Pipe No. 6 (Wajrakarur)

It is situated about 1.5 Km west of the village concealed under a blanket of soil and was discovered by GSI by geophysical survey. It is almost rounded in shape and is 250 m across. It is highly weathered upto 43 m from surface. Pipe rock is highly brecciated and exhibits porphyritic texture. It is made up chiefly of pseudomorphs of olivine in a groundmass of carbonates and serpentine; the accessories are garnet, diopside, spinel and ilmenite. Olivine is generally altered to serpentine and iddingsite. No diamonds have been found in it.

Chelima Dykes

According to Sen and Rao (1970), interesting dykes of rocks transitional between kimberlite and carbonatite occur around chelima (15° 27' : 78° 42'), in Kurnool district, traversing Cumbum Fm (Cuddapah S.G). Several old pits are seen in dykes, which may be ancient workings for diamonds, though there is no past record of diamond mining in this locality.

Petrologically they consist of serpentine, biotite, phlogopite, ilmenite, magnetite, apatite, rutile, perovskite, carbonates, leucoxene, chlorite and quartz. Serpentine appears to be after olivine, which was later replaced by carbonates also. A very minute crystal, <1/100 mm in size, of diamond was observed in one of the thin sections. X-ray fluorescence examination of dyke rock has shown the presence of some trace elements.

Chemical Composition

Anantpur kimberlites show overall similarity in their composition (Paul *et al.* 1975). Silica from 32.25 to 39.04% similar to Panna content, and so is magnesia. The significant difference is in lime which is 8.61 to 20.70% and usually do not show the profusion of calcite veins that are observed in surface rock from Majhgawan. Titania is lower in these samples, varying form 1.25 to 2.45. Total $H_2O + CO_2$ is higher than in other kimberlites.

Chemical composition (in %) of diamond bearing volcanic rocks in Wajrakarur and Lattavaram areas are as below (After G.S.I) :

	SiO_2	TiO_2	Al_2O_3	Fe_2O_3	FeO	MgO	CaO	Na_2O	K_2O	H_2O
(a) Wajrakarur	33.72	2.45	6.53	6.83	4.49	20.22	15.75	0.43	2.04	6.42
(b) Lattavaram	39.04	1.25	4.32	3.28	4.91	28.28	7.92	0.34	1.10	7.04

Trace Elements

Several trace element are abundant in Anantpur kimberlites and Chelima dyke rocks which are widely variable in these kimberlites, and even the ratios do not give any discernable pattern (Gangadharam and Aswathanarayana, 1970).

Total REE abundances in the Anantpur basaltic kimberlites are lower than those of Panna occurrences (Paul *et al.*, 1979). They also determined the noble metal contents of pipe 2. According to them F and Cl abundances are lower in Anantpur kimberlites than in those of Panna. Investigations by Gangadharam and Aswathanarayana (1970) and Paul *et al.* (1977) on Th/U ratios of Anantpur kimberlites and Chelima dyke rocks gave widely divergent results, and hence more work is needed that can be used as indicator of diamondiferous aspect of pipes.

Ages of S-Indian Kimberlites

The whole rock K/Ar ages of Wajrakarur and Lattavaram kimberlite samples (Paul *et al.* 1975) give the following results.

Pipe-1 Wajrakarur	840 ± 33 M.y.
Pipe-3 Lattavaram	956 ± 38 M.y.
Pipe-4 Lattavaram	1023 ± 40 M.y.

Paul (1979) later studied Sr isotopic composition of these kimberlites. The $^{87}Sr/^{86}Sr$ ratios were plotted against $^{87}Rb/^{86}Sr$. It was noticed that Rb/Sr isochron ages obtained for Pipe–2 Wajrakarur is 1350 M.y. and for Pipe-4 Lattavaram is 1505, in both cases higher by about 500 M.y. than those obtained from K/Ar determinations as in the case of Majhgawan kimberlite. Crawford and Compston (1973) obtained $^{87}Sr/^{86}Sr$ ratio for Lattavaram and Muligiripalli, and found that these values fit perfectly the isochron for Majhgawan kimberlite at 1140 M.y. but are far out from the isochron obtained by Paul (1979).

Crawford and Compston (1973) gives an age of 1225 M.y. for Chelima dyke rocks. The difference in the age of Chelima rock and kimberlites (both in Anantpur, A.P. and Panna area) is quite significant.

Banganapalle Conglomerate

Conglomerates associated with Banganapalle Quartzite, forming the lowest formation of Kurnool Group, are known since ancient times to be repository of diamonds and have been exploited extensively. Occurrence of diamonds in this formation was one of the reasons for correlation of Kurnool with the Vindhyans by the earliest geologists (King, 1872).

According to Rajaraman *et al.* (1980) Banganapalle quartzite encloses numerous layers or lenses of conglomerate. In Banganapalle and Yembai areas the thickness of conglomerate is about 1 m or a little more; but at some places it attains a thickness of almost 4 m. Some sections show the presence of more than one conglomerate horizons, mostly lensoid, in the formation. Conglomerate is hard and compact containing subrounded to rounded pebbles of varied coloured banded jaspers and chert up to 2 cm in size. Matrix consists of quartz grains cemented with ferruginous and siliceous material. Another horizon which appears to have been worked by the ancients is a friable ferruginous grit, 10-15 cm thick, which occurs below the conglomerate and rests directly on older rocks.

Banganapalle quartzite formation is widely distributed in Cuddapah Basin mainly in Cuddapah, Kurnool and Krishna districts. Conglomerates occur in several tracts of this formation, in many of which evidence of old workings and excavations for diamonds can be seen. There are still other tracts of conglomerate which are virgin and deserve a closer look.

It may also be stated that the famous Kollur deposits (16° 42' : 80° 01') in Guntur district were not entirely gravel, but the gritty layers in Kurnool quartzite (presumbaly Banganapalle) were also worked about 8 Km E and SE of this locality and in lower hills around Maidarsal peak.

Similar workings occur in the hills near Ustapalle (16° 42' : 80° 08') too. The more extensive alluvial workings here along Krishna river could have derived their diamonds from this horizon. In all Banganapalle tracts the alluvial material is also diamondiferous. Prospecting at Ramallakota (15° 34' : 78° 00') has proved the diamondiferous nature of this conglomerate are quite widespread, and the ancient miners appear to have neglected much of the area.

Younger Conglomerate

Widespread old workings for diamonds occur beween Malavalli (16° 40' : 80° 63') and Gollapalle (16° 43' : 80° 55') in Krishna district in Gollapalli sandstone (Gondwana) and overlying Rajahmundri sandstone (Tertiary) along the contact. This raises an possibility that diamonds could occur in a much younger stratigraphic horizon, and Krishna river gravels could have derived part of their stones from this formation. This also widens the area of search in non traditional locations.

Alluvial Deposits

Gravels found along Krishna and Penner rivers have been the richest repository of diamonds in past, and some of the most famous diamonds in history were mined from them. Production in these deposits completely ceased a century ago, but their most glorious period ended much before. On the basis of past records many of these localities have been identified and efforts are under way to examine them afresh to determine whether they or their extensions have any further potential.

Partiala-Ustapalle area was one of the most productive in old times. It is situated on the left bank of Krishna in the district named after this river. Partiala (18° 39' : 80° 25') was considered eastern limit, while Ustapalle (16° 42' : 80° 09') forms probably western end of diamondiferous gravels. Partiala is about 30 Km from Vijayawada. Although this tract has a long history of diamond mining very little evidence of past activities survive now since much of the tract has been reclaimed for cultivation and ground levelled.

Investigations in this area, including geophysical surveys have not been successful in delineating gravel patches. Treatment of some gravel in Partiala block resulted in the recovery of 3 diamonds weighing 4.27 carats (Deshpande, 1975) confirming this indeed a diamondiferous area.

Main source of diamonds in Kollur area evidently was the conglomerate associated with Banganapalle Quartzite formation and the derived eluvial detritus. A secondary, and apparently also profitable repository, was the alluvial gravel found along the banks of Krishna though to stretch between Pulichinta village (16° 43' : 80° 2') and Kollur.

Ancient diamond workings also exist further upstream of Krishna near Amargiri (16° 04' : 78° 23') up to about 4 Km east of Bollarum (16° 05' : 78° 26') in Mahaboobnagar. Bollarum area is rocky, and there are two parallel lines of ancient workings apparently in old terraces of the river. The ground near Amargiri is flat and gravel occurs in alluvial terraces on river bank. There is a belief that the famous 'Nizam' diamond was obtained from Amargiri workings.

Another important river along whose banks there existed prosperous diamond workings is Panner, which drains the southern part of Cuddapah basin. Old pits have been noted between Chennur (14° 34' : 78° 48') and Doriampalle (14° 32' : 78° 50') in Cuddapah (A.P.), on both the banks of the river.

Other Areas

Jungle Kimberlite

Possibility of occurrence of kimberlite plug in the jungle area was first suggested by Tewari (1971). Subsequently, during geophysical work in 1974 two magnetic anomalies were noted which were geologically investigated and found to be occupied by kimberlitic rock. GSI thereafter, mounted an integrated, multidisciplinary study which resulted in the discovery of the very first unknown diamond bearing area outside those tracts which have been traditionally known to contain diamonds.

Jungel valley is situated about 32 km NW of Obra in Mirzapur (U.P.). As a result of geological and geophysical investigations five elliptical plug-like kimberlite bodies have been located in basic and ultrabasic rocks in the area (Chattopadhyay and Venkataraman, 1977).

The rock is highly weathered and friable in the upper zones; it is dominated by serpentinous clay, tremolite, actinolite, magnetite, ilmenite, calcite and spinel; veins of calcite and asbestos criss-cross the rock. The texture is porphyritic to clastic porphyritic. Pseudomorphs of olivine and pyroxene represented by antigorite and chlorite are well distributed. The groundmass is dominantly ferruginous and glassy. Xenoliths of magnetite rich rock, crystalline limestone and basaltic fragments are common. In boreholes the texture of rock is tuffaceous to agglomeratic.

The percentage of oxides and oxide ratio of this rock are comparable to other kimberlites. Normative compositions show unusually high apatite and illuminite and general lack of clinopyroxene. Ternary plots of oxide relationship fall within the kimberlite field and electron microbrobe studies confirm this identification.

The rocks from Jungel Plug-II yielded small diamonds of gem quality, proving the rock diamondiferous; its commercial viability is doubtful, however, since later efforts of beneficiation have yielded negative result. Investigations are continuing in other plugs as well as to prove whether they are also diamondiferous.

Mahanadi Valley

One of the areas in India from where diamonds were obtained in ancient times is Mahanadi valley in Sambalpur (Orissa) and adjoining territories. Winning of this gemstone seems to have ceased during the middle of 19th century.

Ball (1877) considered that the source of diamonds in Mahanadi were the rocks of the nearby Vindhyan formations. These formations are now referred to Chhattisgarh Super Group. He could not identify the actual bed or beds from which diamonds were derived but on general lithological resemblance of sandstones and shales of Barapahar hills nearby with diamond-bearing beds of Bundelkhand and Banganapalle he opined these rocks as the matrix. He suggested that the search for original source of diamonds be confined, in first instance, to the streams and small rivers which rise in Barapahar hills and join Mahandadi on south.

GSI carried out systematic mapping, sampling of conglomerates and panning of Mahanadi sands in this area and the adjoining Chhattisgarh, but without success.

Chandrapur

One of the ancient areas from where diamonds were mined is Wairagarh (20° 26' : 80° 10') in Chandrapur, Maharashtra. Later attempts to locate diamond bearing horizon here have not met with success. Suspected old workings exist in an autoclastic conglomerate in felspathic quartz schist.

Jharkhand

Although there are no diamond digging areas in Chhota Nagpur territory of Jharkhand now, in past it was a productive diamond region. Diamonds were found in the basins of Sankh and Koel rivers which together become Brahamani river. Investigations were carried out in past years in N-Koel area (Palamu district,), but without success.

Simla Area

Although some ancient Sanskrit texts mention occurrence of diamonds in Himalaya, no definite areas were recognised. In 1870, some small diamonds stated to have been found in a small stream near simla were sent to Geological Museum at Calcutta (Ball, 1881). Nothing however appears to have been heard of this stream or Simla area afterwards in connection with diamonds.

New Diamond Fields

New kimberlite fields have been discovered in Raipur and Bastar districts of (Chhattisgarh), Nuapada and Bargarh districts (Orissa), Narananpet-Maddur-Krishna areas (A.P.) and Richur-Gulbarga districts of Karnataka.

Diamonds are not only prized stones for use in jewellery but are also a valuable industrial material. Demand for both the varieties of diamonds is steadily increasing. Preeminent position of India as a producer of diamonds is no more although we are emerging as an important supplier of finished diamonds on the basis of roughs imported from abroad. Annual production of this stone in India in 2006-07 was 2769 carats.

Most of our known deposits and fields are practically exhausted, and few deposits have come to light during recent years. This state of affairs can be rectified to some extent by adopting a bold and farsighted exploration policy. Moreover, some of the existing deposits could be exploited more intensively by expanding their mining and treatment. This could be accomplished by an ambitious and wide-ranging program.

Discovery of vast Yakutia diamond field in USSR after World War II is a saga in itself; it has shown that perseverance coupled with sound geological judgement can do wonders. In 1980 a large diamond field has been discovered in NW-Australia. Discovery of diamondiferous kimberlite pipes in virgin jungel valley clearly shows that intelligent enterprise is rewarding.

There is still great potential lying hidden in Panna field and its extensions, and in some tracts of A.P. Besides looking for diamonds in traditional areas, we have got to cast our exploration net wider afield. We need to identify new prospective areas to which we may direct our search for additional and new diamond resources, and further exploration that may be undertaken.

Let us first take known diamond regions. Of the two, Panna field has been studied more systematically. But a considerable amount of search still remains to be made in this field and adjoining territories in strike continuation. There is need to cover this entire territory with systematic ground geophysical surveys. Wherever geophysical anomalies are favourable the location should be test drilled for possible concealed pipes and buried river channels that may be carrying diamondiferous gravels. The best results can be obtained where geophysical surveys are integrated with geological and geomorphological studies. Aerial photographs may also be of considerable use in showing up depressions and other features that may be occupied by hidden kimberlite bodies. Geomorphological features may also assist in such search. There are several likely spots for search for concealed diatremes in Panna area, several of which have still to be tested. One of the strong possibilities is Pandwa Fall, and another is chimney-like vertical hole in Dhandraul Quartzite near Tikaria railway station known as Biradh Kund.

Finding a new kimberlite body is not the end of the quest. Whether it is diamondiferous and economically workable has to be tested by actual mining and treatment of the rock. Determining whether a kimberlite contains diamonds or not, or of demarcating pay portions in a productive pipe can be done on the basis of Th/U ratios. Aswathanarayana (1971) has suggested that gross radioactivity measurements for locating kimberlite areas from the air and Th/U ratio measurements for locating diamondiferous zones in kimberlite on ground, could be conveniently made by γ-ray scintillometer for making ground determinations for Th, U and K in diatremes.

The problem of source of diamonds in different conglomerates and gravels in Panna field is not an academic problem alone; if the source could be identified it is bound to reveal many more diamondiferous kimberlites as well as extensions of rich conglomerate horizons. Sedimentological studies of known conglomerate and other sediments and rocks in field may throw light on the provenance of diamond in conglomerate. Since most of the conglomerate horizons are covered by sediments this is no easy task. A beginning could be made, however, by cleaning old, disused pits and exposing the conglomerate and their host formations need to be traced in the tracts beyond traditional limits of Panna field.

Prospect for deep alluvial gravels seems to be limited on Vindhyan plateau, but the river gorges that dissect the scarps hold much promise. For superficial gravels on Kaimur and Rewa sandstones the outlook is fairly bright. Geomorphic and stratigraphic features are good guides for search of these gravels.

A.P. diamond belts have been more extensively exploited, particularly gravel deposits along the river courses. We should be optimistic since judicious exploration is bound to give encouraging results. Geophysical exploration for kimberlites in Anantpur (A.P.), has already resulted in the discovery of several pipes and possibility of locating more is quite bright. It is however necessary to carry out extensive production tests before any pipe is rejected as uneconomic. These surveys need to be extended over wider areas. Possibility of occurrence of diamondiferous gravels in Wajrakarur area should be thoroughly explored.

Conglomerate deposits appear to be the most promising. The old miners did not work them as massively as the gravels because of difficulty in breaking hard rocks to get at them. There are still vast areas of Banganapalle outcrop which deserve careful scrutiny.

There is general belief that gravels along Krishna and Penner rivers are practically exhausted. But there still remain large tracts of potentially productive virgin ground that will amply reward careful exploration. If one looks at the map of old workings along these two river valleys one can observe two facts : firstly there are long stretches of unworked ground in between groups of workings, and secondly, the old diggings appear to be confined to only one bank of rivers, the other being left untouched. A study of geomorphic features of these valleys made on topographic maps have failed to show any character which could have resulted in the deposition of gravels only in traditional areas and not in those that have been left unworked by the ancients. These valleys deserve another careful examination, particularly the virgin tracts and heads of smaller strengths that debouch into main rivers. Geochemical studies of certain trace elements which characterise the soil over kimberlites could also be used in exploration for diatremes.

One of the most spectacular developments in diamond mining has been the discovery of extremely rich deposits along SW coast of Africa in terraces and off-shores. A notable feature of this territory is that there are no rivers draining this part of the desert and no one knows where diamonds have come from. Similar diamond gravel terraces may occur between the mouths of Krishna and Penner rivers along E-coast of India. The likely spots to search would be the gravel beds or loosely cemented shell horizons under dunes and along raised beaches, and a little beyond the shoreline towards sea.

Industrial uses: Diamond is the hardest substance known in world. Its hardness in Moh's scale is 10, but 10 is not a true or qualitative measure of its hardness compared to that of other materials. There is far less difference between hardness 6 (of orthoclase) and 7 (of quartz) than between 9 (of corundum) and 10 (of diamond) in Moh's scale. Therefore, several artificially manufactured substances (like fused zirconia and tantalum carbide, fused alumina and tungsten carbide, silicon carbide and finally boron carbide) have been included to measure the relative hardness of substances like special steels which are harder than corundum and softer than diamond. Because of its extreme hardness, diamond is one of the most important and essential materials in modern industry. Principal industrial uses are in diamond-drill bits used in prospecting and mining; in diamond dies for wire-drawing; in diamond-tooth saws for sawing stone, glass, quartz, metals, etc.; in diamond-tipped tools for truing and dressing abrasive wheels, boring and cutting metals, hard alloys, bakelite, hard rubber and plastics, etc., cutting and drilling glass, porcelain, and engraving gems and scales; and in diamond-precision tools used in automobile and airplane engine parts and scientific instruments. Diamond dust or powder is used for cutting hard gemstones and quartz plates for radio and for polishing gems. Diamond is the only material used for cutting or engraving the diamond itself, and only diamond-tools can cut and shape high-speed tool-steel and hard and tough alloys, used extensively in manufacture of war materials. There is no satisfactory substitute for diamond in this respect. Attempts have been made to produce under conditions of high pressure and temperature, synthetic diamonds large enough to meet the requirements of some industrial uses.

Gem diamond : On the basis of quantity >3/4th of diamonds produced in world are consumed in industrial purposes, but on the basis of value, gem varieties predominate. Diamond has a high refractive index (2.417 for sodium light) and a strong dispersion (0.063) which gives it an unexcelled brilliance or fire when it is cut as a facetted-stone. It has been the most prized of the gems since >2,000 years (references to diamond and other gemstones occur in *Arthashastra* of Kautilya, written between 321 B.C. and 296 B.C.). Price of diamond depends upon its weight, quality, shape and flawlessness. Gem diamonds are transparent and are colourless or show faint shades of different colours. Transparent water-clear diamonds are known as 'of the first water' or 'blue whites'. When yellowish tinge is present, they are termed 'off-colour stones'. Diamonds with green, blue or red shades are rare but are the most valuable gems. Attempts have been made to colour the natural diamonds artificially. Lustre on uncut crystal-faces is greasy adamantine, but on cut faces it is typical adamantine. Diamond found in alluvial deposits usually shows an unattractive oily glassy lustre but for its relative higher specific gravity (3.2-3.5), it is likely to be confused with quartz. Diamond is transparent to X-rays, while lead glass imitations are not transparent to them.

Flawless stones of good colour are employed in gem trade while the off-colour, flawed and defective stones, chips and cuttings as well as small grains and dust are used industrially. Other types of diamonds used in industry are :

(1) *bort* - low grade imperfectly crystallised diamond of yellowish grey to black colour.

(2) *ballas* - very hard, tough, globular mass of imperfect diamond crystals arranged radially.

(3) *carbonado* or *black diamond* - very hard and extremely tough and non-cleavable aggregate of dark grey to black opaque diamond crystals.

India has been a centre of diamond cutting and polishing. This has led India to establish it as a major exporter of 'cut and polished' diamonds (Table 11.5).

Table 11.5: (A) Export of Gems

Year	*Diamonds*		*Coloured gemstones*	*Others**
	$ M	*M carats*	*$ M*	*$ M*
1994-95	4020.91	15.807	141.46	32.85
1995-96	4661.90	19.211	147.32	79.69
1996-97	4235.10	18.883	136.94	138.28
1997-98	4492.66	20.555	132.81	109.27
1998-99	5026.11	26.799	178.63	155.47

(B) Import of Gems

Year	*Rough* *$ M*	*M carats*	*Rough col. gemstones $ M*	*Others* $ M*
1994-95	2792.36	729.08	58.59	16.31
1995-96	3274.41	899.01	71.46	25.68
1996-97	3382.51	1029.01	71.66	22.50
1997-98	3036.15	1066.12	70.80	147.59
1998-99	3343.18	1526.95	45.05	256.87

Source: Gem & Jewellery Export Promotion Council, 2001

Interesting Facts About Diamonds

There are 60 facets of the most expensive substance diamond:

1. Diamonds are hardest substance known to man.
2. They are pure carbon that has crystallised many million years ago.
3. Every diamond in world is unique. No two are the same.
4. Diamonds are weighed in carats—the word comes from the seeds of 'carob' tree because they have a uniform weight, ancient gem dealers used as weights for their scales.
5. A single diamond weighing 2 carats is worth more than twice as much as two one-carat stones of same quality.
6. Only a diamond can cut another diamond.
7. Best colour for a diamond is no colour—the cool white of an icile or a gin and tonic. The rarest stones are this pure white, called 'exceptionally' white.
8. Best way to see a rough diamond's true colour is to view it against a white background.
9. Rough gem-quality diamonds are sorted into >5,000 categories and sold at diamond 'sights' in London ten times a year.
10. A rough diamond looks like a dull glass pebble.
11. Most polished diamonds have 58 facets to bring out their maximum sparkle, fire and brilliance.
12. These facets are ground on to the diamond one by one.
13. Simplest way to clean a diamond is in a small bowl of warm water to which a mild soap has been added.
14. Dimonds can scratch each other. So don't jumble them together in a box.

15. Atmosphere of Venus was first analysed through diamond window of a US spacecraft. Only diamond had the strength and transparency to cope.
16. Until 15th century only kings wore diamonds as a symbol of strength, courage and invincibility.
17. In 15th century, when Archduke Maximilian of Austria gave a diamond engagement ring to Princess Mary of Burgundy, women began to wear diamonds.
18. A diamond of >1 carat in weight is one in a million.
19. A well-cut diamond will reflect the majority of light within the diamond—before reflecting it back out of the top giving the stone its brilliance and fire.
20. Diamonds have a grain like wood, along which cleaving is possible.
21. Diamonds can be sawn by a thin disc coated with oil and diamond dust.
22. Some diamonds can be 'cleaved' by hand with a blade. It has to be done with great skill or thousands of pounds can be lost.
23. A diamond will melt at 6,900° F. This is 2.5 times greater than melting point of steel.
24. Diamonds are very good conductors of heat (four times better than copper). That is why they are so cold to touch.
25. They are poor conductors of electricity.
26. Diamonds were first found in India >2,500 years ago. For centuries that was the only source.
27. In 18th century they were discovered in Brazil.
28. First diamond in S. Africa was discovered near to what is now known as Kimberley in 1866.
29. Diamonds are still mined there and are found in kimberlite which forms a volcanic 'pipe'.
30. In Namibia diamonds are found on Atlantic beaches. They have been washed down by rivers to sea.
31. On average 200 t of kimberlite have to be mined to produce a 1-carat polished diamond.
32. Only 20% of world's rough diamond production can be considered to be of gem quality.

33. The rest is used for industrial purposes—cutting, grinding, drilling or polishing.
34. Largest diamond ever discovered was Cullinan. It originally weighed just under 1.5 pounds.
35. Cullinan was cut into 9 major stones, 2 of which are in the Crown Jewels, and 96 smaller ones.
36. All the rough gem-quality diamonds found in world in one year could be carried by a 2.5 t truck.
37. Diamond-producing countries include Australia, Angola, Botswana, Brazil, Central African Republic, China, Ghana, India, Ivory Coast, Liberia, Russia, Sierre Leone, Tanzania, Venezuela and Zaire.
38. A carat is one-fifth of a gram. Each carat is divided into 100 points. A point is about the weight of three bread crumbs.
39. Word 'diamond' comes from the Greek adamas meaning unconquerable.
40. 'A diamond is forever' is understood and accepted by women throughout the world.
41. Fire is the blaze of rainbow colours into which a diamond breaks up the white light that is reflected out of it.
42. Scintillation is the flash of light that occurs when a diamond is moved, even fleetingly.
43. The simpler and thinner the setting of a ring, the better it will show off the full beauty of the diamond.
44. Early Egyptians believed the *vena amoris* (vein of love) ran directly from the heart to the top of the third finger, left hand. Hence, rings are often worn in this way.
45. Generally speaking, the larger the diamond, the rarer it is.
46. Diamonds are 140 times harder than rubies or sapphires and 180 times harder than emeralds.
47. Koh-i-noor ('Mountain of Light') is world's most famous diamond. Weighing 108.93 carats and oval, it was first mentioned in 1304. It is now among the Crown Jewels.
48. Cupid's arrows were said to be tipped with diamonds.
49. Blue Hope diamond (44.5 carats) is supposed to be unlucky for its owners. It was held responsible for death of Marie Autoinette and Louis XIV. Henry Philip Hope bought it in

London in 1830. All his family died in poverty. It is now in Smithsonian Institute in Washington.

50. Most diamonds have natural inclusions, however infinitesimally small. The fewer the inclusions, the better the diamond's clarity.
51. There are amber, blue, brown, cinnamon, champagne, green, pink, and yellow diamonds. These rare, coloured diamonds are known as 'fancies'.
52. Zsa Zsa Gabor once said 'I never hated a man enough to give him his diamonds back.'
53. Cutting and polishing normally reduces the weight of a rough diamond by more than 50%.
54. Most popular diamond shape is 'round brilliant.' But we can also get pear, oval, marquise, heart, emerald, square and baguette cuts.
55. Legend has it there is an inaccessible valley, situated on Khorasian border in Central Asia which is carpeted with diamonds. It is said to be 'patrolled by birds of prey in air, and guarded by snakes of murderous gaze on ground'.
56. 100 years ago, diamonds were often mounted in silver. Today, gold and platinum (hard yet malleable) are the perfect complement.
57. 'Diamonds are a girl's best friend' was the title of a song in the musical 'Gentlemen Prefer Blondes'.
58. Four characteristics affect a diamond's value-cut, clarity colour and carat-weight-the 4Cs.
59. Greeks believed the fire in a diamond reflected the constant flame of love nothing has changed.

12

Gold Deposits

Gold, the king of metals, has been highly valued by mankind since antiquity for ornaments, as a store of wealth and for coinage. Of all the metals and minerals, gold is easiest to be sold and the market is always open to it. Most of gold produced is now kept by Government banks as bullions in reserve, to guarantee and provide stability for paper currencies and to serve as a medium to settle international trade balances. Considerable gold is used in jewellery, gold plating, gold plates, gold leaves, gold lace, glass and china inlays, gold embroidery, book-binding and tooling.

Gold has always been important to world economy. Potentially, it can be used as an economic weapon. S-Africa produces nearly three times as much gold as anyone else and has over 60% of world's reserves. Many countries hold enormous stocks against shortfalls in supply and fluctuating prices. Figure 12.1 shows production pattern of gold in various countries.

Occurrence. The most commong ore of gold is metallic or native gold (Au, commonly alloyed with Ag) and it is found as scattered grains, scales or plates in rocks of various types and ages, in veins and lodes, or in placers as nuggets, grains, specks or dust. Most of the primary Au deposits occur in the vicinity of acidic igneous intrusions and have been formed by hydro-thermal solutions as replacement of cavity-filling, Au-quartz veins and a few have been formed by contact metasomatism. The commonest minerals associated with Au in veins or lodes are quartz, pyrite, chalcopyrite, arsenopyrite and stibnite. In placer deposits, Au has been derived from pre-existing gold-bearing veins or rocks and concentrated by mechanical action of streams. Au content in a vein or lode is usually expressed in penny weight (dwt) per ton of ore (1 dwt 0.05 fine oz or 1.55 gm); and gold content in a placer deposit is expressed in penny-weight: per cubic yard of gravel or sand.

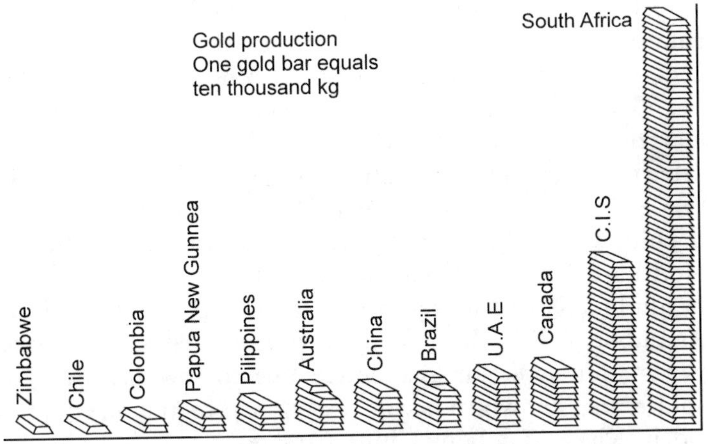

Fig. 12.1: Production Pattern of Gold in Various Countries

Distribution. It is widely distributed in India but not always in workable quantities. In North-India (with exception of Chhota Nagpur region of Jharkhand), Au occurs mainly in detrital or alluvial deposits derived from rocks of diverse geological ages, whereas in S-India (with a few exceptions) it is confined mostly to quartz veins or 'reefs,' associated usually with hornblendic and chloritic schists of Dharwar Super Group. Karnataka is the only Au-producing state in India. Au bearing vein-quartz occurs in two forms, either a dark blue or deep grey semi-translucent variety or an opaque milky white kind. The former is regarded to be of an earlier period and more crushed than the latter and often carries more gold. In both, gold occurs as very minute particles seldom visible to naked eye. The two important gold fields in Karnataka are Kolar Goldfield (KGF) and Hutti Goldfield. KGF has four productive mines—Nundydroog, Champion Reef, Mysore and Ooregaum, all situated along the strike on the same economically important quartz-vein-Champion Reef. Champion Reef mine is one of the deepest mines in world, nearly 2 miles in depth. Ooregaum mine has been closed, and Mysore mine is nearing exhaustion. Proved reserves of Champion Reef mine and Nundydrug mine are large, but the former has reached great depths and mining is becoming increasingly difficult (rock temperatures being >150°F in lowermost working spots); and the grade of ore from the latter mine is now lower than it used to be. Outside KGF gold-bearing quartz reef has also been mined in Bellara area (Tumkur district), and old workings exist in ferruginous quartzites and associated ochres in Chitaldrug and Tumkur districts. Main mine in Hutti Goldfield is Hutti Gold mine. Average tenor of gold ore in KGF is now <6dwt/t and that of Hutti Goldfield <4 dwt/t. Other important gold fields in S-India are Dharwar Goldfield (Karnataka), Wynaad Goldfield (partly in T.N and partly in Kerala) and Ramagiri Goldfield (Anantpur, A.P.). These fields were exploited to some extent in past, but later on abandoned due to low yield of gold. Gold is also known to occur in Coimbatore and Salem (T.N.) and in Khasi and Jaintia hills (Assam).

In N-India, auriferous quartz veins occur at a few localities in Chhota Nagpur region, Jharkhand. Like KGF, both types of vein-quartz occur in this area, but here gold is usually found in the younger opaque white (sometimes spongy and limonitic) quartz of the veins traversing quartzite and chloritic schists. The important fields are Kundra Kocha Goldfield, Singhbhum and Lower Goldfield, Dhanbad, where gold has been mined in past on a small scale. Gold bearing deposit is also reported to occur in Moflang area of Assam.

Alluvial gold is found in sands and gravels of many rivers in India and gold washing (panning) is occasionally carried out at many places by local inhabitants but concentration is poor and yield is therefore very meagre. The notable rivers which carry gold are the tributaries of Brahmaputra (like Subansiri in Assam), Subarnarekha in Singhbhum (Jharkhand), Indus and its tributaries in J&K, Sone, Deo and IB (M.P.), rivers and streams of Koraput and Sambalpur (Orissa), Shimoga (Karnataka), Ambala (Haryana), Garhwal—like Sona nadi, a tributary of Ganga (Uttaranchal) and Purulia (W.B.).

Gold in Archaen-Proterozoic Formations of Central India

Archaen-Proterozoic Fm of Central Indian Shield, in general, and of its Bhandara Cratonic terrain, in particular, are significant in respect of their gold potentialities. Here the present status regarding the known occurrences and incidences of gold in various segments of the Shield has been discussed. Besides, a preliminary evaluation of the prospects and potentialities of gold in various formations, employing the twin parameters of probable age and deposit types, despite several known constraints and gaps of knowledge has been made by Rai (1996).

Deposit-types and their representative occurrences that deserve special mention include Au-mineralisation in Archaen-Proterozoic greenstone belts: Sonakhan-type (in mafic-ultramafic volcanics and intrusives); in chemical metasediments of Archaen-Proterozoic transition period: Sonadehi-type (hosted by Bengpal Group, BIF) and Sakoli-type (hosted by Sakoli metasediments); in Early Proterozoic acid magmatic regimes of island-arc settings: Malanjkhand-type (hosted by Cu-bearing quartz-vein in I-type granitoids); and in Mid-Proterozoic rifts-generated Kotri-Type basins : Gurwandi-type (hosted by felsic volcanics, mainly rhyolite).

Archaen-Proterozoic Fm of Central Indian Shield, covering a vast trapezoid tract between 18°-26°N latitude and 74°-84°E longitude (Fig. 12.2) have in recent years, attracted considerable attention of researchers being the promising repositories of diverse kinds of mineable mineral resources. The resources of primary gold are the foremost among them. However, the prospects and potentialities of various incidences and occurrences of gold from the region can be appraised only preliminarily without much connotations regarding their resource-base.

Prospects and Potentialities

Despite the constraints, evaluation of prospects and potentialities of gold in Archaen-Proterozoic Fm of Central India can be made by classifying the known occurrences and incidences of gold on the basis of twin parameters of probable age and deposit type. This may sound conjectural in several

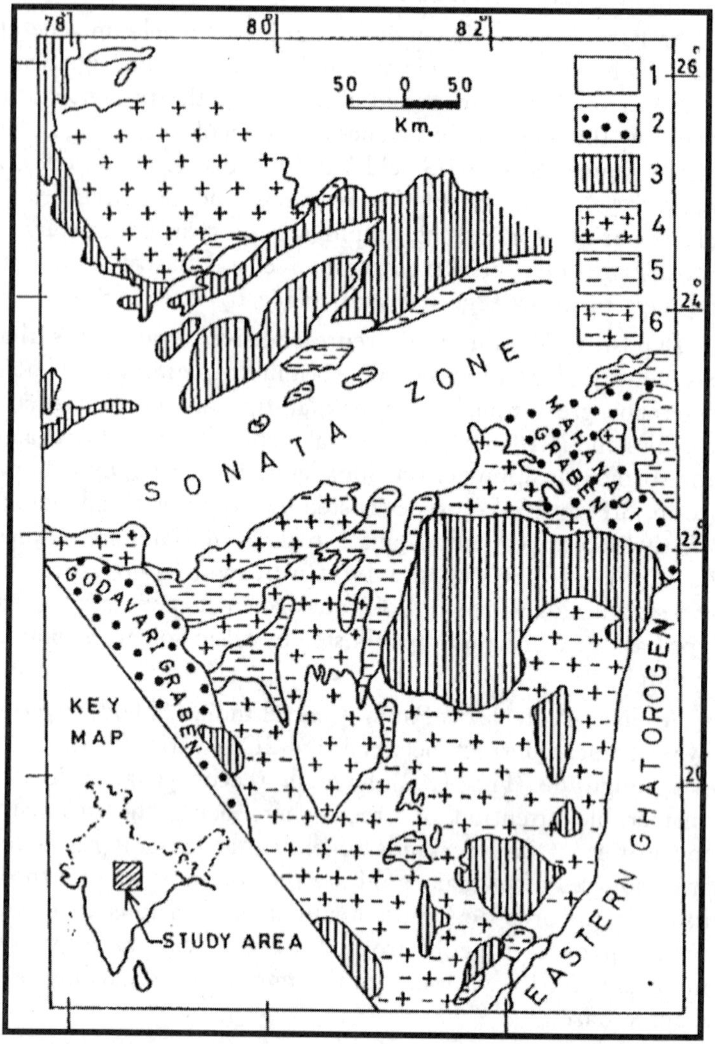

Fig. 12.2: Geological Map of Central Indian Precambrian Shield.

1. Younger cover rocks 2. Gondwana Graben Sediments
3. Chhattisgarh/Vindhyan and equivalent sediments
4. Intrusive Granitoids 5. Sakoli/Sausar metasediments
6. Central Indian granitic complex/granulite patches

individual cases, but evaluation on these lines can lead to a better understanding of the geological controls on distribution of various major-types of Au-deposits/occurrences in the region. This understanding can, in turn, provide a basis for preliminary estimation of the prospects and potentialities of various deposit-types, at least in relative terms.

Sarkar and Padhi (1996) proposed to employ a balanced move between empirical formula and conceptual modelling, on regional scale, of the known mineralisations for prognostication of concealed mineral deposits in pre-cambrians of India. Among various regional concepts employed by them for this purpose, those based upon Au-mineralisations include the following:

(i) Strata-bound Au-mineralisation in BIF, volcanics and volcanoclastics,
(ii) Shear-related Au and base-metal mineralisation in rift basins,
(iii) Concentration of Au-mineralisation in contact zone of volcanics and sedimentary beds,
(iv) Au-concentration in laterite cover of auriferous rocks.

This approach has led to the identification of several gold prospects within rhyolites and associated tuff in the basement of BIF in Kotri rift basin. Besides, some new targets have been identified for the search of gold in red boles in Deccan Trap basalts and in the platform sediments, subjected to alkaline metasomatism.

Central Indian Shield and its Gold Potential

Archaen-Proterozoic Fm of Central Indian Shield (CIS) are divisible geotectonically into three segments, the boundaries of which are demarcated by prominent regional lineaments and/or major tectonic structures. These segments, from south to north, are:

(a) Bhandara Craton
(b) Sone-Narmada-Tapti (SONATA) zone
(c) Bundelkhand Craton

Geological framework of these segments of CIS and major geological processes to which the generation and/or distribution/ redistribution of gold occurring in them may be attributed, are summarised here on the basis of available information which mainly deal with pre-cambrians of India in general and that of CIS or its any specific segment, in particular. Geological set-up and general characteristics of representative incidences/occurrences/prospects of gold occurring in various segments of CIS have also been described.

Bhandara Craton

This Craton having an inverted triangular shape is bounded in S-W by NW-SE trending Godavari graben, in N-E by Mahanadi graben and in north by E-W trending Central Indian Suture Zone (CISZ) that constitutes the S-border of Sone-Narmada Tapti (SONATA) lineament zone north of it.

The terrain is occupied by granite plutons of Dongargarh and Malanjkhand, older gneisses, metasediments and volcanics of Sakoli, Dongargarh and other groups and the granites and pegmatites of Bastar belt. They are overlain by sedimentary sequences of various Proterozoic basins viz. Khairagarh, Chhattisgarh, Indravati and Abhujmath.

Available chrono-stratigraphic and geochronological data regarding the above-stated plutons and sequences is suggestive of following major periods of magmatism and metamorphism during which gold metallization apparently took place in various situations in this most important segment of CIS. These are: (a) Late Archaen period during which the earliest phase of mantle-derived mafic-ultramafic magmatism apparently took place in eugeosynclinal regimes and led to the development of the oldest greenstone-granite complexes e.g. Sonakhan (Raipur district) and Bargaon (Betul district) in the fringe areas of the Craton, (b) Archaen-Proterozoic transition period marking the change over from reducing to oxidising environment resulting in the formation of BIF and sediment-hosted base metal deposits with increase in Pb, (c) Early Proterozoic, around 2.5-2.4 Ga during which the predominant phase of acid magmatism led to the emplacement of granitic plutons of Dongargarh and Malanjkhand and subsequently of the acid effusives of Nandgaon, and (d) around 2.1-2.0 Ga during which a younger phase of basic magmatism, represented by metabasic rocks of Bodal and andesites of Khairagarh basin, took place. Thermal metamorphic imprints of this phase are notable in Amgaon gneisses.

Gold Mineralisation in Archaen Greenstone Belts

Archaen mineralisation is, in general, confined to ensimatic mantle tapping volcanism in submarine environment represented by mafic-ultramafic, komalitic-tholeiitic exhalative activity in Greenstone belts. They bear Au, Pt group metals, Cr, Ni, V-Ti, Fe and Cu mineralisations.

An important geological setting deserving special attention for the search of QPC-types Au or Au-U association is that of basal conglomerates and coarse arenaceous horizons marking the beginning of Archaen sedimentation (2.7 Ga) in intercratonic and peri-cratonic basins. Possibility of occurrences of such horizons at the base of Bengpals in Bastar district and Sonakhan Group in Raipur district should be explored.

Zones of ultrabasic and basic rocks characterising greenstone sequences of this period are likely to host magmatogenic mineralisations basaltophyllic elements with subordinate gold and platinoids. Remobilisation leading to concentration of gold in favourable situations, often in quartz-veins apparently took place much later in such sequences, particular during the emplacement of tonalitic-granodioritic plutons.

The bulk of gold production in world has often been coming from volcano-sedimentary sequences, constituting the main pile of greenstone belts, *e.g.* type deposits of Kalgoorlie in W-Australia, hosted by differentiated tholeiitic basalts and intrusives and of Sigma region in Abitibi greenstone belt of Quebec, Canada. Association of gold in mafic rocks in these and many other Au-deposits including those of Kolar and Hutti belts of India indicates an important link between Au-metallogeny and the products, if not the process of basaltic magmatism.

Sonakhan Schist Belt, Raipur (Chhattisgarh)

Gold mineralisation of Sonakhan-Baghmara area belonging to Sonakhan Schist Belt of Raipur, has assumed significant dimensions after the recognition of the belt as a Late Archaen granite-greenstone complex. Several occurrences and incidences of Au, U and sulphide mineralisation of this belt have been targetted for closer scrutiny by various government agencies.

At Sonakhan, gold has been found concentrated in quartz veins with or without other sulphides such as chalcopyrite, arsenopyrite, pyrite, galena, etc. and in ferruginous chert bands of Baghmara Fm, basal unit of Sonakhan Group that comprises meta-ultramafite, amphibolite, pillowed metabasalt, metamorphosed differentiated basic suite, pyroclastics, acid volcanics and tuffs, BIF, etc. The cycle closes with sulphide-bearing ferruginous chert and after a major stratigraphic break, is followed by deposition of volcano-sedimentary sequence of Sonakhan Group, referred to as Arjuni Fm (Nag *et al.* 1996). Sarang (1995) has reported a base-metal zone in close association with amphibolite schist and leucogranites of Sonakhan Belt in a small restricted area around Bichhia in Raipur district, while Sinha and Hansoti (1995) have recognised an extensively occurring polymetallic sulphide-bearing feldspathic/quartz arenite horizon in Juba Member of Rehatikhol Fm, the basal unit of Proterozoic Singhara Group, which unconformably overlies Sonakhan Group in a small protobasin lying to east of Sonakhan greenstone belt. Sulphide content in these arenites measures upto 10% by volume, comprises principally of pyrite, coffinite and galena and exhibits intimate association with uranium at Juba and Banjhapali. Mineralogy and mode of occurrence of these sulphidic ores bear resemblance to the ores in Sonakhan greenstone belt wherein presence of Au & U in close association has been reported by Das *et al.* (1990).

Detailed investigation on gold occurring in association with base metal sulphides along Barnaidadar sheared auriferous hydrothermal zone has indicated Au values 0.1-0.6 ppm in the bed-rock samples.

Bargaon, Betul District (M.P.)

There is good scope for exploration of gold from unexplored area located in eastern part of an inlier of granite-greenstone association which is exposed near Bargaon in Betul.

The area is dominated by pillowed lava basalts, spilites, acid volcanics, pyritiferous chert and associated metasediments which are intruded by basic-ultrabasic rocks. Mineralisation is essentially polymetallic, comprising mainly Zn-Pb-Cu and is hosted mostly by biotite-talc-anthophyllite schist.

Bimodal volcanism, linear folded disposition of the rocks bordering granitoids, komatitic geochemistry of basalts, low-grade metamorphism and associated volcano-sedimentary assemblage of rocks indicate a greenstone belt-like geological set up, which appears to be favourable to Au-mineralisation akin to Sonakhan type geoenvironment. Indications of a promising base metal mineralisation including a small deposit of zinc located near Bhowratekra in quartz-muscovite schist, tremolite-actionlite schist and chlorite-schist of Bargaon Group have already been recorded by Shrivastava and Chellani (1994). They believe that the presence of large bodies of basic and ultrabasic intrusives in the area is similar to the domination of volcanic flows with gabbroic intrusions in world's largest, single continuous auriferous greenstone belt of Abitibi in Canada reported by Goodwin and Riddler, 1970.

Gold in Archaen-Proterozoic Transition Formations

The change over from Archaen to Proterozoic is a marker horizon with pronounced break at about 2.5 Ga on global scale. This is indicated by several changes in the patterns and styles of magmatism, sedimentation, tectonics, mineralisation, etc. during this transition period. The events of major sequence for gold metallogeny included the change over from reducing to oxidising environment with resultant deposition of iron (BIF in general, and auriferous iron formations of Algoma type in particular) and from ensimatic mantle-derived volcanism to ensialic sedimentary environment which effected a change from volcanogenic mineralisations during Archaen to sediment-hosted strata-bound base metal deposits with increase in their Pb content (e.g. Sakoli mineralisations).

Gold in BIF and associated Volcanics/Volcanoclastics

BIF-hosted gold has been reported from the shield areas of W-Australia, Canada, Zimbabwe and in India from

Dharwarian schist belts of Kolar (Mallappakonda), Chitradurga (Ajjanahalli) and Shimoga (Chinmulgund). Mineralogical investigations in all such occurrences reveal that most of the auriferous BIFs belong to mixed oxide-carbonate or oxide-sulphide facies. High abundance of arsenic and/or sulphur generally characterises such auriferous BIFs. Enrichment of gold values by remobilisation has been noticed in BIFs often in structurally favourable loci, like fold-hinges and shears.

Iron formations of Algoma type from Bailadila region of Bastar dist., Sakoli basin of Maharashtra, and basement of Kotri rift basin have been reported as auriferous.

Auriferous BIFs of Gakoli basin which occur in mafic volcanic domains rich in Zn, Cu, Mn, Fe, W metal ions have led to believe that they were partly generated by hydrothermal exhalate action and partly by chemogenic sedimentation. Au, Ag & Mo contents of these Fe formations may be genetically related to the latter. BIFs from above-stated volcanic domains in areas adjoined Kolari, Bhaona, Kantaon, Nagjhira and Dhabetekri are likely to yield gold bearing Cu, Zn and W ores.

Gold occurrences in Sakoli Basin, Maharashtra

Sakoli Group rocks comprise metasediments with submarine mafic-felsic volcanic flows and felsic, mafic and ultramafic intrusions. Geochemical studies of Sakoli volcanics have, in general indicated an island arc setting.

Gold has been reported in this basin essentially from two different settings : first in association with base metal sulphides, genetically related to mafic volcanism and the second in association with a rare metallogeny of W, Nb, Ta and Sn related mainly to acid magmatism and partly to mafic-felsic volcanism. The two types are exemplified by following important occurrences:

Pular-Parsori-Khapri area: Auriferous Cu-Zn sulphide mineralisation of this area occurs mainly in quartz-chlorite-sericite schist and partly in quartzites belonging to Pular Fm. Several incidences of gold have been observed here in dark bluish grey quartz veins and stringers traversing sulphidic metasediments discordantly as well as concordantly around Kolan, Mukhabardi and Pularparsori (Ray *et. al.*, 1981).

The mineralised area including the zone which is feebly mineralised for copper in Pular Parsori Cu prospect have been found auriferous with Au values ranging from 0.3 to as high as 12.3 g/t indicating highly erratic dispersion of gold.

Silli-Sirasghattola-Matora area: Lithologically, this area is divisible in 3 subparallel sections from south to north. In the area south of Silli-Sirasghat, dominant rock-type is chlorite-sericite schist with andalusite, magnetite and tourmaline. The area to the north of the schist is a sequence of pebbly phyllite-quartzite-BIF. Further north, beyond Nerori, a linear band of strongly intruded quartzite occurs within granitic gneiss country.

Gold reported in stream sediments from Silli-Sirasghattola is possibly derived out of the quartz/quartz-tourmaline veins emplaced in the schist while the one reported from east of Berori is from Matora quartzite body which has subsequently been intruded by quartz-veins. These veins intruding Matora quartzite show finely dispersed pyrite, arsenopyrite and malachite stains indicating some degree of sulphide mineralisation. Presence of disseminated pyrite-arsenopyrite in quartz veins is quite encouraging from point of view of gold mineralisation as quartz veins with similar character are known to be auriferous in the area north of Parsori (Saha and Mohan, 1994).

Bhagri, Bhandara District, Maharastra

Epithermal gold telluride-silver bearing quartz veins, averaging 10.6 ppm gold, have been reported by Dakate and Deshmukh (1972) from near Bhagri village, about 41Km SSW of Sakoli in Bhandara. Mineralisation is hosted by sericite-phyllites, calc-phyllites and quartzites of Precambrian Sakoli group metasediments belonging to green-schist facies.

The mineralised veins, generally comprising quartz and jasper and having variable width from a few cm to tens of mts, appear to be dominantly fill-fractures produced during shearing of host-rocks. The wall-rocks of these veins are marked with prominent hydrothermal alternation which is represented by thick zones of hydromica, alunite and spherosiderite nearer the veins-contact and indistinct propylitic alteration away from it. Extensive development of thick oxidised (geothitic) zone, assaying 5 to 60 ppm gold, marks bright chances of primary native gold at depth in this prospect.

Gold in Early Proterozoic (2.5 Ga) Acid Magmatic Regimes

As elsewhere, early Proterozoic terrains of Bhandara Craton contrast with typical Archaen terrains in terms of the nature of their magmatic and sedimentary assemblages and igneous intrusions. Widespread acid magmatism represented by Dongargarh and Malanjkhand plutons comprising essentially I-type granites, followed by the later phase outpourings of felsic to acid volcanics of Nandgaon Group characterises early Proterozoic terrains of the craton in general. By 2.2 Ga, arc volcanism and younger potassic plutonism had led to the development of island-arc tholeiite suite in large parts of the area. Sulphide mineralisations genetically associated with these acid magmatics of island arc affinity have indicated intimate association with gold in recoverable proportions at many places which include Saletekri area in Balaghat dist and Chauki area in Rajnandgaon, of M.P. As surmised by Hansoti and Sinha (1995), metallization observed in Malanjkhand Complex Sector is of Cu ± Mo + Fe + Ag + Au ± REE ± Zn type whereas in Dongargarh Massif Sector, it is U ± Th ± F ± Fe ± Pb ± Zn ± Cu ± REE ± Nb type.

A younger phase of basic to felsic magmatism represented by metabasics of Bodal area and andesite-rhyolites of Khairagarh Group apparently took place at around 2.2-2.0 Ga in the region. Discovery of a good deposit and several occurrences of uranium in the former should be taken as a positive indication of the prospects of locating auriferous sulphide mineralisations in the same settings. Detection of Fe + U + REE ± Cu ± Ni metallisation in Bortalao sandstone of Dongargarh-Lohara area (Hansoti and Sinha, 1995), located in between Malanjkhand and Chandidongri (granite-hosted) mineralised bodies, renders those prospects all the more promising and significant.

Malanjkhand -Taregaon Porphyry Province (M.P.)

In Malanjkhand-Taregaon Prophyry province of Balaghat district located just south of CIS, possibly at a triple junction, the occurrence of gold has been reported in intimate association with copper ores at 0.2 g/ton an average with proven recoverability as a by-product metal from chalcopyrite concentrates of Cu-Mo ores of Malanjkhand (Sikka, 1989). Concentration of copper sulphides, and perhaps contagiously of associated gold too, is controlled by shear and fracture planes traversing the mineralised quartz-vein and felspathised zones of its host granitoids (Rai and Venkatesh, 1993). Mo-mineralisation, presumably poor in gold, is confined mainly to the granitoids on hanging and foot-wall side of Cu-rich quartz vein. Impoverishment of molybdenite ore in noble metals like Pt & Au or Ni & Co is suggestive of a non-volcanogenic origin of ores. Taregaon occurrence indicating higher values of gold in association with copper in general, needs further evaluation in depth.

The island-arc volcanic setting of this region in an ancient plate tectonic regime, as conclusively deduced from integrated geological, geophysical and geochemical investigations of CRUMANSONATA Project (GSI, 1995), provides a newscope for detailed search and exploration towards locating the loci of economic-grade gold mineralisation. Many occurrences and surface shows of copper ores located along or close to NNE-SSW trending lineament from Malanjkhand in North to Nandora in South, appear to be significant from this viewpoint.

Gold Mineralisation in Middle Proterozoic Formations

The end of late Archaen to early Proterozoic : orogenics was in general, marked with the development of several intracratonic basins of failed-rift type in cratonic areas. Being fault-controlled, these graben-like basins generally assumed elongated shape. They were the sites of extensive volcanic activity in the initial stages and of long-continued fluvial sedimentation under progressively developing oxidising environments and prolonged crustal stability largely during 1.6 to 0.9 Ga. The beginning of sedimentation in some of these basins has been marked by auriferous quartz pebble conglomerate beds of the kind seen in Sidhi (M.P.). In general, such basins have been the hosts of several important stratiform and strata-bound mineralisations which include auriferous iron formations as well as sulphides with or without gold. Potential of volcanogenic gold in these associations has been found to be promising and worth examining from exploitation point of view.

The most important rift-generated basin of Craton from the above-stated point of view is located along NNE-SSW trending, 250 Km long and 30 Km wide Kotri Rift Zone (Sarkar *et al.* 1990), litho-assemblage of which comprises the basal (Nandgaon Group) formations of acid volcanics, volcano-sedimentary units and basic volcanics which are, in turn, overlain unconformably by conglomerate—arkose assemblage (of Khairagarh Group). Gold has been reported from a hydrothermally altered, 40 m thick zone along the contact of rhyolite with tuff, over a length of 2 Km in Gurwandi area of this basin. Primary dispersion pattern in sheared rhyolite band of volcano-sedimenatry sequence of Pachangi Fm of Kotri Super Group in this area has shown correlatable Au-As values with gold in 0.1 to 7.0 ppm range (Patel, 1994).

Gurwandi area, Kotri Linear Belt, Bastar (Chhattisgarh)

The acid and basic volcanic rock-fill along Kotri Linear belt near Gurwandi is well known for the panning of placer gold from recent gravels by local people for several generations. A zone of deformed rhyolite located west of Gurwandi, about 55 Km WSW of Bhanupratappur, constitutes the first known occurrence of primary source—rock of gold in this belt (Sarkar, 1995).

Two distinct lithological associations have been noted in the area: one is an acid volcanic and sedimentary rock association while the other is a gabbro-basalt (mafic igneous) suite. A number of gabbro bodies of this suite show intrusive relationship with acid volcanics. In a rhyolite band about 60 m wide and 500m long, 0.5 Km WSW of Gurwandi, Au-mineralisation has been found associated with disseminations and veinlets of sulphides mainly pyrite, arsenopyrite and pyrrhotite. The veinlets range from <1 cm to 5 cm in length and are often marked with fine-grained yellowish green chloritic alteration along their wall rocks. Sulphide ore minerals occur often in thin discrete bands generally conformable with the flow laminae in rhyolitic host. The ore minerals are also deformed alongwith the host rhyolite. Au-values in deformed rhyolite and tuff range from 0.1 to 1.7 ppm (average 0.5 ppm) while their arsenic contents have been noted 100 to 16,000 ppm range with moderately good positive correlation. This has confirmed that arsenic can be used as a reliable pathfinder in any exploration programs of gold in this region in future.

According to Hansoti and Sinha (1996), U, REE, Fe and even Au enrichment observed in Nandgaon Group volcanics, can be ascribed to rift-related volcanism along Kotri lineament zone.

Gold in Late Proterozoic Formations

Widespread magmatic activity represented by alkaline to peralkaline (alaskitic) granites, syenites, carbonatites, and pegmatites apparently marked the early stage of Late Proterozoic at about 1.2Ga. Generally evolved nature of anticipated magma sources of these rocks points to their potential as a possible factor in gold enrichment in favorable situations.

Development of broad epicontinental seas over large parts of peneplained craton during the large span of this period subsequently led to the accumulation of vast thicknesses of platformal sediments in Chhatisgarh and Indravati basins. No gold minneralisation has been reported from these basins so far.

Sone-Narmada-Tapti Lineament (Sonata) Zone

This ENE-WSW trending belt comprising a wide variety of rock-types ranging in age from Archaen to Recent, separates two distinct pre-cambrian crustal blocks—Bhandara Craton, a constituent of Southern Protocontinent in South and Bundelkhand Craton of Bundelkhand Protocontinent in north. CIS constitutes its tectonic limit all along Satpura Orogenic Belt in South. Pre-cambrian supracrustals of Mahakoshal Group, on the other hand, occur all along the zone's interface with Bundelkhand Craton in north.

The zone was the site of extensive sedimentation and volcanism since early Proterozoic, extending well into middle Proterozoic. Sedimentary sequences of Sausar fold belt and Sidhi-Mahakoshal greenstone belt and other equivalent rock groups were deposited in this elongate and wide trough and, as a result of intense lateral compression, were intensely folded, faulted and subjected to low-angle thrusting on an extensive scale in upper Proterozoic period (GSI, 1995).

Mahakoshal belt of Sone Valley, established as a late Archaen—early Proterozoic linear ensialic greenstone belt in the light of extensive investigations under CRUMANSONATA Project (GSI, 1995), is represented by metabasics, tuffs, orthoquartzite, chert, dolomite, carbonaceous phyllite and BIF, beside ultramafic-mafic intrusions and granite basement rocks. Mafic to felsic volcanics belonging to middle sequences of this Belt have been found associated with a polymetallic mineralisation comprising Cu, Zn & Au.

The earliest phase of sedimentation is marked with deposition of auriferous quartz pebble conglomerates at certain locations in Sidhi district. Subsequent Au-mineralisation has been reported from volcanic sequences of Chitrangi Fm and also from the quartz veins in western part around Sleemanabad in Jabalpur and Barmanghat near Kareli in Narsinghpur, M.P.

Gold Mineralisation of Sudda area, Sidhi Dist. (M.P.) and Gulaldih Area, Sonebhadra Dist. (U.P.)

Potential occurrences of auriferous quartz-veins have been reported from the pelitic sequence of Parsori-Fm of Bijawar Group in Sudda area of Sidhi, and phyllites of Agori Fm of the same Group near Amaranian village of Mirzapur, and from the same sequence in Gulaldih area of adjoining Sonebhadra district.

Gold mineralisation in quartz-veins of all these occurrences is often associated with disseminations of primary sulphides viz. Pyrite, galena, arsenopyrite and pyrrhotite. X-ray studies of the alteration products of auriferous sulphidic rocks in Gulaldih area have revealed that its lemon-coloured part contains mansfieldite in major and minetite + vanadinite in small amounts, while the orange-coloured portion comprises goethite in major and minetite + vanadinite + hematite + arseniosiderite in small amounts. In whole-rock samples, scorodite has been found to be the predominant supergene mineral, resulting from the oxidation of arsenopyrite and other arsenic-bearing minerals and therefore, can be used as an informative pathfinder element for gold.

Amaranian occurrences have assayed 0.8 to 3.8 gm/ton of Au which exhibits sympathetic variation with As, Pb, Hg and Sb contents, in general. Auriferous quartz veins from Sudda area in Sidhi district have also indicated similar relationship of As and Pb with gold contents. Analysis of volcanic rocks of Chitrangi Fm have reportedly yielded gold values from 0.1 to 0.2 ppm (GSI, 1995).

Bundelkhand Craton

This sickle-shaped cratonic block occurring north of SONATA Zone, comprises Bundelkhand granite complex and overlying sedimentary sequences of Bijawar and Vindhyan Super Group. A narrow strip of Mahakoshal Greenstone belt occurs along its southern boundary in S-E.

Occurrences and incidences of gold in association with sulphides have been reported from this cratonic block principally in quartz-veins traversing Bundelkhand granites and also the sedimentary, volcano-sedimentary or volcanic sequences of Bijawar and Gwalior groups overlying the basement rocks of Bundelkhand Granite Complex.

Gold in Cupriferous Quartz-Veins Traversing Bundelkhand Granite

Gold in intimate association with sulphide minerals is likely to occur in some of the cupriferous quartz-veins seen along several shear-zones in granitic terrain or along the zones of granite mylonite traversing the basement gneisses. Some of such areas include those from Salaiya-Jamtoli tract in Chhatarpur; Antri and Lekhnauti-Bageli-Gijora areas in Gwalior and Bahadurpur and Baldeogarh in Tikamgarh, (M.P.). Chaurasia *et. al.* (1994) reported <1 ppm of gold in medium-grained pink granites exposed along the NE-SW trending Dhanupura-Khorroli lineament in Chhatarpur, M.P.

Gold in Veins Traversing Sedimentary Terrains

Occurrence of QPC with U & Au at the base of Bijawars at Mehrauni in U.P. has been reported (Bogdanov *et. al.*, 1979).

Besides, the mineralised quartz-veins containing base-metal sulphides, occasionally auriferous, are also found in Bijawar and Gwalior Group rocks. Known locations of this type include Tigoria in Sagar Dinara in Sheopuri and Sendba in Datia (M.P.). Possibility of occurrence of auriferous magnetitic zones or horizons in Fe formations occurring in the upper parts of volcano-sedimentary sequences in these basins also deserves due consideration.

Important Gold Field

Kolar goldfield

The field is situated in the centre of NS-elongated belt (80 Km long and 1.2 Km to 6.5 Km wide) of Dharwar schists on a plateau 840 m above mean sea level. Here 26 or more lodes comprising gold-quartz lodes, gold-quartz-sulphide and graphitic lodes, have been worked on. The lodes are localised along or in vicinity of stratigraphic contacts of granular; fibrous and massive amphibolites with the schistose amphibolites of schist belt. There is a remarkable structural control of the lodes along NNW-SSE axis of the major crossfold in NS trending synclinal structure of Kolar schist belt. The lodes are made up of 'en echelon' quartz vein zones in altered and mineralised wall rock. Besides native gold, the ore minerals in lodes include chiefly pyrrhotite, arsenopyrite, galena, pyrite, chalcopyrite, scheelite and magnetite. Quartz lodes contain only sporadic sulphides and other ore minerals whereas sulphide-bearing reefs show a high proportion of sulphides occurring as layers and veins parallel to quartz veins and stringers.

Gold Mining

Modern mining for gold has been in operation here for past 100 years, and the most productive Champion Reef has been mined over a strike of >8 Km and down to an average depth of >2700m. Champion reef does not form one continuous quartz-filled fissure but is a system of individual fissures which peter out both along the strike and in depth, and is further dislocated by major faults, dykes and other disturbances. The width of lode-filling of Champion reef arises from a mere stringer to >4 m in the folded areas, the average being about 1 m.

Access to the reefs in KGF is by crosscuts from major vertical shafts placed either in the hanging or foot-wall side. The shafts has a depth of 2 Km and provides access to S-section of Champion reef mines, one of the deepest single wind in the world. It serves as the main downcast airway into the southern portion of the mine. From the bottom of the main vertical shaft, main haulage ways lead to a number of auxiliary shafts which make access possible to the lowest horizons. The levels on the reef are driven at 30 m intervals but below 90th level the interval has been reduced to 22.5 m. Cut and fill method of mining is practised here. For rock breaking underhand, overhand, flat back and rill methods are used in the stopes, depending upon the nature of walls and inclination of the reefs.

As mining in this field has extended to great depths (up to 3000 m), there are certain special problems connected with hot environment conditions. For example, high temperatures (about 150°F in the lowest working horizons) resulting in the rise of temperature of the ventilating air descending downcast which is also augmented due to the auto compression in deep downcast shaft; and the heavy rock pressures due to weight of the superincumbent rock (the vertical stress being about 6 t/inch^2 in solid rock at deepest level, but actually increasing to 2 or 3 times this value near excavations). The ventilating air is conducted from the surface through major vertical shafts and is then led as rapidly as possible to the deepest areas of mines. There from it finds its way back to the surface through the slopes where most of the productive men are employed. The air is forced through these circuits by powerful fans which are sited at strategic points of the mines. Before leaving the surface the air is cooled and considerable part of its moistue is removed in main air cooling plants sited at main downcast shafts of the field and sometimes also in secondary cooling plants sited at a deeper horizon. In the warmer parts of mines, piped water is too hot for drinking and cool water is lowered to such horizons from the surface or from one of the water cooling plants installed at convenient sites. Water is normally drunk with admixed salt in order to prevent muscular cramp resulting from saline losses due to excessive perspiration. As a result of very high vertical stress at deeper levels and sometimes also due to high lateral stress at shallower levels, 'rock burst' in which the highly stressed rock fails with explosive violence is frequent. It may be either local or may result in catastrophic damage to large sections of the mine. Researches have been carried out to investigate the nature of this rock pressure and the means of controlling it by predicting the incidents.

Extraction of gold from the gangue, mostly quartz, is affected by crushing the ore in jaw crushers and then in stamp batteries along with water. The pulp is allowed to flow over amalgamation tables covered with blankets. The free gold in the pulp is caught on the blankets from where it is washed and recovered. The pulp is pumped to classifier cones in which the finer particles are separated from the coarser ones. The coarse portion from the classifying cones is fed into tube mills and the tailings from tube mills is also passed over blankets before returning to the classifying plants. The concentrate is agitated with strong cyanide solution which dissolves the remaining gold in the pulp. After about 12 hours agitation, the solution is filtered and the filtrate is passed through Zn extractor boxes where gold is precipitated. This precipitate is roasted with fluxes and smelted in oil-fired furnaces to produce bullion, usually cast into bars. The recovery amounts to about 98% and the tailings contain 3-4 grains of gold/t. The fineness of gold from the smelter is about 910 parts/1000, which is further refined at Bombay Mint to 995 parts/1000.

Reserve and Production. Total resource of gold (ore) in India is 390.29 MT and of the metal is 49081t (2005-06).

The reserve, production and resources of gold are shown in Table 12.1. (A-B):

Table 12.1 (A): Reserve & Production of Gold in India

Year	Reserve		Production
	Recoverable (Lakh ton)	*Metal content (000'Kg)*	*(000'Kg)*
1998-99	177.9	6.8	2.7
1999-2000	—	—	2.6
2000-2001	—	—	7.6
2001-2002	177.9	6.8	10.2
2004-2005	–	–	3.5
2005-2006	–	–	3.5
2006-2007	–	–	2.9

Table 12.1(B): Gold Resources (2005-06)

Category	*Total*	*Reserve*	*Remaining*
(A) Ore (MT)	390.29	19.25	371.03
(B) Metal (t)	490.81	85.12	405.69

Production of gold has been mainly from Karnataka. Production has however declined considerably. So in June, 2000 Board for Industrial and Financial Reconstruction gave the verdict to wind up Bharat Gold Mines Ltd. but has been challenged.

Properties and Uses. The principal properties which make gold useful in industry are its malleability, ductility, decorative colour, resistance to corrosion, high electrical conductivity and reflective power and therapeutic effects of some of its salts. Pure gold is the most malleable and ductile metal. It can be drawn into very thin and long wires (1 gm can give 2 miles long wire) and it can be beaten into very thin leaves (each gold leaf may be 1/300,000 of an inch in thickness). Gold is too soft for most industrial uses, but it readily alloys with a number of metals which give it strength and resistance to wear and also make the natural colour of gold vary. Red gold is obtained by addition to Cu and Ag, white gold by addition of either Ni-Zn-Cu or Pt-Pd-Ag green gold by adding Ag-Cu-Cd, blue gold by adding Fe and steel and purple gold by adding Al. These golds are used mainly in ornamentation.

The common jewellery gold is an alloy of pure gold with Cu, Ag or Ni and its purity or fineness is designated in carats (1 carat means 1 part of gold in 24 parts). Gold for coinage is usually alloyed with Cu (9 parts of Au to 1 part of Cu). Industrial uses of gold alloys are in dentistry, chemical plants, laboratory apparatus, thermocouples, watches, radio-conductors, sound-transmitting apparatus, X-ray equipment, photography, theraphy, etc.

13

Iron Deposits

Iron-Ores in India

India is endowed with vast quantities of iron-ore in several places in the Peninsula. It had an iron-making industry since at least the second millennium B.C. when Indo-Aryan civilisation flourished and extended into the region of Babylon, Persia and Levant. There are references to iron in Rig-Veda which is about 4000 years old. It seems that knowledge of iron-making spread into the near East from India and that Kassites in Babylon, Mittanis in Syria and Hittites in Asia Minor were all familiar with iron-making. This knowledge was handed down in India from generation to generation till Industrial Revolution of W-Europe led to the marketing of cheap iron in India by British merchants. For many centuries, till about 1850 A.D., Indian iron ore was in great demand in the markets of Levant, where it was sold in the form of damascus swords, weapons and armours.

Iron articles readily undergo rusting in humid tropical climates of India. But a few examples of non-rusting wrought-iron have come to light in some places. Archaeological Survey of India found a variety of Prehistoric iron weapons such as daggers, swords, tridents, arrows, etc., in the ancient burial sites of Adittanallur, Tinnevelly (T.N.). An iron spear probably of 4th or 5th century B.C. was discovered at Piprahwa in Basti district (U.P.). Of about the same age are some iron clamps discovered in the excavations of Bodh Gaya temple in Bihar. Figure 13.1 shows the distribution of iron in India.

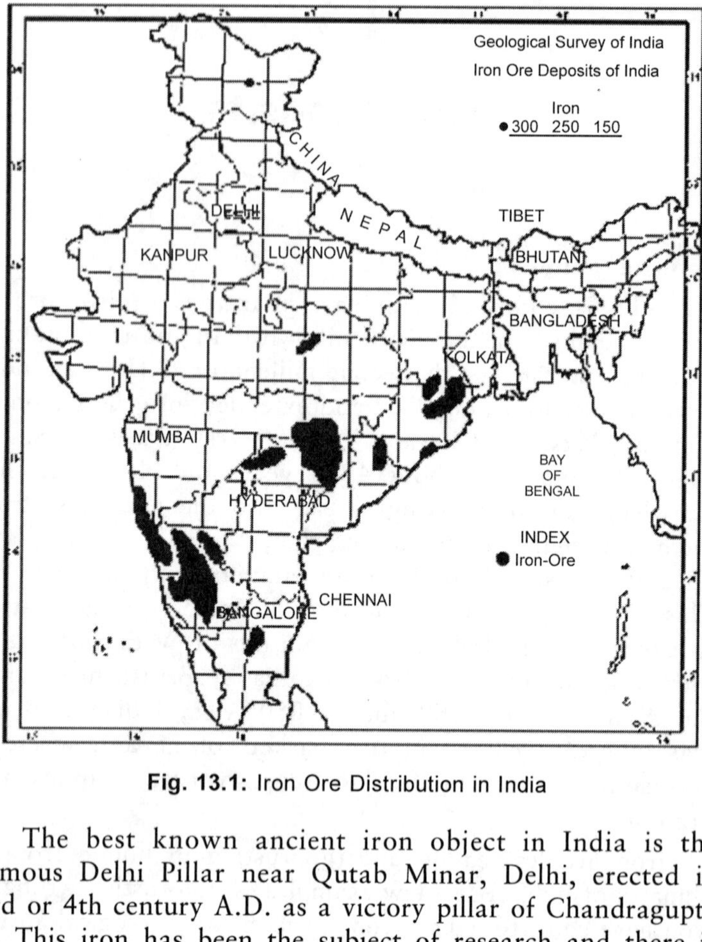

Fig. 13.1: Iron Ore Distribution in India

The best known ancient iron object in India is the famous Delhi Pillar near Qutab Minar, Delhi, erected in 3rd or 4th century A.D. as a victory pillar of Chandragupta II. This iron has been the subject of research and there is much speculation as to what makes it non-rusting. The analysis given by Robert Hadfield, the great British metallurgist (1925), is as in Table 13.1.

The specimen inspected by Hadfield was free from slag and other inclusions. It consisted of large grains of ferrite with very little cementite. He stated that the iron was much purer than anything which was produced on a large scale during his time.

Table 13.1: Analysis of Ancient Delhi Pillar (Hadfield, 1925)

Elements	%
C	0.080
Si	0.046
S	0.006
P	0.114
N	0.032
Mn	—
Cu	0.034
Fe	99.720
Total	100.032
Specific gravity	7.81

Iron Deposits: Indian Fe-ore deposits of Precambrian (BHQ-Banded Hematite Quartzite and BHJ—Banded Hematite Jasper) are strata bound and stratiform. Besides this there are two other categories of deposits which are roughly strata bound and sometimes stratiform. These are :

(a) **Bog iron ore.** These look like black band and are not very popular in India.

(b) **Iron stone type.** These also are not very popular in India and are found as siderite ($FeCO_3$) ores.

In India only BIF (Banded Iron Formation)—BHQ and BHJ is important. In Brazil, Precambrian iron formation is called Itabarite. In W-Australia it is called Taconites (Precambrian age). Here, they are also called Jaspillite. But the most important iron formation are of Precambrian in Africa.

BIF is formed in marine environment containing 15 to 25-30% Fe and contain amporphous variety (or crystalline variety of silica like chert, japser) and are bedded. Iron formation with 15% Fe is called original rock or mother rock for hematite in India. It is also called Protore (initial ore or embryonic ore). In India BHJ is not used as ore because other high grade ores containing as high as 65% Fe are available.

Previously, it was thought that there is only one horizon of iron deposits of Precambrian (Early middle Precambrian). Recently, it has been found that there are >1 (3 or 4) horizons of Precambrian iron ore which are even of later Precambrian.

Genetic Classification of Iron-Ore

Genetically, Indian Iron-ore deposits can be divided into four main groups (i) banded precambrian iron-ore formations — banded hematite-quartzite, banded hematite-jaspar, banded hematite-chert, banded magnetite-quartzite, banded magnetite-hematite schist (itabarite); (ii) titaniferous and vanadiferous magnetite and apatite-magnetite, (iii) sideritic and limonitic bog iron deposits, and (iv) highly ferruginous lateritic iron ore.

On physical characteristics, hematite iron-ores can be classified mainly under two types, massive and laminated. Variants of these two types include massive laminated under the first; porous, laminated, biscuity, etc. under second type; and lateritic variety under both types. There is also a variety of powdery ore known as blue dust.

Commercially, Indian iron ores are usually classified (i) by size: lumps, rubbles and fines, (ii) by treatment: raw, washed and concentrates, and (iii) by chemical grade: bessemer and non-bessemar, high phosphorous or low phosphorous, limy; aluminous, siliceous or mangniferous, Indian ores are usually high in alumina, giving an unfavourable alumina-silica ratio.

Mode of Occurrence

BIF are stratified and layered. They occur as layers of iron ore (hematite) or layers of geothite varying from few cm to metres (even 2 to 3 m) or may be as paper thin lamellies. Along with this there are cryptocrystalline silica layer (chert and jasper) alternating with iron layer of varying thickness. Sometimes iron layer is dominant while sometimes silica layer may be of more thickness. These bands of Fe & SiO_2 may suffer leaching by the solution due to the presence of cavities in the bands which are secondary pores or secondary cavities. Besides, these bands may show different degree of deformation like tilting or syn-sedimentary deformation or penecontem poraneous deformation that is deformation taking place in the basin during sedimentation and produce different structures like convolute laminations which are characteristics of syn-sedimentary deformation. These structures include diagenetic folds or slump structures etc. The layers may show local deformation and depression or wrapping structures. There may be some local dislocation by some undefined faulting. The layers also may show boudinage structure which is a purely sedimentary structure of local deformation in Fe-ore formation. There may be other different sedimentary structure due to shallow depositional environments like mud cracks or shrinkage or cross bedding of local scale. Ripple marks and graded bedding are not very common. One may get brecciation where breccias from different sources are present in these BIF. These breccias may be from volcanoes, tectonic activities or by solution activities of solidified BIF which gets fragmented and gives breccias. There may be some hexagonal or rhombohedral joints, huge folding (isoclinal fold, synclines, recumbent fold), low angle faults (thrusting), granulation due to faulting or tectonic effects. There are some BIF in Chhattisgarh which do not show any tectonic structures like folding and faulting.

Mineralogy

Mineralogy of BIF is simple. They occur as Fe-oxide with some water as hydroxide. In nature they occur as (a) Hematite with compositions α-Fe_2O_3 and γ-Fe_2O_3 (b) Geothite with same composition as that of hematite but with more H_2O-FeO. OH (c) Limonite in which water content varies and accordingly its colour varies. Besides they contain other forms of Fe-ores like geothite having needle form and lepidocrocite showing acicular form of Fe-oxide.

In India Precambrian BIF occurs in Chhattisgarh, T.N., Maharashtra, Karnataka. Iron ores of Jharkhand, Orissa contain magnetite and not hematite. Besides, martite—a mineral which is neither pure hematite nor magnetite (but partly hematite and partly magnetite) occurs as iron ore. Sometimes magnetite is formed by the process of magnetization of hematite (conversion of hematite to magnetite) is found in BIF. Igneous instrusions of dolerite or amphibolite may heat BIF when hematite reacts with silica and gives iron silicate minerals. If rise in temperature and pressure is less, low grade metamorphism takes place and minnesotoite (an iron bearing talc), a low grade metamorphism mineral is formed. This is not common in India. Other iron silicate mineral is grunerite which is common in Jharkhand, Uttaranchal, A.P., Karnataka. There are also other iron rich amphiboles, rare earth minerals and gangue minerals associated with these BIF. Manganese is invariably associated with BIF. In case, where the ores have undergone metamorphism one may get chlorite, mica as gangue minerals. There are also other oxide iron ore minerals but they are not found in BIF.

Ore Minerals. The chief minerals that are worked as ores of iron are given in Table 13.2.

Table 13.2: Principal Iron Ores

Ore	*Composition*	*Fe%*	*Type*
1. Magnetite (Magnetic) ore	Fe_3O_4	72.4	
2. Hematite or red ore	Fe_2O_3	70.0	Fe-oxide
3. Limonite or brown ore	$2Fe_2O_3 \cdot 3H_2O$	59.8	
4. Geothite	$Fe_2O_3 \cdot H_2O$	62.9	
5. Siderite or spathic ore	$FeCO_3$	48.2	
6. Ankerite	$FeCO_3$		Fe-Carbonate
7. Pyrite	FeS_2		Fe-Sulphide
8. Lepidochrosite			
9. Minnesotoite			
10. Grunerite			
11. Chemosite			Fe-Silicate

Depositional Environment of Formation: Characteristic mineralogical, petrological texture and features of BIF :

1. Mineralogy is comparatively simple, *i.e.* only simple oxides with silica or chert. Detritals of zinc, tourmaline, rutile, or feldspars etc, are almost absent.
2. The junction between iron and silica layer is often sharp. Replacement characteristics are found only in special cases. No transitional or gradational contact indicating any replacement, but direct deposition.
3. Features associated with shallow water deposition like mud cracks, cross lamination etc. are observed.

4. They are associated with tuff or pyroclastic rocks in some sub-marine rocks. At Bailadila, Dhanjori lava, Dalma lava rocks are associated with BIF, particularly in Jharkhand and Karnataka. They may be above or below lava rocks.
5. In India mostly iron oxide facies is present. There is no carbonate facies of iron like other world deposits. Hence, siderite and ankerite etc. are not present. Also, there is no iron silicate deposits (that is silicate facies of iron) like chemosite (iron with aluminium), thuringite, minnesotoite (iron bearing talc), grunerite (rich in Fe^{+3}). There is also no sulphide facies, *i.e.* black chert, pyretiferous chert, iron sulphide called covatite except in Orissa.
6. There are >1 horizons of BIF in India. This may not be true for other world Precambrian deposits. For example, in Jharkhand and Cuddapah there are BIF of pre-kolhans while there are also 1 or 2 formations which are post-kolhans in Orissa.

Mode of origin of BIF in India

1. Precambrian iron sediments were deposited in shallow, well-irrated, well-oxygenated, agitated water within the depth less than a few 100 ft in continental shelf over an extensive area. It means that they deposited in shallow sea having current in oxygenated condition in epicontinental sea or epiric sea or shallow intracratonic areas or in Precambrian shield areas of India.

2. Land mass or the source area or provenance which was supplying iron and silica alternately might have been a peneplained surface with sluggish streams flowing into epicontinental basin. So the clastic components are rare. The rivers carried only iron and silica in solution.

3. Climate in the provenance might have been variable with tropical humid summer and high rain and dry winter (six months each). These seasonal variations of climate might have been responsible for deposition of silica during winter and iron during tropical summer period. This variation explains why the thin layering or lamination was formed during winter and summer.

Silica might have leached out from the layer at one place and deposited at another place as secondary SiO_2. So that silica layer of one place becomes thin and of other place becomes thick. This may be the case with iron also. There might be replacement of silica by iron by the process called secondary ferruginization of chert. This explains the thick layer of iron. When silica replaces iron the process is called secondary silicification. The layers vary greatly in thickness. There may be cherts with relicts of iron oxide or iron oxide with relicts of silica. (Fig. 13.2a & b)

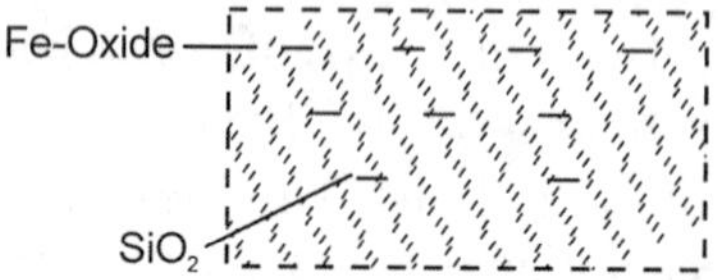

Fig. 13.2a: Fe-oxide with relicts of SiO_2

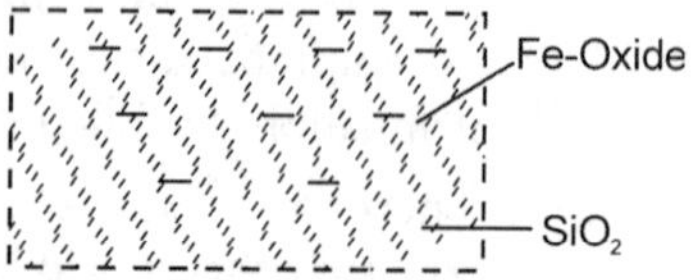

Fig. 13.2b: SiO_2 layer with relicts of Fe-oxide

Role of Bacteria in BIF

Relicts of bacterial and algal cells have been found in Precambrian rocks. Sargur Group rocks have yielded the relicts of micro-organisms. There might have been some iron bacteria which helped in precipitation of iron called bacterial precipitation of iron.

There is also volcanic or submarine volcanic origin of BIF. Some believe that BIF is not sedimentary, but have resulted by secondary replacement of the tuffaceous rocks by iron and silica giving rise to iron and silica layer respectively. This is called 'tuff-theory' of formation of BIF.

Source, Transport and Deposition of Iron

Source of iron may be : (a) continental i.e. the crust (b) marine (c) sub-marine (d) submarine hot springs or volcanoes.

Continental Source

Crust contains 100s of iron bearing minerals e.g. pyroxene, amphibole, olivine, mica etc. All types of igneous rocks contain iron (Fe^{+2} or Fe^{+3}). Ultrabasic and basic igneous rocks contain 8-12% Fe while acid igneous rocks contain 4-5% iron. Sedimentary rocks do not contain much iron. Only limestone contains 5% iron and others like sandstone, shale, dolomite, ferruginous shale etc contain some iron.

Metamorphic rocks like schist, gneisse, marble, quartzite contain iron. Thus, there is no lack of iron in crust. So, crust supplied all the iron of BIF. But why crust yielded so much iron only in Precambrian and not in Postcambrian age? Some BIF of Devonian called Devonian iron formation has also been found, but not in India.

This facies variation can be explained on the basis of temperature, pressure, Eh, pH, CO_2, oxygen and sulpher contents (Fig. 13.3).

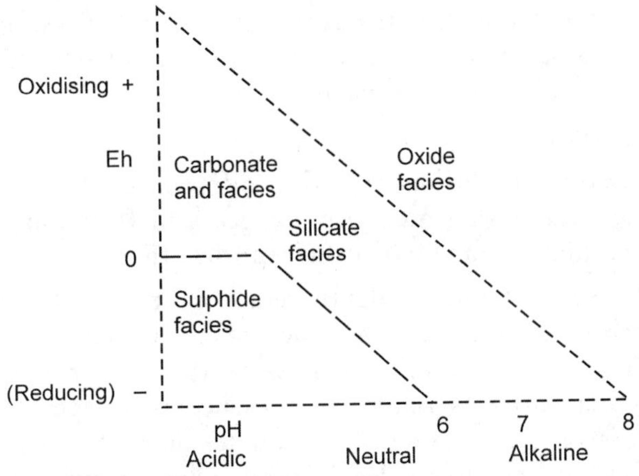

Fig. 13.3: Stability diagram of iron deposits.

Most of the iron deposits of India are oxide deposits. Why we do not get iron deposits of other facies (carbonate, silicate, sulphide) in India can be explained in the following way:

1. Indian deposits were formed in shallow water with oxidising conditions in wide area with circulation of water.
2. There are no relicts of carbonate or silicate or sulphide as original rocks.

Volcanic source or sub-marine volcanic springs act as a source of iron and silica. Heavy metals and fumes come from great depth. The solutions are rich in heavy metals. This theory is called "Hot Brine Theory". The question arises that what the condition was that never repeated later. The answer is that there must have been some special conditions like —

1. Atmosphere was more reducing and iron was more soluble. Iron is more soluble in Fe^{+2} condition.
2. There might have been organic life (like premordial form of life e.g. bacterial) which affected separation of silica and iron besides leaching.
3. It was a period of intense thermal activity or volcanism or deformation. After Cambrian very less deformation or thermal activity took place. (exception is formation of Himalaya, Alpine, and Rocky mountains which formed later in the geological history). These activities might have liberated iron and silica or have helped in the liberation of Fe and SiO_2.

Marine Source

Different conditions prevail at different depths in oceans. In lakes and fresh water also we get iron formation. There are four marine facies of iron deposits. (Fig. 13.4)

The CO_2 of middle depth removes Fe as $Fe(HCO_3)_2$ in the carbonate zone and this zone takes the maximum iron. This $Fe(HCO_3)_2$ may go to shallow depth where it precipitates and is deposited as oxides in presence of oxygen. But the source is in deeper zone of carbonate facies. Some iron like $Fe(HCO_3)_2$ may go to the deeper zone and precipitate as sulphide deposits (FeS_2).

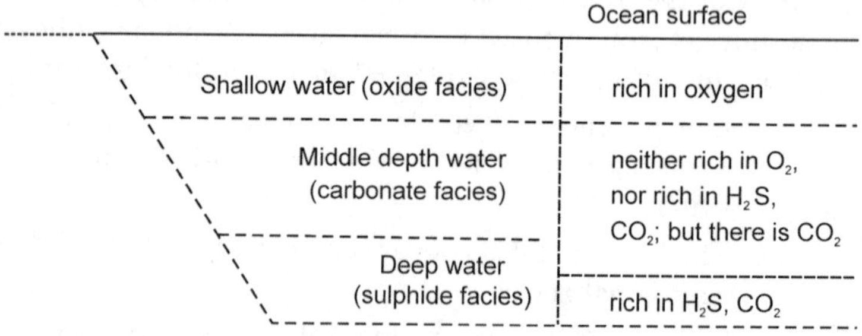

Fig. 13.4: Different Fe-Facies/Zones according to depths

Transport

Iron is transported in both ferrous (Fe^{+2}) and ferric (Fe^{+3}) form. They may be transported as Fe $(HCO_3)_2$, $FeCl_2$, $FeCO_3$ or Fe_2O_3 which are insoluble. In Fe^{+3} form iron is transported/ carried as gel as coating on mud particles. Environment may be acidic, basic or neutral, but it is carried in solution in reducing condition and later precipitates.

Types of Deposits

Rich hematite iron ore may develop from BIF by the process of supergene enrichment. Other types of iron ores (other than BIF) are:

1. Hydrothermal
2. Early magmatic
3. Late magmatic
4. Metamorphic
5. Volcanic
6. Metasomatic

Fe-ore deposits of India fall into 3 types: igneous, sedimentary and residual :

1. Titaniferous magnetite (vanadiferous and chromiferous) associated with basic igneous rocks.
2. Apatite-magnetite rocks found in Cu- belt of Singhbhum which are considered to be of hydrothermal origin.
3. BHQ of Precambrian age; metamorphic equivalents of these are BMQ.
4. Sedimentary ore deposits as layers and lenses of siderite in Gondwana strata. These are generally altered to limonite near the surface.
5. Hematite in fissures and faults.
6. Lateritic material derived from surface alteration of Fe-bearing rocks under humid tropical conditions. Though they are widespread, they are generally of quite low grade. There are several deposits in many parts of the country which, in the aggregate, yield an enormous tonnage.

Geological Distribution

Geologically, ores are found in rocks of different ages, though Precambrians are by far the most important as they comprise all the great hematite and magnetite deposits of India (Table 13.3).

Table 13.3: Geological Distribution of Indian Iron-Ores (after Krishnan, Jhingran & Sinha)

Formations	*Nature of Ore*	*States*	*Occurrence*
(1) ***Pre-Cambrian***			
Basic and ultrabasic rock.	Titaniferous and vanadiferous magnetite	Jharkhand Orissa	SE Singhbhum Mayurbhanj Southern districts.
Granodiorite and granophyre in Singhbhum	Apatite-magnetite rock	Jharkhand Orissa	Singhbhum Bonai, Keonjhar and Mayurbhanj.
		Karnataka	Shimoga, Chikmangalore, Sandur etc.
		Maharashtra	Goa, Ratnagiri.
		M.P.& Chhattisgarh	Bastar, Chanda, Durg and Jabalpur
Banded Iron formation (Metamorphosed)	Magnetite-quartzite	T.N.	Salem and Tiruchirapalli.
		A.P. Karnataka H.P.	Guntur Shimoga etc. Mandi.
(2) ***Cuddapah***	Ferruginous quartzite	M.P.	Gwalior, Rewa, Indore, Bijawar,etc.
(3) ***Gondwana***		A.P.	Cuddapah and Kurnool.
Barakar-Mahadeva	Ironstone	W.B. Jharkhand	Birbhum. Auranga Coalfield.
Ironstone shale	Ironstone and siderite	W.B.	Raniganj Coalfield.
(4) ***Triassic***	Hematite and limonite		
(5) ***Jurassic***			
Rajmahal Trap (Inter-trappean bed)	Ironstone	Jharkhand W.B.	Rajmahal Birbhum
(6) ***Miocene and Eocene***	Ironstone	T.N.	Travancore, Malabar etc. NE districts.
		Uttaranchal	Almora, Nainital.
Quaternary	Laterite	Other States	Derived from many formations including Deccan Trap.

Precambrian

Iron-ore series of Jharkhand and Orissa, Bailadila series of Chhattisgarh, Dharwar formations of Goa, Dharwar and Mysore areas include all the richest hematite deposits of India. The deposits occupy the top of hills which are made up of BMQ from which the ores have been derived by a process of enrichment, by removal of silica and other impurities in solution. The best ores are found at and near the surface, and contain 60 to 69% Fe. The ore bodies grade downwards into lower grade ores and into the original bed-rock.

Metamorphism in some areas converted hematite into magnetite and jasper into quartz. Some ore bodies found in Salem (T.N.) and Guntur (A.P.) are BMQ of this type.

Precambrian ultramafic intrusions in some parts of Jharkhand (Singhbhum) and Karnataka are associated with lenses of magnetite ore. Magnetite is titaniferous (up to 21% TiO_2) and often contains other elements such as V (Singhbhum) or Cr (T.N.). These ores are not much used because of the difficulties of eliminating or of easily utilising Ti content.

Precambrian rocks in some areas, *e.g.*, Kerala, Nilgiris, E-Ghats, etc., are often covered by a thickness of several metres of ferruginous laterite. These cannot be used now as richer hematite ores are available.

Cuddapah and Vindhyan

These formations contain scattered deposits of hematite, generally of small sizes. In some cases hematite has been concentrated from shale or quartzite and forms lenses and irregular bodies.

Gondwana

Lower Gondwana, particularly Barakar and Barren Measures, contain beds, lenses and pockets of siderite and limonite. Ironstone beds (mainly limonite) of Barren Measures in Raniganj coalfield were worked for several years and used in blast furnaces at Kulti near Asansol (W.B.).

Raniganj-Kamthi group of Lower Gondwana also contains limonitic ores but they are not important. Intertrappeans of Rajmahal hills contain thin beds and concretions of ironstone which were formerly used in indigenous furnaces.

Deccan Trap

Deccan Trap in many places is covered by fairly thick caps of ferruginous laterite, though iron content of these laterites is only 25 to 30%. This is not used at present but may become important in future. Some of these laterites are of the nature of bauxite.

Tertiary

Lower Siwaliks of Uttaranchal and Assam, Tipam Series of Assam and Miocene sandstone of A.P., T.N. and Kerala contain thin beds and concretions of ironstone which were formerly used as ore, but may be ruled out as unimportant at present.

Geographic Distribution

Assam

Clay-ironstones are found as thin beds and nodules in coal measures of Eocene age and in Tipam series of Miocene age. Some magnetite, weathered out of granite, was worked during 19th century in Khasi and Jaintia hills of Assam. These deposits are of no importance today.

West Bengal

Birbhum district contains deposits of iron-ores derived from various formations. Magnetite was obtained from weathered metamorphics; limonite from sandstones of Gondwana, thin beds of limonite and hematite in laterite associated with Rajmahal Trap. These were worked about two centuries ago for manufacture of iron locally.

Raniganj coalfield in Burdwan district contains bands of iron-stone in ironstone shale group of Damuda system (Lower Gondwana). Ironstone bands are found in an area over 40 Km long and 1.5–3 Km wide. At a depth of 6 to 9m, ironstone gives place to carbonate of iron. These deposits were used for several years by Bengal Iron Company (present IISCO). Ironstones were replaced by rich hematites of Singhbhum district in about 1915.

Analyses of ore show that the average is about 42-45% Fe; 15-20% SiO_2; 0.5-1.3% P and 0.9-3% Mn (Walker, 1914).

Jharkhand

Palamu district : Magnetite deposits

Two groups of deposits occur near Daltonganj, one near Gore (23°38' : 83°58') and the other near Biwabathan (23°55' : 84°03'). Magnetite beds are underlain here by tremolite-schist derived from the metamorphism of limestones. Granite is found to underlie the deposits in central part of the hill while dolerite is seen on NW side of the hill. The deposits extend over a length of 914 m with a width of 320 m. The ore varies from massive magnetite to magnetite-schist.

Singhbhum District : Apatite-magnetite deposits

Cu-belt of Singhbhum, which occupies the major shear zone separating northern from southern Singhbhum is characterised by Cu-lodes as well as lenses and tabular bodies of apatite-magnetite rocks. They are considered to be of hydrothermal origin. The largest lens is 274 m long and 18m thick at its maximum. Granitic rocks of the area are thought to have been responsible for the mineralisation of shear zone. Apatite is generally the dominant mineral but may vary from 20 to 70% of the rock. These deposits were regularly worked between 1906 and 1914 by Bengal Iron Company for mixing with the limonitic ores for producing high phosphorus pig iron. Some deposits were also worked in a small way for sometime for their apatite contents.

Singhbhum and Mayurbhanj districts : Titaniferous magnetite

Near the border of SE-Singhbhum and Mayurbhanj, there are a few fairly important lenses and veins as well as irregular patches of titaniferous magnetite at Kudada (22°42' : 86°12'), Dublabera (22°29' : 86°17'), Kumardubi (22°17' : 86°19') and other places. Magnetite is a product of segregation from gabbroid and ultramafic igneous rocks which occur in the area. It is associated also with serpentine, steatite, epidote, etc. Magnetite shows microscopic intergrowth with ilmenite and coulsonite, the latter being an oxide mineral of vanadium. Hematite is seen to replace magnetite to some extent. The ores are undoubtedly related to ultramafic rocks. Magnetite always contains a small amount of vanadium oxide which amounts generally <2% but rises rarely up to 7%. Titanium content of magnetite varies from 10 to 24%. Analyses of magnetite ore are given in Table 13.4. Similar deposits are also found in Simlipal area of Mayurbhanj. It is considered that the deposits may contain a few million tons of titaniferous and vanadiferous magnetites.

Table 13.4: Analysis of Titaniferous Magnetite Singhbhum (Jharkhand), Mayurbhanj (Orissa) and Channagiri (Karnataka)

Oxides	*Kumardhubi-Singhbhum*	*Channagiri*
SiO_2	0.57	0.88
TiO_2	13.84	11.60
Al_2O_2	2.83	1.79
Fe_2O_3	72.17	—
FeO	8.07	56.82
Cr_2O_3	—	3.09
MnO	0.14	0.48
CaO	trace	0.72
MgO	0.20	1.58
H_2O	1.33	1.23
P_2O_5	0.97	—
V_2O_5	0.59	—
Total :	**100.71**	**78.19**

Jharkhand and Orissa

Hematite ores (Lake Superior type)

Numerous large deposits of high grade hematite were discovered in the early years of last century by TISCO and GSI in southern Singhbhum, Keonjhar and Bonai. The largest group of deposits occurs between Manoharpur in north and Keonjhargarh in south. Similar deposits are also found in Mayurbhanj district, some distance east of this belt. Some deposits are well known as they have been worked for several years : Gorumahisani Pahar, Sulaipet-Badampahar and Simlipal hills in Mayurbhanj; Pansira, Buda, Noamundi and Gua in Singhbhum; Kiriburu and other deposits along a 48 km ridge on the borders of Bonai and Keonjhar; Thakurani, Joda Banspani and other deposits in Keonjhar. Mayurbhanj deposits, particularly Gorumahisani (22°19' : 86°18'), were the first to be exploited by TISCO when they established their Iron & Steel works at Jamshedpur in 1906. This hill is made up of BHQ associated with quartzite, epidiorite and grunerite-bearing rocks which have been derived from the metamorphism of banded iron-ores. BHQ occupies eastern part of the hill and is capped by ore bodies in north and east. Both massive and lateritic hematite is found amongst the ores. Two other deposits, *viz.* Sulaipet and Badampahar have also been worked for several years.

The most spectacular group of deposits is on the hill range forming the border of Singhbhum-Keonjhar-Bonai for a distance of 48 Km in a NNE-SSW direction. This region exposes more or less parallel, isoclinally folded rocks composed of BHQ associated with phyllite, conglomerate, arkose and some basic lava, all these constituting Iron-ore series.

The rocks are folded with their axis in a NNE-SSW direction. Iron-ore series in southern Singhbhum is intruded by Singhbhum granite whose age is now given as about 2,000 M.y. Geological succession in southern Singhbhum is given in Table 13.5.

Table 13.5: Geological Succession in South Singhbhum and Keonjhar (After Dunn, J.A. and Dey, A.K.)

	Kolhan series
	—Unconformity—
	Newer dolerite (?)
	Singhbhum granite
	Arkasani granophyre
	Chota Nagpur granite gneiss
Dhanjori series	Dalma and Dhanjori lava
	Dhanjori sandstone, conglomerate etc.
	—Unconformity—
	Upper shale and volcanics
Iron-ore series	Banded hematite-quartzite
	Lower shales (local limestone)
	Sandstone, conglomerate
	—Unconformity—
Older metamorphics	Mica-sillimanite, hornblende, chlorite and quartz-schist, quartzite etc.

Sarkar and Saha (1962) have revised the above succession on the basis of isotopic age determinations. They place an orogeny (Fe-ore orogeny) and Singhbhum granite (2038 M.y.) just above Fe-ore series; Kolhan series (1584 M.y.) follows Singhbhum granite after a hiatus; then comes Dhanjori series followed by Singhbhum orogeny (905 M.y.); then come the ultrabasics of Jojohatu, Arkasani granophyre and granite and finally Newer Dolerite. Confirmation of age data is now required.

BHQ consists of alternating thin bands of hematite, chalcedony or jasper generally having a thickness of about 2mm. It is possible that in some places thick beds of practically pure hematite (several metres thick) may have been deposited. In general, hematite-quartzites of Singhbhum-Keonjhar-Bonai area are practically free from detrital matter, indicating that the deposition must have taken place at some distance from the shore beyond the zone of clastics. The rocks are similar to jaspilite of USA and itabirite of Brazil. The colour of different bands varies from red or dark brown to grey, white and lavender. The bands often show much contortion and minute folding and faulting which are considered to have been developed during the process of replacement accompanied by slumping. Silica in jasper bands is generally chalcedonic but sometimes microcrystalline. The banding is often rather irregular. Real thickness of BHQ Fm has not been determined in several exposures in the area though it was estimated to be about 914 m in Bonai and 304 m in the ridge between Singhbhum and Keonjhar. The apparent thickness is, however, quite deceptive because of repeated folding.

Under the microscope, siliceous bands exhibit grains of quartz with undulatory extinction. Small crystals of martite or hematite may be found amongst silica grains and also in hematite bands. Some magnetite is also known to occur. The colour of jasper is controlled by the abundance and distribution of minute flakes of blood-red hematite. Rhomb-shaped crystals of silica as well as of siderite have occasionally been noted. It appears therefore that part of iron might have been deposited as carbonate in some places. Some of the bands show microspherulitic structures which may be attributed to the shrinkage of original colloidal Fe-hydroxide after deposition.

BHQ generally changes into hard massive iron-ore at and near the surface. The change from ore to BHQ may occur at any depth below ground level, usually between 45 m and 61 m and also laterally. 'Horses' of BHQ are sometimes found in the ore. There are often fairly large lenticular patches composed of friable, crystalline to dusty hematite showing banded structures. These lenses are inferred to be due to the complete removal of silica from the original rock without the space occupied by silica being filled by iron-ore later. The rock also changes sometimes into a laminated biscuit-like ore or into shaly looking ore which however may be quite rich in iron. These are not necessarily due to the enrichment of shale but may merely be very fine grained laminated hematite with some impurities.

A study of ore bodies indicates that they are due to the enrichment of BHQ by Fe-hydroxide under suitable conditions of topography and drainage at some former age. Enrichment has been due not only to the removal of SiO_2 but also to the simultaneous introduction of Fe in solution. It is only this way that very hard massive Fe-ores could have been produced at and near the surface. Other types of ore, namely, laminated and shaly ore, are generally not quite so rich as massive ore and may represent only partial replacement. Many features of folding, crumpling and faulting in ores and in hematite-quartzites should be attributed to slumping and movement during the process of replacement. Individual layers must at that time have been sufficiently plastic to undergo deformation on a minute scale.

Size of ore bodies varies considerably. Individual hills often contain several MT of ore as far as could be judged from surface observations though data are available from drilling in a few deposits. Jones (1934) listed >135 ore bodies in this region and estimated a total of nearly 2800 MT. Percival (1947) revised the above estimate to 8128 MT of good quality ore. In addition to these high grade ores, there are certainly large quantities of low grade ores containing 50 to 60% Fe. Analyses show average quality and range of composition of ores (Table 13.6).

Table 13.6: Composition of Fe-ore (Hematite-Quartzite)

Element	*%*	*Range*
Fe	65.06	60—69
SIO_2	1.47	1.0—40
Al_2O_2	1.47	1.5—3.0
P	0.048	0.040—0.090
S	0.017	0.014—0.050
Mn	0.022	nil—0.03

Orissa

Iron-ore deposits at Daitery (21°06' : 85°40'), Tomaka (21°05' : 85°56') and Kansa (21°04' : 85°51) all in Cuttack district, revealed that they contain Fe-ores of high grade and are available within 45 m of the surface.

Maharashtra and Goa

Some deposits are known in Goa as well as in the adjoining Ratnagiri district. They are all associated with Dharwar Fm and belong to Lake Superior type.

At Bicholim (15°30' : 73°55') there are two ore bands making up a syncline which can be seen as parallel ridges 400m apart and about 7 Km long. Width of the ore bands varies from 30 to 100 m. The surface contains some laterite ore. Main ore bodies are hematite with a little magnetite. These pass downward into bedded, friable hematite schist.

There are two other deposits one at Sirigao, 40 Km up Mandvi river and at Kosti just south of Calai station about 60 Km from Mormugao. These deposits were developed by Japanese for export to Japan but after Goa was taken over by India, the mining activities have suffered retardation.

In Ratnagiri (Maharashtra) some exposures of BHQ enriched to ore are found at Kankauli (16°16' : 73°45'), Vagda (16°14' : 73°45'), Kasal, Katta (16°05' : 73°40') and Redi (15°45' : 73°44'). The last deposit forms a ridge running EW for nearly 2 Km and rises to a height of 70 m.

Madhya Pradesh and Chhattisgarh

Rich deposits of BHQ are found in Bastar, Durg and Chanda districts. Other deposits are known in Jabalpur, Gwalior and few other areas.

Bastar

Deposits of large size are found in Precambrian rocks which have been called Bailadila Fe-ore series comprising BHQ associated with quartzite, phyllite, micaceous phyllite etc.

The deposits form 2 parallel ridges which may be the limbs of a syncline running from north to south. Thickness of BHQ formation is probably of the order of 300 to 450 m. The formation is capped by rich Fe-ores with some lateritic Fe-ore at the surface. There are also large quantities of poorer ore containing about 50 to 55% Fe.

Another group of deposits occurs at Rowghat. These deposits are close to Antagarh-Narayanpur road and occur on high hills about 450 m above the general level of the country. Six major deposits are known. These ores are also of the same type as those of Bailadila and contain over 60% Fe.

In Kanker area, near Pareekaro (20°23' : 81°04'), there is a ridge about 5 Km long containing hard hematite ore which is of the same character as that in the other deposits of Bastar. Deposits are hard ore and scaly micaceous ore.

Durg

Ores of this district have been known for over 75 years as they were originally examined by TISCO. They occur on Dhalli and Rajhara hills (20°34' : 81°04') forming a zig-zag range which attains a height of 12 m above the plains. The length of the ridge is about 32 Km. High grade ores of the same charcter as Bastar ores occur in a few places. Estimates made by GSI (70 yrs ago) are of the order of 115 MT. Deposits are now being exploited for feeding Bhilai Steel Works near Raipur.

Chanda

There are 5 or 6 deposits in this area at Lohara (20°23' : 79°44'), Pipalgaon (20°33' : 79°29'), Asola (20°14' : 79°46'), Dewalgaon (2o°23' : 79°59') etc. Of these, only Lohara deposit is large.

Jabalpur

In NE part of this district, formations formerly described as Bijawar series (Precambrian) contain a few deposits of hematite associated with ferruginous shale and quartzite. Some of them contain appreciable quantities of manganese. In general, however, ores are poorer than those of Orissa and Bastar and contain anything from 4 to 60% Fe. Deposits occur at Majgaon, Agaria (23°23' : 80°11'), Saroli (23°24' : 80°13'), Jauli (23°23' : 80°15'), Lora (23°30' : 80°11'), Gosalpur (23°24' : 80°05'), Kanhwara hills (23°55' : 80°28'), Imalia (23°46' : 80°24') etc. Estimates made about a century ago indicated about 50 MT in Agaria hill, 3.4 MT in Saroli, and some 50 MT in Kanhwara hills. Table 13.7 gives typical analysis of these ores.

Table 13.7: Analysis of Ores of Jabalpur District

Elements	%age (avg)
Fe	24.4—64.6
SiO_2	3.7—19.6
S	0.02—0.03
P	0.02—0.38
Mn	12.3—21.4 (may be lower upto 0.36)

There are also small deposits of red and brown hematite occurring in limestone and quartzite of Bijawar Fm in Hoshangabad, Nimar and Narsinghpur districts, but they are of little importance in modern iron smelting.

Himachal Pradesh

In Mandi area of H.P., magnetite-hematite ores occur in quartzite and slate of Simla slate Fm. The rocks strike NW-SE with steep dip. They are of bedded character and are conformable to the slate. The ore body extends over a length of 18 km with a gap, the northernmost point being at about (31°38' : 77°05').

Ores consist of banded magnetite-schists and some micaceous hematite associated with quartzite. The ore body varies in thickness from 30 to 180 m. Ore is friable and will easily be amenable to concentration by electromagnet. Analysis of magnetite-schist and hematite-schist is given in Table 13.8. Magnetite-schist can be concentrated by electromagnet to yield ores containing >60% Fe.

Table 13.8: Analysis of Ores from Mandi (H.P.) and Salem (T.N.)

Oxides	*Mandi Average (Magnetite schist-Hematite schist)*	*Salem Range*
Fe_2O_3	38.5	31.2 — 50.8
FeO	10.1	4.1 — 15.7
SiO_2	47.2	4.12 — 56.2
TiO_2	0.4	Nil — 2.7
Al_2O_3	1.6	0.2 — 2.7
CaO	0.8	0.1 — 1.5
MgO	1.3	0.1 — 2.6
P	0.088	0.017 — 0.19
S	0.007	Nil — 0.02
	34.75	25.05 — 47.3

Andhra Pradesh

Iron-ores occur in Guntur, Cuddapah and Kurnool districts and in a few places in Adilabad, Karimnagar and Nizamabad districts.

Guntur

In Guntur area there are magnetite-quartzite bands in Konijedu hills, Ongole hills and Pernametta. Another group of deposits occurs in Gundlakamma valley at Burepalle (15°41' : 79°59') and also near Vemparala (15°54' : 79°58'). Ores are of banded sedimentary type and have been generally metamorphosed into magnetite-quartzites. The reserves are found mainly in Guntur group and in Gundlakamma group. Ores, being magnetite-schists, contain about 35% Fe and about 40% SiO_2. Some other deposits of the same type of lower grade are also present in the same area.

Cuddapah

There are two small deposits of hematite which appear to be due to the enrichment of the iron content of Pulivendla quartzite of Cuddapah system at Chabali and Pagdalapalle (14°27' : 78°37'). Though the ores exposed at the surface are quite rich and contain around 60% Fe.

Kurnool

A fault zone through Lower Cuddapah formations traverses the country between Veldurti (15°33' : 77°56') and Ramallakota (15°34' : 78°01'). Fault zone trending ENE -WSE direction has a steep dip towards NW. Ore bodies form small hillocks and extend intermittently over a length of 8-10 Km. Ore material is hematite which is somewhat siliceous. There are about a dozen individual ore bodies in this zone. Few deposits have been investigated in Adilabad, Karimnagar and Nizamabad districts at Chityal (19°03' : 78°47'), Kalleda (19°08' : 78°53') and Amberpet (18°84' : 78°11'). All these occur in Dharwar Fm and are of banded hematite type. Ores generally contain 45 to 50% Fe and varying amounts of SiO_2. The ores need to be beneficiated before use.

Tamil Nadu

Though Fe-ores have been recorded from several districts, important occurrence is in Salem and Tiruchirapalli districts where they occur in a series of hills. Ores of Dharwar age and of banded hematite type have been converted into BMQ by metamorphism. Nine groups of deposits have been described in this area. Ores are of bedded character and form prominent hills, the ultimate reserves may be much larger. Magnetite-quartzites contain some hematite and amphibole which have been produced during metamorphism. Being magnetite-quartzites, ores show an average Fe content of only 35 to 40% but they are amenable to crushing and to electromagnetic concentrations.

Karnataka

There are several deposits in this state. Ores include (a) banded ferruginous types which give rise to magnetite when metamorphosed, and (b) titaniferous magnetites of igneous origin. Quantitatively the most important are of BHQ type which occur in Bellary (Sandur), Shimoga, Chitaldurg and other districts.

Sandur

Ores in this area form a part of Dharwar succession consisting of phyllite, banded quartzite and amphibolite and green schist. Rich ores have been derived, as in Orissa, from the enrichment of BHQ. There are also fairly important Mn-deposits in this area derived from the concentration and enrichment of Mn contained in the shale of Dharwar Fm. Fe ore bodies form ridges whose crest and upper part are composed of rich, hard and massive hematite. There are also shaly and laminated ores as in Orissa. There are 6 groups of deposits here which are high grade ore.

Ores similar to those of Sandur occur in Chikmagalur, Chitradurga and Shimoga districts. Amongst these, deposits at Kemmangundi have been worked for several years for feeding Mysore Iron & Steel Works. Magnetite ores derived from BHQ type through metamorphism also occur here.

There are few exposures of ultramafic rocks in Central and Southern Karnataka. Some of these, especially in Tumkur district, are associated with bodies of titaniferous magnetites. These contain 50-60% Fe, around 10% TiO_2 and 2-4% each of SiO_2 and Al_2O_3. Some of the ores also contain a little Cr (but rarely any vanadium). Reserves in few deposits which have been investigated are of titaniferous magnetites.

Rajasthan

There are few small deposits in Jaipur, Udaipur, Alwar and Bundi areas. Jaipur deposits occur in mica-schist as hematite bands. A few irregular masses were also found intercalated with Rialo limestones. In Udaipur rather poor quality ferruginous breccias occur near Jahazpur and some bands of hematite-schist amidst hornblende-schist of Aravalli near Thana. In Alwar district hematite ores are found in hornstone-breccia of Delhi system near Kushalgarh (27°25' : 76°29') and micaceous hematite-schist in Alwar quartzite near Tehla (27°14' : 76°28').

Some brecciated limestone of Aravalli age contains pockets of hematite near Rajgarh (27°14' : 76°41').

Punjab

Magnetite-schist and magnetite-quartzite are found at few places in Kangra district as at Dharman (32°04' : 74°49'). The bands continue into Mandi over a length of about 48 Km. Other deposits composed of lenses of magnetite with some hematite occur in Boileauganj quartzite near Nahan. These are also probably fairly extensive. Two or three bands of magnetite with some hematite of a total thickness of about 7.6 m and extending over a distance of 4 Km are found near Chhapri (27°56' : 76°07'). The bands occur in marble and schist of Aravalli system.

Uttaranchal

Nainital and Almora districts were scenes of iron smelting activity over a century ago. There are several deposits of hematite mainly in Siwaliks near Ramgarh (29°26' : 79°35'), Natuakhan, Dechauri, Kala Dhungri (29°17' : 79°22') etc. Quality of ore is apparently poor averaging 30-50% Fe but fairly large quantities, perhaps of the order of several million tons, may be present.

Kashmir

Of several deposits, those of Matah (33°08' : 74°44') and Khandli (33°26' : 74°28') may be mentioned. Matah deposits constitute a body of siliceous hematite associated with ferruginous chert having a thickness of 2.1 m in Great limestone series. Khandli deposit occurs at the base of quartzites overlying Agglomeratic Slate series of Permo-Carboniferous age. There are two beds of hematite here, the lower averaging a thickness of about 2.4 m and the upper at about 60 cm. Fig. 13.5 shows major iron ore mining areas of India.

Reserves of Iron-Ore

India possesses large reserves (about 24,535 MT) of good grade Fe-ore (with 60-63% or more). Total recoverable reserves of iron ore in the country are about 14,630 MT of hematite and 10,619MT of magnetite (2005-06). Resources of very high grade ore are limited and are restricted mainly in Bailadila sector of (Chhattisgarh) and to a lesser extent in Bellary-Hospet area (Karnataka) and Barajamda (Jharkhand and Orissa). Hematite resources are located in Orissa, Jharkhand, Chhattisgarh, Karnataka, Goa, Maharashtra, A.P. and Rajasthan.

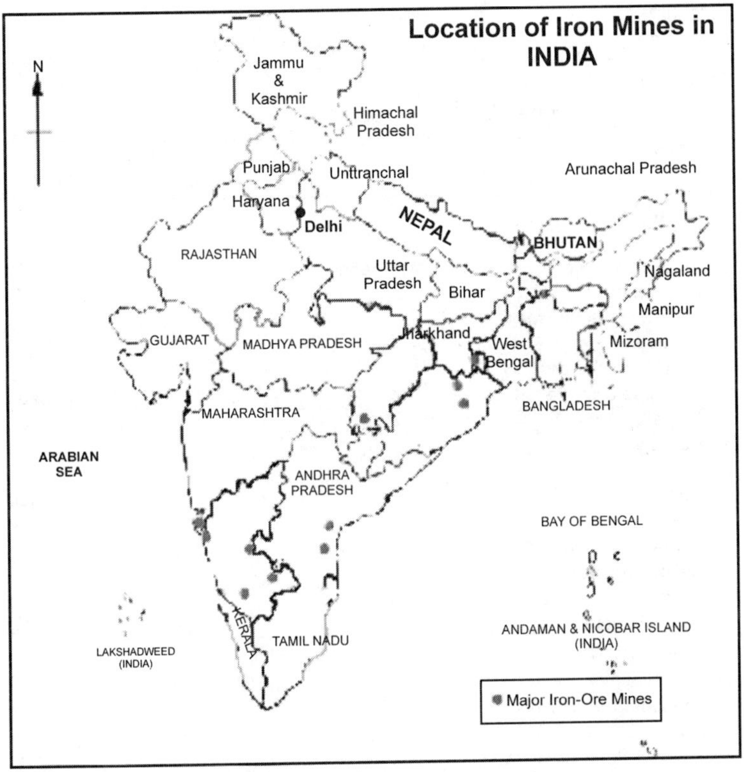

Fig. 13.5: Major Iron-ore Mining Areas

Magnetite resources are located in Karnataka, A.P., Goa, Kerala, Jharkhand, Rajasthan, Assam, Nagaland, Meghalaya, Maharashtra, Orissa and T.N. Table 13.9 presents an account of Fe-ore reserves and production in India. In comparison to statistics of 1970s as shown in Table 13.10 there has been tremendous improvement in recoverable reserves of Fe-ore in country.

Table 13.9: Fe-Ore Reserve and Production in India

Year	*Reserve (Recoverable)* Hematite	*Magnetite*	*Production* Iron Ore
1990-91	—	—	5.6
1998-99	9,602	3,408	7.2
1999-2000	—	—	7.5
2000-2001	—	—	7.9
2001-2002	10,052	3,408	8.3

Source : IBM

Table 13.10: Reserves of Iron-Ore (MT)

States	*Proved and Indicated*	*Possible*
I. Hematite Ores		
Jharkhand and Orissa		
Singhbhum, Keonjhar, Bonai, Mayurbhanj	2,740	8,000
Barajamda		
M.P. and Chhattisgarh		
Bailadila	610	3,600
Rowghat-Kanker	745	
Jabalpur-Chanda	120	3,400
Maharashtra		
Ratnagiri	5	
Goa	285	
Durg	115	500
A.P.	40	
Karnataka including Sandur	900	2,300
Bellary-Hospet		
Kashmir, Rajasthan, Punjab, and Uttaranchal	20	20
Total Hematite Ores :	5,580	17,830
II. Magnetite Ores		
A.P. : Guntur	300	600
Jharkhand and Orissa	5	
H.P.	60	60
T.N. : Salem-Tiruchirapalli	305	1,000
Karnataka	215	500
Total Magnetite Ores	885	2,160
III. Limonite and Spathic Ores		
W.B.	500	2,000
Grand Total	**6,965**	**21,990**

At present reserves of hematite is 14630 MT of which 95% (13916MT) is in Orissa and Jharkhand and that of magnetite stands at 10619 MT with 99% (10560 MT) in Karnataka (74%), A.P.(14%) and Jharkhand (11%).

Statistical Approach to Ore Genesis

Sandur Schist Belt in Bellary District of Karnataka has a reserve of 1250 MT of high grade iron ores and also contains 14.65 MT of Mn-ores. Iron ore deposits are associated with Precambrian Dharwarian rocks and at Donimalai there are 9 major deposits with a known reserve of 150 MT of chiefly hematitic ores of +64% Fe.

Ore horizon is hosted in BIF and which is also the protore. Much of the surrounding rocks in this region are silicified along with major occurrences of secondary chert bodies. Ores have undergone considerable metamorphism but they along with BIF still retains primary sedimentary and diagenetic textures which are overprinted with the later supergene enrichment process. Ores have been classified as lump and fines but are sub classified to 6 types: very hard ore (type 1), hard laminated ore (2), medium hard laminated ore (3), soft laminated ore (4), friable ore(5) and blue dust (6).

Ore mineralogy consists of Magnetite (Fe_3O_4, FeO, Fe_2O_3, or X $\{Fe^{+3}{}_8\}$ Y $\{Fe^{+2}{}_8\ Fe^{+3}{}_8\}O_{32}$ which is inverse spinel structure where the ferrous in octahedral position is oxidized, hematite alpha (Fe_2O_3, goethite (a-FeO OH), lepidocrocite (g-FeO OH). Major portion of magnetite had been selectively oxidized by the process of low temperature martitization, which affected the grains randomly, but mainly along peripheral regions rather than along cleavages which are indicative of preferably high temperature process as commonly seen in igneous rocks. Resultant mineral is termed as martite due to its origin from preexisting magnetite, which either imparts its isometric shape to hematite or is present as unaltered inclusions.

Genetically, lithologic control of mineralisation is reflected by total confinement of ore bodies within BIF, even though BIF are flanked on both sides by shales. Broadly, hardness of ores decreases downward but with consequent increase in grade. The bottommost ores are dominated by type 6 blue dust variety with Fe>68%. Texturally, ores exhibit a variety of replacement texture where oxides are both martitised as well as replaced in turn by hydroxides. Near the top colloform goethite often coexists with colloidal hematite.

Classical theory of origin of iron ores by replacement or ferruginisation of prexisting shales has been changed to the widely accepted theory of supergene process of ore formation from BIF. For well banded and/or finely laminated ores, iron in ferrous state from BIF were leached by a near surface GW possibly of acidic nature but definitely deficient in dissolved oxygen i.e. of low Eh and pH. The leached solution containing iron then either precipitated iron due to solubility product been exceeded or were oxidized by SW containing dissolved atmospheric oxygen that peculated down and on admixing raised the Eh to positive state while the pH may not have been significantly changed. New pH and Eh ranges of 4-7 and –4 to +4 being the stability range of hematite at near NTP conditions, supergene form of iron ores were therefore hematite rather than magnetite. Magnetite however is presently observed on in BIF and in some biscuity iron ores that has a different mode of origin. Leaching of silica by alkaline GW resulted in formation of biscuity iron ores where original iron oxide mesobands were left intact and on coalescing formed iron ores of this type cemented by residual silica. Total removal of silica from ores resulted in the formation of blue dust type of ores.

For statistical modeling of genesis ores were geochemically divided to lump and fines. For both ores Fe shows high negative correlation with SiO_2 and Al_2O_3. Positive correlation between SiO_2 and Al_2O_3 is significantly higher at 0.47 for fines and 0.36 for lump. However Al_2O_3 correlation with Fe is at –0.80 for lump but significantly lower at –0.80 for lump but significantly lower at –0.67 for fines. Multivariate linear regression equations using Al_2O_3 and SiO_2 as independent variable to predict iron content is

For lump $Fe = 67.20 - 0.87\ Al_2O_3 - 0.23\ SiO_2$

For fine $Fe = 68.38 - 0.94\ Al_2O_3 - 0.50\ SiO_2$

Positive correlation between Fe and other two oxides can be interpreted as either iron was removed independent of other elements which were stable or that during removal of silica, alumina was also affected. Geochemical similarity of the two oxide group is along with antipathy relation to iron is well known. Further the near similar regression equation indicated that the variation was linear and ultimately the two resultant ores have a co-affinity independent of the processes of formation. Low coefficient of determination ranging from 0.20 to 0.63 indicates that only bivariate regression equations cannot satisfactorily predict iron content for which alumina and silica together are required and only a first degree equation is sufficient for predicting iron content.

Grades. Indian commerical grades of hematite ore :

(A) Based on iron content are :

1. *low grade* (54-57% Fe)
2. *normal grade* (57-60% Fe)
3. *high grade* (>60% Fe)

Low grade ores are generally blended with high grade ones.

(B) Depending upon size, ores are :

1. *fines* (–10 mm),
2. *nuts* (+10 to 40 mm)
3. *lumps* (+40 mm).

Maximum size of lumps generally accepted by buyers is 200 mm.

Resultantly, presently India ranks among the largest producers of iron ore in world as shown in Table 13.11.

Table 13.11: World Ranking in Fe-Ore Production

Country	*Production (MT)*
Brazil	18.3
China	16.3
Australia	14.2
Russia	10.6
India	6.8
USA	5.9
UKraine	4.5
S. Africa	3.1
Canada	2.7
Venezuela	2.1
Sweden	1.8
France	0.15

(C) According to percentage and type of impurities are :

1. *bessemer* (containing not more than 0.045% P)
2. *low-phosphorous/nonbessemer* (0.045% to 0.180% P)

3. *high-phosphorous* (>0.180% P)
4. *manganiferous* (containing >2% Mn)
5. *siliceous* (containing >18% SiO_2)

In addition iron ore may be described as high sulphur, titaniferous, etc. Specifications are rather rigid within the above named grades and maximum impurity limits are always considered.

Types of ore which are found in India associated with BHQ formations are: (1) *massive ore*, (2) *laminated ore*, (3) *shaly ore*, (4) *flaky ore*, (5) *powdery (biscuit) ore* and (6) *lateritic ore.*

Uses : Iron ore is basic metallic raw material and iron the backbone of modern civilization. Industrial progress and military potential of a country can be judged by its per capita consumption of iron metal. High standard of living strained by advanced countries of world has been largely based on Fe-ore and coking coal, the two essential raw materials for manufacture of iron and steel. The factors which have made iron the most essential and basic industrial metal are that its ore deposits are wide spread and extensive to supply large tonnage at low cost; it has got high tensile strength which can be considerably increased by alloying it with other metals and non-metals to produce different kinds of steel and it has the property of magnetism which enables it to be used in generation of electricity.

Hematite is the main ore of iron in India. For the production of 1t of pig iron about 1.5 to 1.7t of iron ore (lumpy), 0.9 to 1t of hard coke and 0.5 to 0.7t of limestone (flux) are required and to produce 1t of ingot steel, generally 0.8t of pig iron, nearly 20% of steel scrap (home scrap, locally available) and 10 to 12% iron ore are required. In addition to the main use of iron ore minerals for production of iron and steel, magnetite is used in heavy-media coal washeries, micaceous hematite is used in paint manufacture and coating for electric welding rods, soft impure varieties of hematite (red ochre) and limonite (yellow ochre) are used in mineral paints and sideritic ore (after being calcined) is used for generation of hydrogen by steel-iron-contact process.

14

Manganese Deposits

Systematic record of mining of Mn ore is traced back to 1891 in Srikakulam district (A.P.), 1899 in Nagpur district (Maharashtra) and 1902 in Jhabua district (Chhattisgarh, earlier M.P.). From 1907 to 1928, India had been almost a leading producer of Mn-ore in world, but since 1929 this lead has been taken over by Russia and Brazil. Several mines of Mn-ore have been opened up in Maharashtra, M.P., Chhattisgarh, Bihar, Jharkand, Orissa, Gujarat and Karnataka.

A major part (about 3/4) of Mn-ore produced is exported. Seven plants producing standard grade of ferro-manganese have been established. Mn-ore mines are situated mostly in Karnataka, M.P., Orissa, A.P. and Jharkhand.

There are several types of Mn-deposits like bog iron type deposits, lake deposits etc. But these types are not important from Indian point of view. In India there are marine Mn-deposits. Like iron depsoits there are facies of Mn-deposits. There are three types of Mn-deposits:

(1) Orthoquartzite or Sandstone type (Precambrian age)

(2) Carbonate type

(3) Sedimentary exhalated or submarine volcanic type.

Richest Mn-deposits are more or less associated with BIF of pre-Cambrian. In India there are several types of deposits. They may be sedimentary or non-sedimentary deposits mostly of pre-Cambrian age.

Mineralogy

Manganese and iron behave very similarly. Mn is transitional element in periodic table and has more variable valency state (Mn^{+2}, Mn^{+3} and Mn^{+4}). Mn^{+4} state is common in Indian deposits. Ionic radius of Mn varies in different states. It varies from 0.97 in Mn^{+2} to 0.57 or 0.6 in Mn^{+4}. There are several metals like Fe, Ti, Al, Zn, V, Cd which have their ionic radius within the range of 0.097 to 0.57. Hence, Mn forms isomorphous compound series with many of these metals.

In nature Mn is found in several forms such as oxides, silicates, carbonates, sulphides etc. (Table 14.1).

Table 14.1: Principal Manganese Ores

Ore Minerals	*Composition*	*Mn%*	*Properties*
1. Pyrolusite	MnO_2	63.2	finely crystalline form.
2. Polianite	MnO_2		Coarsely crystalline (needle like crystals)
3. Psilomelane	Mn-oxide of uncertain composition containing water and varying amounts of oxides of Ba, K and Na	45.60	botryoidal or Stalactitic form, amorphous nature
4. Soft impure Psilomelane with some Pyrolusite (wad)	Soft and impure variety of Mn-ore Colloidal Mn has ability to absorb impurities		
5. Manganite	Mn_2O_2. H_2O	62.4	alters to pyrolusite
6. Braunite	$3Mn_2O_3$. MnSiO2	62	
7. Hollandite	MnO_2 with some Ba and Ferric oxide (Fe_2O_3), like psilomelane but anhydrous	45-46	
8. Sitaparite	$(Mn, Fe)_2O_3$ with CaO	44	magnetic
9. Vredenburgite	a mixture of Jacobsite $(Mn,Fe)_2O_4$, inter grown with Hausmonite (Mn_2O_4)	72	magnetic
10. Rhodochrosite	$MnCO_3$	47.7	
11. Rhodonite	$MnSiO_2$	41.9	
12. Spessartite	Mn – garnet 3MnO. Al_2O_3. $3SiO_2$	33.3	

Mn-Sulphide Ores (MnS and MnS_2). These are less common than iron sulphides. Psilomelane, braunite and pyrolusite are the common ores in India. Indian commercial ore of Mn is found as a mixture of psilomelane and braunite. In Nature there are parallel facies of sedimentary Mn-formation like oxide, carbonate, silicate and sulphide facies.

Texture and Structure

They are mainly primary, sedimentary or sedimentary exhalative (sub-marine) deposits. Mn-deposits are very fine grained. Some of them are even amorphous and also colloidal. They do not have crystalline structure. They have been usually transported to a longer distance than the other detrital compounds. Because of their colloidal nature they adsorb some impurities. These deposits on the whole are bedded, stratiform or layered and invariably strata bound as in the case of Fe-ores.

Occurrence. Mn-ore deposits are of five geological types:

1. ***Hydrothermal deposits*** : These are formed by magmatic hot water solutions.
2. ***Sedimentary deposits***: Such deposits are formed in seas or lakes by chemical and mechanical action (Mn-minerals being deposited from solutions derived from pre-existing Mn-bearing source rocks or other deposits.
3. ***Residual deposits***: These deposits are formed by the leaching out of gangue minerals from pre-existing Mn-bearing rocks or deposits, leaving a residual concentration of Mn-Minerals.
4. ***Metasomatic replacement deposits***: These are formed by UG water containing Mn-solutions.
5. ***Metamorphosed deposits***: These deposits are formed by intense alteration by heat and pressure of any of the above type of deposits, generally the sedimentary ones.

Hydrothermal deposits are not important but such deposits might have provided Mn to other deposits. Sedimentary and residual deposits are most important economically and most of the metamorphosed deposits were originally sedimentary.

Genetic Classification

Indian Mn-ore deposits can be classified as follows:

Epigenetic

1. Regionally metamorphosed Mn–sediments. These deposits are of 2 types : (a) associated with pelitic and psammitic rocks and very often with Mn-silicate rocks (gondites) *e.g.* Mn-ore bodies associated with Mansar Fm of Sauser Group of M.P. and Maharashtra; with Gangapur Group of Orissa; and with Aravalli Group of M.P. and Gujarat. (b) associated with marbles and Calc-silicates, *e.g.* Mn-ore bodies in Lohangi marbles of Sausar Group of M.P. and Maharashtra and in calc-silicate rocks of Khondalite Group of Srikakulum distt. (A.P.). The latter ore bodies were formerly considered by Fermor to have been formed by supergene alteration of Mn-Silicate in kodurites. Kodurite ore bodies are now believed to be of metasedimentary nature and regionally metamorphosed syngenetic sediment.

2. Thermally metamorphosed Mn sediments. These are contact metamorphic deposits formed by thermal effect of granitic intrusion in pre-existing Mn-sediments associated with impure limestones e.g. Goldongri deposit, Panchmahal (Gujarat).

Syngenetic

3. Residually concentrated (*in situ*) ore bodies. These occur in parent rocks (lateritic) including supergene oxidation products of primary ore bodies or Mn-silicate rocks, *e.g.* Mn-ore deposits in Goa (in part), Singhbhum-Keonjhar-Bonai area (Jharkhand and Orissa) superficial oxidation products of regionally metamorphosed ores and gondites.

4. Replacement and Cavity filling deposits. These are formed by meteoric water by precipitation from 'gel' or solution, e.g. Singhbhum-Keonjhar-Bonai area (Jharkhand and Orissa), Sandur-Shimoga-Chitaldurg area (Karnataka), Banswara & Udaipur distt (Rajasthan).

Sources of Manganese

1. Continental Crust. All the igneous, sedimentary and most of the metamorphic rocks contain Mn, but its content is less than iron. Mn-Fe ratio is 1:40 or even 1:60 but Mn is present in all the crustal rocks-igneous, sedimentary and metamorphic. Thus, primary source of Mn is the continental crust as that of iron. All the common rock forming minerals *viz.*, Pyroxene, Amphibole, Garnet, Mica, etc. contain Mn while ultrabasic rocks do not contain Mn.

2. Marine Source. Moderately deep water of Oceans are source of Mn. Iron gets dissolved as Co_3 in Co_2 zone in marine basins. It is called Halmyrolysis, a marine process of origin of iron. A similar process is applicable in case of Mn also.

3. Hot submarine thermal springs. These springs errupt materials on sea floor. Like iron, Mn is also derived from these eruptions. Source of Mn are sub-marine exhalations, thermal springs, volcanic activities and hydrothermal activities. Red Sea has a sub-marine hot spot giving out heavy metals and manganese. This theory is called 'hot brine theory' of ore genesis. The ore charged materials come from deeper part of ocean and get deposited on ocean surface. In some parts of Pacific and California hot exhalations are coming out from the bottom of sea and depositing Ni, Co, Pb, Cu, Mn etc. Origin of deep sea Mn-nodules can be explained by this theory. These are common in Pacific and Indian Ocean and not in Atlantic Ocean.

Transport of Manganese

Mn is transported like iron compounds. These are transported in form of solution as bicarbonates or chlorides or sulphates or clay pellets or organic compound like humic acid or humates or suspension in form of fine particles.

Deposition

The depositional process of Mn is similar to that of iron bicarbonate solutions or chloride solutions which are formed under acid conditions and precipitate Mn in the form of oxide in oxygen rich environment and also in slightly alkaline condition (pH>6). Mn-sulphide is formed only in very reducing condition, and is rare unlike Fe-sulphide which is much more common. Similar to iron precipitation, bacteria plays an important role in precipitation of Mn, e.g. Leptothrax bacteria. Algae also helps in precipitation of Mn Pebbles and cobbles of rivers bed has dark colour coating of Mn due to algal activity in stream water.

Thus Fe and Mn has same source and similar mode of transportation and deposition. So most of the Fe-deposits are associated with Mn deposits. Such type of deposits are known as ferruginous Mn-ore deposits or manganiferous Fe-deposits which are quite common. Lateritoid ore (Mn + Fe) are found in Jharkhand. Ferruginous Mn-laterites are also found there.

Separation of Fe and Mn: Most of Fe and Mn ores are associated. In nature, both Fe-ore without Mn and Mn-ore without Fe are known. In India there are examples of both such deposits. The question arises when, how and why they (Fe and Mn) get separated then? Separation may take place either in the provenance or during transportation or deposition or it may be diagenetic. Natural rocks are weathered by sea water at normal temperature and pressure if subjected for 4 to 15 days. The solution after analysis indicates that ratio of Fe and Mn is 10:1 as compared to that of the original ratio 50:1, that is, originally Fe which was 50 times more than Mn has reduced to 10 times. This means that Fe has precipitated during transportation. This indicates that Mn has increased or more Mn has gone in to solution.

Separation may also take place under different Eh and pH conditions. Under constant pH, and same acidity and alkalinity, Fe is removed in solution at a low Eh (*i.e.*, Fe is leached out and not Mn) or if Eh condition remains same, Fe is leached out in more acid condition and Mn will leach out later in alkaline condition. This selective separation may be responsible for separation of Mn and Fe in sedimentary cycle. Hewetts experimental work has shown that there is selective leaching at different Eh and pH conditions.

Separation may also take place during transport. Experiments show that Mn is more stable than Fe compounds. Hence, Fe-compounds precipitate earlier than Mn-compounds which precipitate later and farther away from the source. Thus Fe separates out at shallower part while Mn at deeper part. In such cases upper part of deeper Mn ores can be found associated with Fe rich material. In some cases iron is not found at shallow depths, which may be due to biogenic separation of Fe and Mn. Here, Mn bacteria causes Mn to precipitate because it needs manganese for its metabolic activities. This can be exemplified by precipitation of manganese in water pipes. Analysis of water carried through pipes reveal that it contains more Fe than Mn because of precipitation of Mn along the walls of pipes. This is due to biogenic separation during transport. Separation of Mn and Fe may take place during deposition also. Russian scientists believe that separation may occur during the diagenetic stage before lithification. Fe and Mn both precipitate in the basin but Mn which is more soluble than iron, is carried farther away from iron in the form of solution.

Indian iron formation which is known as BIF (pre-Cambrian), contains very little Mn and other detrital contents/sediments. Similarly Mn-deposits of India contain very little Fe. Thus, there has been natural separation in Indian Mn-deposits.

Genesis of Mn-deposits

1. Orthoquartzite type or sandstone type. In this type of deposits Mn-ores are associated with sandstone and clays. Thus, these are partly arenaceous and partly argillaceous. Sometimes there may be K-bearing clays like chlorite, glauconite, etc. indicating precipitation in marine environment.

2. Carbonate type. These Mn-deposits are associated with limestone and dolomites etc.

3. Sub-marine volcanic type. (Sedimentary-exhalated type). These deposits are mainly associated with tuffs and pyroclasts.

Indian Mn-deposits are mostly layered or bedded in nature and are associated with Fe. Manganese is deposited as chemical precipitates of Mn oxide. Sometimes Mn oxide layers contain silica or silicates. These layers are found to be present in alternation with detrital sediments carried from land which deposited in shallow sea of pre-Cambrian.

Besides these main types of Mn-ore deposits, there are some other types of deposits also.

Plateform type deposits: These are associated with carbonates like limestones, dolomite. These may or may not be associated with tuff rocks. These are sometimes geosynclinal (Eugeosynclinal—with volcanism and miogeosynclinal—without volcanism) in nature, *e.g.* Lohangi marble of pre-Cambrian, in M.P. In India, generally Mn-deposits are not associated with tuff rocks or pyroclastics. Such type of deposits are found in Africa. Thus, submarine exhalative Mn-deposits are generally absent in India.

Hydrothermal deposits: In India there are some deposits of Mn-formed from submarine hydrothermal solution or submarine streams from the bottom of sea which deposits or precipitates Mn on ocean floor. Volcanic materials come up with Fe and Mn. Fe precipitates at deeper depth while Mn at shallow depth. Here, separation takes place during ascent of hydrothermal solution on the theory that Fe precipitates first and Mn later. Such deposits are found in Brazil, Africa, former USSR, Morocco.

Manganese nodules. These are formed both at shallow as well as at deeper part of Oceans. These consist of mainly Mn, Fe, Cr, Ni, and Co.

In deeper part of sea, Mn-nodules are present because of separation or precipitation of Fe at shallow part and Mn at deeper part.

Other process may be that they are formed from thermal springs rich in metallic constituents like Mn and Fe. Mn precipitates at deeper part. There are also evidences that Mn-deposits have formed at greater depth and later carried to shallower depth. These deposits occur as Mn-nodules (and not like beds) because of bacterial action. Mn deposition occurs around some nucleus. Such deposits are found in Arabian, Indian and Red Sea where base metals or heavy metals are present at greater depth.

Indian Distribution

The most important Indian Mn-ore deposits occur as bedded sedimentary (metamorphosed) deposits associated with Gondite series and Kodurite series of Archean (pre-Cambrian).

Mn-ore deposits are found associated with Gondite series of M.P. (Balaghat, Chindwara) and Chhattisgarh (Jhabua), Maharashtra (Bhandara and Nagpur), Orissa (Gangpur in Sundergarh). Mn-ore deposits are found associated with Kodurite series of Orissa (Ganjam and Koraput), A.P. (Srikakulum and Vishakhapatanam).

Gondite is mainly composed of quartz and spessartite (Mn-garnet) with varying amounts of rhodonite (Mn-pyroxene) and Mn-amphiboles. Spessartite-rock, rhodonite rock and rhodonite-quartz rocks are also common. Gondite rocks were formed by intense metamorphism of manganiferous sediments (probably oxides). Sands and clays on metamorphism were changed into quartzite, phyllites and schists and purer manganiferous sediments into Mn-ores and mixtures into Gondite rocks.

Kodurite rocks consist of feldspar, spessartite andradite garnet, rhodonite, quartz and apatite in varying amounts. They may vary from acid (quartz-orthoclase rocks) through basic (kodurite) to ultrabasic (Mn garnet rock or Mn-pyroxene rock) and are believed to be hybrid rocks produced by assimilation of Mn-ore and Mn-silicate rocks by an acid igneous magma.

Ore bodies of Gondite series persist to a considerable depth from the surface while those of Kodurite series are generally irregular, but sometimes follow the original bedding.

Lateritoid Deposit

These are next in importance to bedded ores Lateritoid deposits of Mn-ore or surface enrichments of lateritic character which occur as segregations and are evidently derived from underlying Archean manganiferous rocks or deposits or are associated with Fe-ores. These ores also contain Fe and may vary from Mn-ores to Fe-ores. Indian distribution of Lateritoid deposits are shown in Table 14.2.

Table 14.2: Distribution of Latritoid Deposits

State	*District*
Jharkhand	Singhbhum
Maharashtra	Ratnagiri
Karanataka	Bellary, N Kanara, Tumkur, Chitaldurg and Shimoga
Orissa	Keonjhar
Rajasthan	Banswara, Bhilwara, Jaipur and Pali
W.B.	Midnapur
Goa [important producer—Mn-Ore]	–

Secondary Manganese Oxides: These are less important and occur as veins, nodules, lenticles or irregular replacement masses in phyllites, schists, limestones and quartzites in parts of Jharkhand, Goa, Karnataka, Orissa and Rajasthan.

Such deposits may have been formed by segregation of Mn from adjacent rocks or the source may be deep seated connected with igneous activity. Float ores some times occurring as beds of boulders, derived from *in situ* beds are also worked at some places.

Ores of M.P. and Maharashtra have high Mn (>46%) which are suitable for manufacture of ferro-manganese alloy. Some battery grade ore (MnO_2) is also found. Bhandara and Ratnagiri deposits are of low grade (20-30% Mn). Jharkhand and Orissa ores vary from high grade to low grade (Mn 38-54%) and metallurgical and battery grades both are found. In Karnataka, Shimoga and N Kanara district ores vary from high to low grade (33 or 34% to 51 or 52% Mn) and sometimes suitable for manufacture of ferro-manganese. In A.P. Srikakulam district ores are low to medium grade (28-30 % Mn). Average grade of Mn-ores produced from Gujarat contains 44.50% Mn, from Rajasthan 36-40% Mn and from Goa (two grades 10-35% and 35-48% Mn).

Lateritoid Manganese

Mn deposits of India are either metamorphosed to various grades and syngenetically associated with Gondite (M.P., Maharashtra) and with Kodurite (A.P. and Orissa) or are unmetamorphosed epigenetically associated with shales of volcanic affiliation mainly in Jamda Koira Valley (JKV) of

Jharkhand and Orissa. Minor deposits are also found in Goa and other places.

Jamda Koira Valley Deposits: These overlie a younger sequence (1.9-2.1 Ga) of BIF and belong to the broad Iron Ore Group of pre-Cambrian Age. Numerous deposits are scattered all over JKV where ore bodies occur commonly as stringers, veins, pockets but rarely as massive or conformable beds or horizons. They exhibit no structural conformity with regional or even the local host rock structures both at macro and micro levels, which leads one to conclude that structural control was not responsible for ore emplacement or mineralisation.

Ore mineralogy consists of low temperature higher oxides represented by primary pyrolusite, psilomelane and cryptomelane along with rare nsutite. Prominent ore texture is colloform banding with very localized evidences of minor recrystallisation. They show typical botryoidal structure while internal texture is mainly oolitic. Braunite of secondary origin showing well crystalline nature is found either in between colloidal band or is more common in vugs or fracture zones. Associated metallic gangue consist of minor hematite which often shows colloidal form or massive aggregate and goethite that acts both as intergranular cement and surface coating. Non-metallic gangue is chiefly made up of alumina in form of clay, which is more often in admixed form rather than in structural form. Silica is always present but in the form of quartz rather than in any silicate matrix a concretionary form of ore is called WAD, which is a admixture of Mn-ores and clay often cemented with ferruginous material. WAD is frequently rejected as primary ore but in recent times many have found some limited use of these WADs. Sulphur content of primary ores are below industry level but from many deposits phosphorous in objectionable amount has been reported.

The bulk of deposits belongs to metallurgical grade with MnO_2 varying from 35% (low grade), 45% (medium grade) and >55% (high grade) with corresponding decrease in Fe content but SiO_2 and Al_2O_3 almost remains constant and their total decreases in a small manner. Chemical grade of ores with MnO_2 >85% is confined only to a few deposits where it is mostly found as massive beds or very thick bands of secondary origin. Pre requirement of nsutite for utilization of Mn-ores in dry cell battery industry prevents use of these ores but in recent times a few deposits from this area and a few places in Orissa which were found to contain appreciable amount of nsutite makes them suitable.

Submarine volcanic origin of syngenetic nature was advocated due to the intimate association of ore bodies with shale formation or confinements within various shale horizons. However, absence of structural conformity prompted a different genetic theory. Near surface confinement along with erratic and sporadic occurrence coupled with highly variable dimensions of individual ore bodies seems to indicate a definitely epigenetic emplacement though the ores may be endogenic in nature. Field studies indicate that extensive weathering coupled with deep seated and possibly acidic GW may have caused removal of Mn from manganiferous shales that were later randomly deposited giving rise to presently observed ore bodies. From borehole data it appears that whitish clay beds, which are extensively weathered shale or pyroclastic tuff acts as an barrier to circulating ore bearing solutions as very rarely ores are found below clay horizons. Some people are of the opinion that possibly ore bearing solution also deposited the ores at culmination and depression points of regional structure but to prove this a very detailed structural-geological work is required in this valley.

Many a workers have postulated a biogenic origin or precipitation of Mn by biota based on the presence of stromatolyte seen in the cross cutting chert bands. However, other than colloform banding seen in ores, evidences of biological activities are absent. Direct precipitation from submarine volcanism is also negated in view of absence of ores either as a stratigraphic horizon or strata bound nature.

Mining of Manganese Deposits : Majority of Mn-mines in India are worked by OC method as most of the deposits occur very close to surface. In shallow surface, superficial and lateritoid deposits (where ore occurs as boulders, float ore or as small segregated masses overlain usually by soil, laterite and rare rocks like quartzite, shale or limestone), OB is first removed and ore thus exposed is excavated and screened. Large pieces of ore free from contamination with OB are picked up by hand and stacked separately. Smaller pieces associated with spurious matter consisting of laterite, ochre, lithomarge, iron hydroxides, country rock fragments and secondary quartz are subjected to jigging (in Japlin jiggs) where heavy manganese ore pieces concentrate at the bottom of screen while lighter impurities come to the top and are scraped off and thrown away.

OC method of working Mn-ore deposits which occur as reefs or beds exposed on hills consists of deep pits or quarries following primary beds of Mn-ore as in M.P., Gujarat and Maharashtra. It varies from removal of upper parts and sides of hill ridges, as in some of the mines in M.P. to deep quarries as at Kandri in M.P. or Garividi and Garbham workings in Visakhapattanam (A.P.). Most of these OC mines are worked manually. Shot holes are made in ore and surrounding hard rock by means of crow bars and broken ore is carried out of the quarry in baskets. The height of benches varies from 1.5 to 9m according to the hardness of ground. Pneumatic Jack hammers and in some of the larger mines mechanical shovels, dumpers and other earth moving machines are used. With use of self-acting haulages, ores from hill slopes are brought down to ground level.

There are few UG mines, such as Ukwa and Bharweli mines in Balaghat district (M.P.). Mansar mines in Nagpur district (Maharashtra) and Shivrajpur mines in Gujarat. These mines use either the cut and fill method of stoping or the square set method. Development work, *i.e.*, drives, cross cuts, raises and winzes driven prior to actual mining by these methods proceeds as for methods of stoping. Cut and fill method employs a sequence in which the ore is removed in a series of parallel slices and as each slice is removed, a layer of waste fill is placed in the stope, leaving sufficient working head room in which to drill the ore. The fill is usually broken rock debris rejected from the screens and jigs. Square set method of stoping is applicable in mining those deposits in which ore is structurally weak or surrounding rock is fractured, faulted and is made very weak by alteration. Here, ore is mined in blocks, almost of the same size—1.5m × 1.5m × 2m to 1.8m × 1.8m × 2.4 m. As soon as one or more blocks have been removed, a timber frame (square set) is erected in open space with blocks or braces to support the timbers next to ore and walls. The stope is started from a rise timbered with square sets.

In some mines, rejected and small size ore is thrown along with waste rock, while at other mines such ores, which had been dumped separately and are called black dump, are now being worked by simple manual quarrying. Use of explosives is not required as dumped material is quite loose and can easily be dug out by picks and shovels. Ore is concentrated by jigging. Manually operated Joplin Jig is generally used for cleaning and grading the ore. Washing is done to reduce Fe content of Peroxide grade ore. Mechanical washers are used in some M.P. and Orissa mines and rumblers are employed in some mines in N-Kanara (Maharashtra) for removing laterite coatings.

A heavy media separation plant is used to treat the reclaimed old dumps. The plant uses ferro-silicon (sp gr. 7.1) kept in suspension in water by churning to maintain a sp.gr. of pseudo-liquid at 3. Gangue material floats, while ore, which had a sp.gr. of 3.5 sinks. A plant of capacity of 70 tons/hour requires 40,000 gallons of water in circulation per day.

Total reserves of all grade of Mn-ores in India are estimated at about 167 MT of which Maharashtra-M.P.-Chhattisgarh belt contains 144 MT and Jharkhand-Orissa belt 22 MT (Table 14.3). The plants producing standard grade ferromanganese ore are situated in A.P., Karnataka, Orissa and Maharashtra. National Metallurgical Laboratory & IBM are carrying on series of investigations for beneficiation of low grade ores (siliceous, ferruginous, garnetiferous and complex ores) to render them suitable for manufacture of ferromanganese. A major part of production is exported as ore; but with development of ferromanganese industry and increase in production of steel, there has been a great increase in consumption of Mn-ores in the country in steel industry as well as for manufacture of dry batteries and manganese salts. India ranked third in world production of ore in 1980s. It was an important item of our exports. Its present ranking is as shown in Table 14.4.

Table 14.3: Reserve of Mn Ore

Type	*Reserve (MT)*
Proved	40
Probable	49
Possible	78
Total	**16**

Table 14.4: Major Mn Producers

	Countries	*Production (MT)*
1.	China	7.00
2.	Ukrain	3.00
3.	S. Africa	3.11
4.	Brazil	2.00
5.	Australia	2.01
6.	Gaivon	1.90
7.	India	1.80
8.	Mexico	0.53
	World (Total)	**22.70**

Chemical composition of Mn-ores of important deposits in various state has been given in Table 14.5.

Table 14.5: Chemical Composition of Mn-Ores

Area	*Mn%*	*Fe%*	*SiO_2%*	*P%*	*Production ('000 t)*
I. A.P.					93
Srikakulam	32.21 to 49.05	4.8 to 15.70	1.1 to 10.3	0.13 to 0.48	
II. Jharkhand					—
Chaibasa and Singhbhum	47.77	—	4.63	0.34	
III. Maharashtra					297
Kandri	44 to 52.5	5 to 7	11 t0 18	0.18 to 0.25	
Mansar	49 to 52	6.7	7 to 12	0.06 to 0.26	
IV. Gujarat					—
Shivrajpur, Bamankua and Pani	31 to 55.53	2.08 to 7.5	2.12 to 19.1	0.14 to 0.35	

V. M.P.					384
Chikla (Bhandara)	47.3 to 51.0	6.5 to 8.0	8.0 to 16.0	0.075 to 0.2	
Dongri Buzurg (Bhandara)	51.25	6.5	—	0.275	
Tirodi (Balaghat)	48.01 to 52.32	5.79 to 0.05	3.6 to 8.9	0.08 to 0.16	
Bharweli (Balaghat)	50	5.0	5.0	0.07	
VI. Chhattisgarh					312
Kajili Dongri (Jhabua)	47.77 to 48.49	9.60 to 10.4	6.6 to 8.6	0.165 to 0.172	
VII. Karnataka					494
Shimoga	19.9 to 41.04	—	—	—	
Sandur (Bellary)	21.51 to 61.6	—	0.41 to 3.66	0.02 to 0.25	
North Kanara :					
Chipoli Nagari	15.79 to 53.78	3.88 to 28.21	1.38 to 12.3	0.023 to 0.114	
Usoda	—	—	—	—	
VIII. Orissa					0.65
Kutingi	44.34 to 55.42	1.12 to 5.02	—	Nil to 0.02	
Ghoriajor (Sundargarh)	45.58 to 54.13	2.60 to 7.92	2.6 to 11.2	0.06 to 0.15	
Patna	38.0 to 54.15	3.1 to 18	Nil to 8.6	0.224 to 0.377	
Keonjhar	31.07 to 57.4	1.9 to 20.53	0.99 to 3.1	0.036 to 0.103	
All India Total					1600

Production of Mn: Production of Mn Ore in India is as shown in Table 14.6.

Total 14.6: Production of Mn Ore

Year	*Production (MT)*
1990-91	1.5
1997-98	1.6
1998-99	1.5
1999-2000	1.6
2000-2001	1.6
2001-2002	1.5

Industrial Uses of Mn-Ores: Mn-ores are an essential constituent of steel in which its percentage of presence varies from 0.5 to 0.75 (12 to 15 lbs in every ton of steel). It has no satisfactory substitute for it in making steel and may well be described as workhouse of steel making. Steels used for crushers and well drill bits have 12% or more Mn and are extremely tough, because Mn is stronger, harder and able to resist abrasion. Mn is used in steel in form of Fe-Mn alloys ferromanganese (containing >75% Mn) and spiegeleisen (containing >75% Mn) and manganiferous pig-iron (containing 10.30% Mn), usually containing small amounts of C and Si. These alloys are made either in blast furnace or in electric furnace. Main purpose of using or adding Mn is to:

(1) Remove oxygen from molten steel
(2) Combine with sulphur and so improve the workability of steel
(3) Restore to steel the necessary carbon, and
(4) Increase the strength, toughness and wearing qualities of special steel.

In pig iron production, Mn (0.4 to 1%) is added in form of Mn-ore to remove bad effects of sulphur. Mn-metal (prepared commercially by electric furnace, thermal reduction of oxide or electro deposition from $MnSO_4$ solution is used as a deoxidiser and to improve the physical properties like strength, ductility etc. of non-ferrous metals and alloys (with Ni, Cu, Al, Mg, Pb, Zn and Sn).

Non-metallurgical uses of Mn-ore include use in (1) dry-cell batteries (MnO_2 as a depolariser); (2) manufacture of driers for paints and varnishes (3) ceramics as colouring agent (4) glass to impart colour when used in large quantity, otherwise as a decolourizer (5) production of permanganate (MnO_4) and other Mn-Chemical (6) agriculture.

Grade : According to composition of manganiferous ores in regard to proportion of Mn to Fe, it is customary to use the terms shown in Table 14.7.

Table 14.7: Gradation Based on Fe and Mn%

	Ore	*Mn%*	*Fe%*
(a)	Mn-ores	>40-63	0-1
(b)	Ferruginous Mn-Ore	10-40	10-30
(c)	Manganiferous iron ore	5-10	30-65
(d)	Iron ores	<5	>65

In general, manganiferous iron ores are used for making manganiferous pig iron; ferruginous manganese ores for production of spiegeleisen; and Mn-ores for production of ferro-manganese alloys and chemicals.

For the purpose of trade, Indian Mn-ores are classified in to the grades shown in Table 14.8.

Table 14.8: Commercial Grades of Mn-ores

	Grade	*MnO_2%*
(a)	Battery	80-86
(b)	Peroxide	78 (Fe 4%)
(c)	High	46-48
(d)	Medium	42-44
(e)	Low	38-40
(f)	Ferruginous	30-35

Mn-ores have also been further subdivided in 3 grades as shown in Table 14.9.

Table 14.9: Mn% Based Gradation of Mn

Grade	Mn%
First	48% and over
Second	45-47
Third or Low	40-44

Specification

(A) For ferromanganese Alloy making — Mn-ore should contain:

(a) Mn	48% (minimum)	(d) Al_2O_3	6%
(b) Fe	7% (minimum)	(e) P	0.15%
(c) SiO_2	8% (minimum)	(f) Zn, Pb	0.1%

There should be low percent of Cu, Ba, Pb; and Mn and Fe ratio should not be less than 7:1.

(B) Spiegeleisen — Mn ores should contain:

Mn 20-40%

Fe not more than 2.7 times of Mn

Presence of S and P are objectionable and are considered as impurities.

(C) **For Dry batteries** — In batteries mainly Pyrolusite is used. MnO_2 should be atleast 75%. Ores should be moderately hard, somewhat porous, and free from Cu, Ni, Co, As, Pb and Sb compounds. Fe variety is less objectionable than Fe due to its relative insolubility and Pyrolusite containing 3-4% Fe_2O_3 can be used.

(D) **For Chemicals** — Ore should be high in grade than that used in steel industry and it should contain

MnO	75%	SiO_2	6%
Fe	1.5%	Cu	0.02%
Al_2O_3	1%		

(E) **For Glass industry** — Ores should be high in grade with MnO_2 (85-90%) and Fe_2O_3 (0.75% Maximum). Total insoluble material should not be >2%.

15

Tin Deposits

Tin is one of the scarce metals. Cassiterite (tinstone, SnO, with Sn 78.7%) is the only tin ore of commercial importance. Sn is also recovered from stannite (tin pyrites, Cu_2S FeS. Sn_2S; with Sn 27.6%) and other rare Sn-bearing minerals when these are associated with cassiterite. Cassiterite is usually deep brown or black, sometimes green or red, has white streak and adamantine lustre. Its hardness (6½ - 7) and sp. gr. (7) helps in the concentration of the ore in placer deposits. Fused with a mixture of soda, sulphur and powdered charcoal, cassiterite gives a metallic globule, and when pulverised cassiterite is treated with dil. HCl and Zn, it forms a coating of metalic tin which may be polished on rubbing.

Occurrence: Tin deposits are usually associated with acid igneous rocks, such as granites, pegmatites, aplites, rhyolites and quarta porphyries. Cassiterite occurs in granite, greisen (granite in which feldsper is replaced by muscovite and quartz, by vapours), pegmatite, aplite, or magmatic quartz veins; but by weathering of these rocks, the mineral, being heavy and resistant is concentrated in placers as sand with various sizes of grains of cassiterite (from size of pea down to the finest flour). Cassiterite concentrate obtained from placers is purer than that from veins. In acidic rocks and veins it is commonly associated with quartz, muscovite and lepidolite, topaz, tourmaline, apatite, fluorite and tungsten minerals. Sn content of lode deposits is generally 1 to 8% and that of placer deposits 0.4 to 5 lbs per cubic yard.

Distribution: Chhattisgarh is the main producer of tin metal in India. Table 15.1 shows statewise areas of country having tin deposits. Cassiterite has been found in Jharkhand and Bihar (Purgo, Pihra, Simratand and Chappatand in

Hazaribagh district as crystals or grains in granite, lepidolite-granite and granulite; at Dhamras Pahar and Chakrabandh in veins in Gaya district and in veins and pegmatite at Silli and Paharsingh in Ranchi district); Gujarat (Hassainipura in

Table 15.1: Tin Deposits of India

	State	*District/Area*
1.	Chhattisgarh	Govindpal-Tongpal area (Bastar Dist.)
2.	Orissa	Mundguda area (Koraput Dist.)
3.	Bihar	Gaya
4.	Rajasthan	Bhilwara, Jhunjhunu
5.	W.B.	Purulia
6.	Haryana	Bhiwani
7.	H.P.	Sirmur
8.	Karnataka	Gulbarga
9.	U.P.	Mirzapur
10.	Jharkhand	Hazaribag, Ranchi
11.	Gujarat	Palanpur, Banaskantha

Palanpur, Banaskantha district, in tourmaline pegmatite); Karnataka (Dharwar district in stream sand); and Rajasthan (Paroli and Johra silli in Bhilwara district in pegmatite). None of the deposits has proved to be economically workable, but there has been an occasional output of tin ore from pegmatite and associated alluvial placer in the hills north of Paharsingh in Ranchi district. There is occurrence of Sn-ore associated with wolfram near Chakrabandh area in Gaya district. The deposit is located in an area about 4 Km long and 365m wide in the hilly ranges locally known as Dhankanahwa and Dhanras. Cassiterite occurs at Dhankanahwa in garnet-tourmaline veins and is also distributed in aplite, pegmatite and quartz veins. SnO_2 content of ore varies from 0.20 to 0.25%. Cassiterite-bearing pegmatite of Paroli, north of Shahpura in Bhilwara district of Rajasthan needs detailed prospecting. In Johra Silli area of the same district Sn bearing lodes have been proved for about 1.6 km. along the strike and about 12.19 m in depth along the strike but they are very irregular in shape and size and on an average the ore assayed 0.25% in metal. At present, India is dependent entirely on imports of tin. India ranks nowhere in reference of tin production. Table 15.2 shows principal tin producing countries of world. In India total reserves of tin are placed at 29 MT, most of which are in Bastar district, Chhattisgarh.

Table 15.2: Major Tin Producing Countries

Country	*Production (in '000 Metric t)*
China	60.0
Indonesia	40.3
Peru	27.2
Brazil	20.4
Australia	9.9
Malaysia	5.2
Thailand	2.7

Industrial Uses: Cassiterite is reduced to Sn-metal by heating with carbon at 1200° to 1300°C. Tin is soft, malleable and can be rolled into sheets 1/1000th of cm in thickness. The most important use of tin is for plating metals, specially in manufacture of *tinplate*—thin mild sheet steel, coated with commercially pure tin and used for manufacture of cans and 'tins', food containers, etc. and *terneplate* thin sheets of iron coated with an alloy of Sn and Pb and used for roofing (as a substitute for galvanised iron) and in oil tanks and oil containers. Other important use of tin is in making alloys like bell metal, gun-metal and bronze (Cu-Sn alloys); babbitt metal (Pb-Sn bearing alloy); brittanial metal (Sn-Sb-Cu alloy, sometimes with some Zn), and phosphor-bronze. (Sn-phosphate and Cu). Tin is also extensively used in making solders (alloys of Sn and Pb) and tin foils and collapsible tubes. Tin compounds (chlorides) are used in textile industry in dyeing and calico printing and for weighting fabrics, particularly silk and wool. Tin oxide is used in ceramic industry as an pacifying agent for white enamels, glazes and glassware and as a polishing agent.

16

Tungsten Deposits

Tungsten is the strongest and most refractory of all metals and is resistant at ordinary temperatures to all acids. It is largely because of its very high melting point (6150°F) that tools of tungsten steel retain a tough cutting edge. It is elastic, ductile, has high tensile strength and can be drawn in to very fine wires, which retain their rigidity at higher temperatures more than any other metal. Tungsten is so ductile that 10 gm of it can be drawn into a wire 4 Km in length. Less than 2t of tungsten metal can supply filaments for 100 million electric lamps.

Tungsten ore deposit of Agargaon (Nagpur district) was discovered in 1907, Rewat hill near Degana (Jodhpur district) in 1913, and Kalimati near Tatanagar (Jharkhand) in 1916. Tungsten ore deposit around Chhendapathar in Bankura district (W.B.) was worked for the first time in 1942-43 and 46.44 (t) of ore was produced. There was no output from Rewat hill after 1955 till 1965 when mining was restarted. India has to import tungsten concentrates and tungsten metal each. Chhendapathar deposit of W.B. and Agargaon (Nagpur) deposits of tungsten ore has also been investigated.

Occurrence: The common ore minerals of tungsten are wolframite or wolfram, (Fe, Mn) WO_4 (containing about 76.5% WO_2 or 51% W) and scheelite ($CaWO_4$; 80.6% WO_2 or 64% W). Wolframite, is a tungstate of Fe and Mn, while Scheelite is a tungstate of Ca and is of a white colour. Its hardness is 4.5 to 5. It is also less heavy than Wolframite, its sp.gr. being only about 6. Wolframite is distinguished by its brownish black colour and streak, prismatic or bladed crystals, one perfect cleavage, submetallic lustre (metallic on cleavage faces), high sp. gr. (7.2-7.5), and being attracted by electromagnet. Scheelite is white or pale yellow, has high sp. gr. (about 5) and gives a characteristic blush-white fluorescence under UV rays. Wolframite often occurs in pegmatites and quartz veins containing Sn-ore and traversing granite rocks and in greisen and intruded metamorphosed sedimentaries. W-deposits may be classified into 5-types-pegmatites, fissure veins, replacement lode, contact metamorphic deposits and placer deposits. The usual associated minerals are quartz, cassiterite, tourmaline, fluorite; topaz, beryl, etc. The tenor of W-ores is usually <1% and W concentrates obtained contain about 60% WO_2. If present, As, S, Sb Bi, P are objectionable impurities.

Distribution : In India, wolfram deposits are restricted in size but are capable of yielding small production of ore. Area-wise description of Indian states having Tungsten deposits is given in Table 16.1. There are two workable deposits one in Rewat hill near Degana in Nagaur district (Rajasthan) and the other in the Chendapathar area in Bankura district (W.B.). On Rewat hills, wolfram occurs in veins traversing both granite and the intruded Aravalli phyllites and is associated with quartz, fluorite, biotite and small quantities of pyrite, chalcopyrite and phosphates (triplite and libethenite). The veins are usually thin. The mineral also occurs as eluvial deposits around the foot of the hill. In Chendapathar, wolfram associated with small quantities of malachite, azurite, magnetite, etc. and occasionally with scheelite occurs in quartz veins traversing phyllites and quartzites. Small amounts of wolfram and scheelite are also found further north near Porahpahar. Scattered deposits of wolfram occur in quartz veins at Agargaon in Nagpur district (Maharashtra); at Kalimati in pockets associated with quartzites in Singhbhum district (Jharkhand); at Jher and Palla in quartz bodies in Ahmedabad district (Gujarat) and at Kadavur in Tiruchirapalli district (T.N.). Scheelite is reported to occur in quartz veins of Kangundi area east of Bisanatharn in Chittoor district (A.P.). Wolfram is also reported from Dhanras hill associated with ferruginous material and tin ore in Gaya district (Bihar) and KGF (Karnataka).

Table 16.1: Indian Distribution of Tungsten Deposits

State	*District/Area*
1. Rajasthan	Degana, Sirohi
2. W.B.	Chendapathar (Bankura)
3. Maharashtra	Bhandara, Nagpur
4. Jharkhand	Singhbhum
5. Gujarat	Ahmedabad
6. T.N.	Tiruchirapalli
7. A.P.	Chittoor
8. Kerala	Palghat
9. Karnataka	Kolar Gold Field

Reserve: As per UNFC, total resources of tungsten ore in the country have been estimated at 87.39 MT with a WO_3 content of 142094t. All these resources are placed under Remaining Resource category.

Industrial uses: The chief uses of tungsten ore are for making: (1) *ferroalloys* (mostly in the form-ferro-tungsten containing 75-80% W, for tungsten steels—magnetic steels, dic-steels, hot-work steels, high-speed steels, tool steels etc. (2) *non-ferrous alloys* (cutting tool alloys, alloys for resistance welding, alloys for electric contacts, jet-engine alloys, alloys for resistance wire, high density alloys, etc. (3) carbides (sintered, fused or crystals, with or without other carbides); (4) *metallic tungsten* (wire, rod, sheet and powder) and (5) *tungsten chemicals*.

Tungsten steel are used for various cutting tools and materials subjected to hard wear. The self-hardening steels contain 2-12% W and high-speed cutting steels which retain their hardness at red heat commonly contain 14-18% W (with about 4% Cr and 1-3% V). Mo may replace some W and some contain Co also. The greatest use of high-speed steels is for high-speed drills, milling cutters and hacksaw blades. W enters into the composition of steels for armour plates, gun barrels, armour-piercing shells and other military equipment and is of vital strategic importance in times of war. It is often said, "The more the tension, too more the tungsten". Non-ferrous W-alloys (of Co, Cr, Ni, Mo etc.) retain their hardness at red heat are very resistant to abrasion and corrosion and are extensively used in parts of aircraft engines and chemical plants and other machineries. The best known non-ferrous alloys are 'stellites' which contain 5-10% W, 45-65% Co and 25-30% Cr; they give extremely smooth and high-reflecting polished surface and are said to retain a higher hardness a very high temperatures than the best high-speed steel. Tungsten Carbide (WC or W_2C) is much harder than the hardest steel. It is next to diamond as cutting agent and with metallic cobalt as a binder or matrix, is used for tips of rock drills, wire-drawing dies, cutting glass and porcelain etc. The most important uses of metallic W (as wire, sheets or rods) are in electric lamps; electronic equipment; X-ray tubes; fluorescent and projector lamps; radar and television apparatus; radio valves; cathode-ray tubes; spark-plugs; ignition and other contacts; welding electrodes; for shielding radioactive material, etc. W-metal powder is now the raw material for the manufacture of tungsten carbides. Amongst W-compounds, sodium tungstate is used as a mordant in calico-printing and silk-dyeing and in the preparation of phospho-tungstic acid used in printing inks; and tungsten oxide is used in ceramic industry for producing yellow tint in glass and porcelain, and for making decorative pigments known as 'tungsten-bronzes' (golden yellow, purple red, violet and blue in colour according to the proportion of tungsten present).

Due to the replacement of the diamond bits by tungsten carbide bits in drilling practice and the expansion of drilling program in the mineral fields including oil fields, the prospects for a tungsten carbide industry in India are very bright. The output of W-ore produced are partially consumed in this industry (manufacturing of W-carbide tools). The ore concentrate contains on average of 65% WO_2.

17

Copper Deposits

Copper is the most significant base metal which is stepping ahead keeping pace with this 'era of Information Technology'. India has the mining history for Cu for almost 150 years. Copper ore is reported to have been worked in 1862 between Rajdah and Badia (Singhbhum district) and the ore was carried to Calcutta for smelting. The Rakha mine was developed and copper metal was produced locally in 1919; but the mine was closed in 1923 after working up to a depth of 240 m. Production of Cu started from Mosaboni mine in 1928 when the copper smelter was set up at Maubhandar and went into operation. Since then the Mosaboni mine has been the regular producer of Cu ore and Cu metal. As a result of intensive prospecting by Cu metal. As a result of intensive prospecting by GSI and IBM in the copper belt of Rajasthan, another copper field came into existence at Khetri–Kolihan and mining operations started under Hindustan Copper Ltd (HCL). India is deficient in copper, but with the expansion of the capacity of Maubhandar Smelter, the establishment of electrolytic refinery, the establishment of a mill, smelter and electrolytic refinery at Khetri by HCL and the beneficiation plant for Rakha Mine to produce Cu, Mo & Cu-Ni concentrates able to compensate somewhat this deficiency. Other public sector copper exploration and development projects include Agnigundala (A.P.), Ingladhal (Karnataka), Dariba (Rajasthan) and Mamandur (T.N.) Ore minerals of Cu (Composition and Cu%) are shown in Table 17.1.

Table 17.1: Ore Minerals of Copper

Ore	*Composition*	*Cu%*	*State of occurrence*
1. Native copper	Cu	100	Sulphides
2. Chalcopyrite (copper pyrite)	$Cu\ Fe\ S_2$	34.5	
3. Bornite (or erubescite)	$Cu_2\ FeS_4$	63.3	
4. Covelite	CuS	66.4	
5. Chalcocite	Cu_2S	79.8	
6. Cuprite	Cu_2O	88.8	Oxide
7. Malachite	$CuCO_3.\ Cu\ (OH)_2$	57.3	Carbonate
8. Azurite	$2CuCO_3.\ Cu\ (OH)_2$	51.1	
9. Chrysocola	$CuSiO_3\ H_2O$	36	Silicate

In addition to these there are several complex ores (mostly sulph-antimonites and sulph-arsenties) usually found associated with Pb-Zn-Hg ores from which Cu is recovered as a byproduct.

Gangue minerals of Cu-ores

Quartz, calcite, dolomite, siderite, rhodochrosite and baryte are the main gangue minerals. Cu-pyrite is the most common primary ore of Cu (which is characterised by golden yellow colour and differs from gold in being much lighter, brittle and soluble in HNO_3). The primary Cu-ores are usually altered on the surface to oxidised Cu-ores distinguished by their characteristic colour bright green (malachite), blue (azurite) and bluish green (chrysocolla). The presence of these minerals in the outcrop gossan or the colouring of the rocks at the surface by them is indicative of primary Cu-ore below.

Grade of ore: The average percentage of Cu in an ore which can be mined economically depends upon the type of ore and the nature of its occurrence. Few Cu-deposits below a grade of 2% Cu are capable of being worked profitably by UG mining but presence of metals like Au and Ag make them payable. If very large tonnage can be handled daily, ores carrying even 1% Cu may be workable at a profit. Most Cu-ores are milled and concentrated before they can be smelted economically. Thus, grade of ore as low as 0.6 to 0.8% Cu is now economically worked. In India, the Cu-ore which has been found in workable quantities is only Chalcopyrite. The two most producing areas (1) Singhbhum (Jharkhand) has ore of grade 2.0 to 2.44% Cu and (2) Khetri Cu-belt, Rajasthan has ore of grade 0.8 to 2.50% Cu Kolihan section have indicated an average of about 6% of Cu.

Occurrence

(1) **Cavity filling or replacement deposit :** Most of the primary Cu-sulphide deposits (and the only important native Cu-deposit of Michigan, USA) are of this type. They have been formed by hydrothermal solutions, emanating form intrusions of acidic igneous rocks such as granites, granodiorite and Quartz-monzonites.

(2) **Magmatic Concentration or by Contact Metasomatism:** Only a few deposits have been formed by this type.

(3) **Oxidised ores and secondary Cu sulphide ores:** Some primary deposits have undergone superficial oxidation giving rise to deposits of oxidised ores oxides, carbonates and silicates, etc. whereas some have undergone supergene enrichment by downward circulating waters resulting in the formation of deposits of secondary Cu-Sulphide ores.

According to their form, the Cu-deposits may be classified as:

(a) ***Disseminated Ores*** — In these cu-minerals are dispersed through considerable part of the rock body (igneous, sedimentary, metamorphic).

(b) ***Massive irregular or lenticular ore bodies*** — These are formed by more or less complete replacement of large bodies of rocks.

(c) ***Vein deposits or loades*** — In these the Cu-minerals are formed by infilling of open fissures or by replacing narrow porous shear channels.

(d) ***Deposits following certain stratigraphic beds*** — These may be termed stratiform or stratabound deposit. Some people regard them as a special form of disseminated deposits.

The Cu-deposits of Singhbhum are of type (c) and those of Khetri and Dariba Cu-belt may be classified under types (c) & (d).

Distribution: Important Cu-deposits of India are shown in Table 17.2.

Table 17.2: State-wise Distribution of Important Deposits

	State		*District*
1.	A.P.	-	(Garimanipenta) Nellore and (Gani) Kurnool district
2.	Assam	-	
3.	Manipur	-	
4.	W.B.	-	Darjeeling
5.	Jharkhand	-	Hazaribagh, Singhbhum
6.	Gujarat	-	Chota Udaipur, Danta
7.	J.K.	-	Kashmir Valley, Jammu
8.	M.P.	-	Jabalpur
9.	T.N.	-	S Arcot
10.	Karnataka	-	Mysore, Chitaldurg
11.	Rajasthan	-	Khetri-Dariba Cu-belt (Jaipur and Alwar)
12.	Uttaranchal	-	Garhwal, Almora district (Kumaun)
13.	Haryana	-	Mahendragarh
14.	Chhattisgarh	-	Bastar
15.	Orissa	-	Mayurbhanj
16.	Sikkim	-	
17.	Meghalaya	-	
18.	Maharashtra	-	

The important Cu-belts developed so far in India are in Jharkhand and Rajasthan. Largest resources of Cu is in Rajasthan having 669 MT of ore.

Table 17.3 shows major Cu producing states of India.

Table 17.3: Major Cu Producing States (1997-98)

	State	*Production ('000 t)*
1.	Chhattisgarh (earlier M.P.)	2082
2.	Rajasthan	1471
3.	Jharkhand (earlier Bihar)	945
4.	Sikkim	0.8

1. Singhbhum Cu-belt of Jharkhand. It is 160 Km long belt. The Cu-ores of Singhbhum are genetically related to the intrusion of a soda-granite and occur as veins in the granite (occasionally altered to biotite—chlorite schist) and the neighbouring mica-schists, Quartzite and epidiorites or hornblende and schists. These veins are best developed along a zone of overthrust where they form well defined lodes. The individual lodes consist of one or more veins of solid sulphide average thickness being 1.8 m. On either side of this the sheared country rock is partially replaced by sulphides.

The main sulphides are chalcopyrite, pyrrhotite and Fe-pyrite with minute quantities of pentlandite (Fe, Ni) S_2, violarite $(Ni, Fe)_2$ S_2 and millerite (NiS).

Gangue minerals in ore: Quartz, chlorite, biotite, tourmaline, magnetite and apatite are the main gangue minerals. At the surface, the sulphides have been oxidised and deposits of malachite, azurite, chrysocolla, cuprite and native Cu are found at the surface.

2. Khetri Cu-belt of Rajasthan. It is about 64 Km long in Jaipur district. The belt has three richly minerlised sections-Madhan, Kolihan and Akhwali sections. The Cu-ore bodies occur in phyllites, slates, schists of the Ajabgarh series (Delhi Group) as irregular stringers, fillings of schistose planes and fractures and dissemination in the host rock.

Mineralogically, the ore bodies consist of chalcopyrite; pyrrhotite, pyrite and bornite as primary sulphides and malachite and azurite as supergene minerals. Besides traces of Co and Hg have been found in Cu-ore.

Cu-ores are also found in Dariba area of Alwar district where Cu-lodes occur usually in sheared calc-phyllites and consist of chalcopyrite, pyrrhotite, pyrite and bornite in lenticular veins.

The mineralisation in Rajashtan Cu-belt is epigenetic and seem to have occurred under Mesothermal conditions from post Delhi (Erinpura) granite magma.

Reserve and Production: Category-wise reserves and production of Cu are shown in Table 17.4.(A-B).

Table 17.4(A): Reserve and Production of Cu in India

	Year		
	1990-91	1998-99	2001-02
1. Reserve (Ore) MT			
(a) Recoverable (Cu-content)	—	416.8 (4.37)	537.86 (5.3)
(b) Conditional	—	647.1 (1.89)	721.00 (3.16)
(c) Prospective	—	—	0.6
2. Cu-Metal Production ('000 t)	5255 (ore)	199	164

Table 17.4(B): Cu-Resources

Category	*Total*	*Reserve*	*Remaining*
a-Ore	1.39BT	369.49MT	1.02BT
b-Metal (000't)	11418	4383.97	7033.75

Uses: Cu is the most important non-ferrous metal having the widest application in industry and among all metals it is only surpassed by Fe in that respect. Cu was the earliest metal used widely by man in the form of Cu-bearing bronze for his weapons, tools and utensils during the bronze age. The metal is at present of great industrial importance due to its high electrical and thermal conductivity; high ductility and malleability; moderate tensile strength; good resistance to corrosion by air, water and many chemicals (except acids and ammonia); and its ability to take brilliant polish and to form many useful alloys like brass (Cu-Zn alloy), Ni-Ag (Cu-Zn-Ni alloy), bronze (Cu-Sn alloy with some Zn), phosphor-bronze (bronze with some phosphorus), etc. Cu has the highest electrical conductivity than any metal except Ag and is therefore almost indispensable in the electrical industry. Cu is fabricated in many forms such as wires, bars, rods, slabs, bullets, sheets, strips, tubes, pipes etc. and is used in the manufacture of generators, motors, electric locomotives, switch boards, telephone, telegraph and radio equipments, light and power lines, light bulbs and sockets and other equipments. Cu-alloys are widely used in building construction, automobile and airplane manufacture, naval and other ship construction, railway equipment, chemical plants, war ammunition, a wide variety of manufactured products, machinery and also in coinage. Cu-powder is used in powder metallurgy for making porous bronze filter discs and (with graphite) for making self-oiling bearings. Cu-powder and Cu bronze powder (produced from Cu-Zn-Al alloys) are used as pigments in the paint industry. Cu-compounds, mostly Cu-sulphate, cuprous oxide, Cu acetate and Cu naphthenate are used in the preparation of insecticides and fungicides to protect timber, textile fabrics, canvas tents and ropers etc. High grade Cu carbonate minerals, malachite and azurite are used for brilliant green and blue paints but the cost is relatively high.

Present Status: Copper is considered to be the most important metal next to steel and aluminium. It is a material of strategic advantage. Combination of its excellent properties such as high electrical and thermal conductivity, high strength, malleability, easy workability, corrosion resistance, precision castability, ability to produce alloys with other metals in attractive range of colours has made Cu a strategic metal for the engineering industry. Domestic demand is overriding supply in spite of commissioning of two new smelters—Birla Copper and Sterlite Industries and proposed expansion to increase capacity by 1.5 times in each smelter. The domestic supply was only 59% of the demand in 1998-99. (Fig. 17.1)

Apart form Hindustan Copper (a public sector enterprise), there are 3 new private sector players in this field.

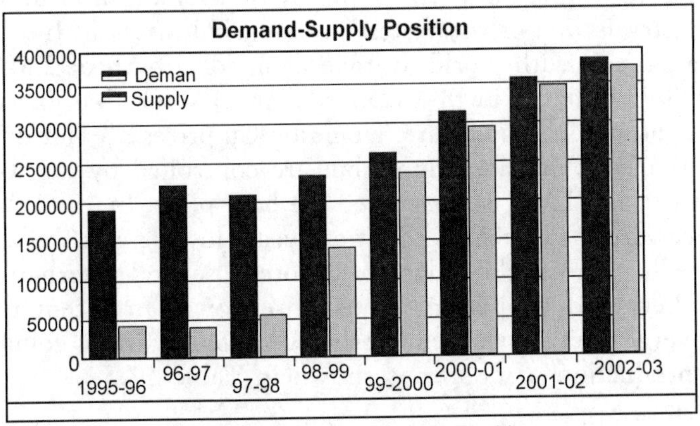

Fig. 17.1: Demand-Supply Position

Sterlite (imported concentrate based) was commissioned in 1997-98 with a capacity of 60,000 tpa. The company expanded the capacity to 1.5 lakh tons in 2000-01. Birla Copper (imported concentrate based) commissioned its unit in 1998-99 with a capacity of 100,000 tpa. It has expanded the capacity to 1.5 lakh tons in 2000-01.

HCL is also planning expansion in its Khetri smelter from 34,000t to 60,000t. Continuously increased growth rate in India emanates on account of several factors, one of them being increased thrust on infrastructure, consumer goods, telecommunication and power sectors and progressive replacement of Al by Cu in its various applications. The new revolution, optical fibre also, instead of replacing Cu has increased its demand as communities are getting wired fast through networks involving LAN, WAN, PCs and other hardware.

In time ahead, Cu use in electronics and telecom, consumer durables and handicrafts will increase. Fast developing sectors such as telecom will lead. But the share of wire application will remain at 50-52%. Engineering sector will find increased utility for copper in various areas such as precision equipment, ship building, automobiles and special alloys. Fig. 17.2, 17.3 and 17.4 show consumption pattern of Cu in various sectors.

Future of Cu is secure in India and country is emerging as a potential market. However, it would take at least 4-5 years for India to become totally self-reliant in Cu. This industry is one of the fastest growing industries in India. In comparison with world average demand growth of 2.5% in 1998-99 Indian growth was more than 11% which is expected to continue in near future. Domestic Cu prices are not driven by domestic demand-supply but are controlled by the world market and LME is considered the base price. In 1999, LME prices dropped by 22% to hit a 6 years low. It was the worst time for the world industry. It forced several international smelters and copper mines to close down. In recent times however prices have been improving. State of various countries in production of copper is shown in Table 17.5.

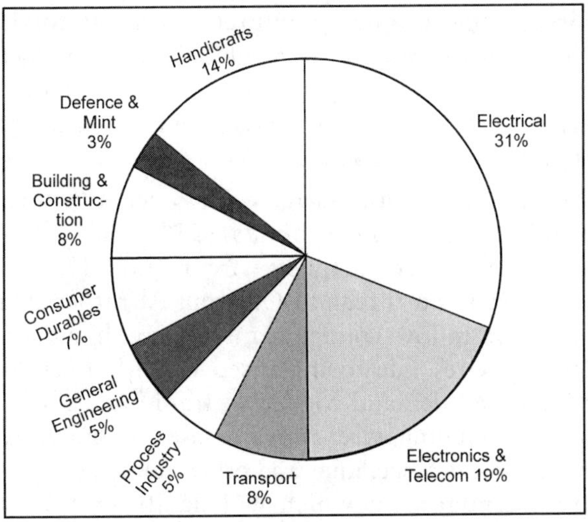

Fig. 17.2: Consumption Pattern

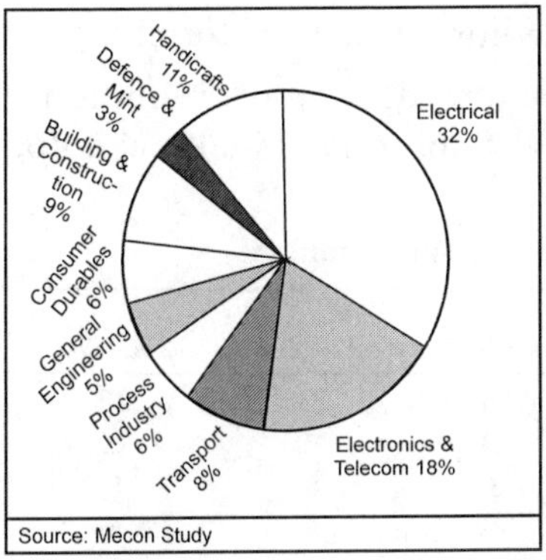

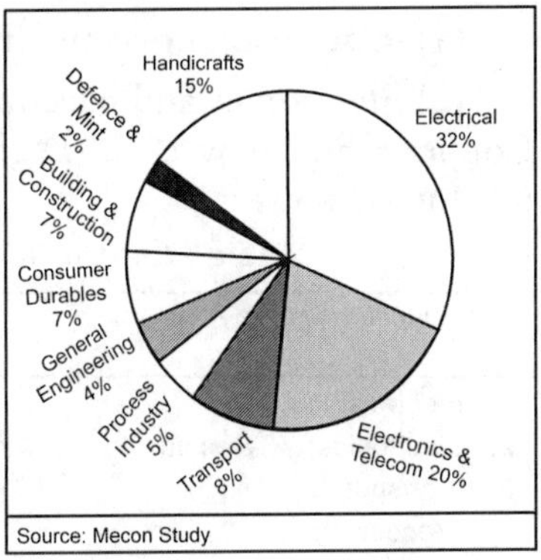

Fig. 17.3 & 4: Consumption Pattern: 2002-03 and 2004-05

Table 17.5: Major Cu Producing Countries in MT (UNO)

Country	*Production*	*Country*	*Production*
1. Chile	44.8	7. Peru	4.9
2. USA	19.8	8. Jambia	3.4
3. Indonesia	8.7	9. Mexico	2.3
4. Canada	6.3	10. New Guinea	2.1
5. Australia	6.2	11. S Africa	1.7
6. Poland	4.6	12. India	0.6

Copper, being the best non-precious metal conductor of electricity, sets the standard to which other conductors are compared. In addition, copper's exceptional strength, ductility and resistance to corrosion and creeping make it the preferred and safest electric conducter.

There is a preferential usage of Cu over Al in the field of home appliances/consumer goods (25% Cu, 6% Al) and building/infrastructure, magnet wire market (52% Cu and 7% Al). There is no threat to Cu from Al rather Al is being replaced by Cu in low voltage transformers, home appliances and building wires due to increased benefits over Al. There will be increased demand for Cu in traditional areas such as power cables, winding wire, alloys, house and building, wiring, consumer goods, defence, mint and other industrial uses. Besides, there are potential areas which will significantly increase Cu demand (Table 17.6). Industry-wise potential areas are given as under :

(1) **Electrical:** Low and medium Kva transformer.

(2) **Electronics and telecom :** Introduction of revolutionary Copper Chip, new DSL (Digital Subscriber Line) technology for Internet users.

Table 17.6: Cu-Consumption Pattern (%)

Industry/Sector	*Year*		
	1994-95	*1999-2000*	*2004-05*
1. Electrical	34	31	32
2. Electronics & Telecom	18	19	20
3. Transport	8	8	8
4. Processing	6	5	5
5. General Engineering	5	5	4
6. Consumer durables	6	7	7
7. Building & Construction	9	8	7
8. Defence & Mint	3	3	2
9. Handicrafts	11	14	15

18

Lead Deposits

Little information is available about the working of lead-zinc ore deposit of Zawar after mining activity had ceased there in 1812-13. During the last century, owing to the shortage of lead and zinc in India as a result of the loss of Burma, Government decided to reopen Zawar mines and GSI carried out exploratory work from 1942 to 1945. Meanwhile in 1942, a deposit of lead ore at Banjari, about 100 km from Jaipur (Rajasthan) was established and a plant was set up in 1947 at Tundoo (Jharia coalfield) with the help of Burmese engineers who had fled to India during Japanese invasion. They set up an ore dressing plant in 1950 to treat run-of-mines ores for preparing lead and zinc concentrates, the former being sent to Tundoo for smelting and the latter shipped to Japan for treatment (due to absence of any zinc smelter in India), on an understanding that half their metallic content of zinc is returned to India.

The recovery of silver at Tundoo was started in 1954. Now Zawar mines are nationalised and are being worked by HZL (a Govt. of India undertaking) and a zinc smelter has been installed at Debari near Udaipur to produce electrolytic zinc in addition to Pb, silver and cadmium. Another smelter has been installed near Cochin (Kerala), and at Vishakhapatnam (Andhra). They smelt the concentrates of zinc imported from abroad. Steps are also being taken to develop Pb deposits of Bandalamottu (Guntur dist, A.P.). India is deficient in Pb and Zn. Important ores of lead are as shown in Table 18.1.

Table 18.1: Ore Minerals of Lead

Ors	*Composition*	*Pb%*	*Colour*
1. Galena	PbS	86.6	black (lead gray)
2. Cerrussite	$PbCO_3$	77.5	white
3. Anglesite	$PbSO_4$	68.3	white

Cerrussite and Anglesite are formed from the alteration of galena. Galena has typical high S.G. (7.5) and low hardness (2.5). Crystallised Cerrusite and any ferrite have adamantine luster and are heavier than other carbonate and sulphate minerals.

Occurrence

Galena (PbS) usually occurs associated with Zinc ore (sphalerite) as veins and massive or tabular lodes, usually in limestones or dolomites, formed by cavity-filling or replacement by low-temperature hydrothermal solutions. Lead ores squarely contain 32 to 12% lead and most of them carry silver and many carry gold. Fine grained galena is more commonly silver bearing than coarse grained. Minerals associated with lead ores are sphalerite (zinc blende), pyrite, chalcopyrite, quartz, siderite, calcite and barite.

Distribution

Lead ores are distributed in the following states of India as shown in Table 18.2.

The other places in which lead ore has been reported are districts of Bans Kantha in Gujarat, S. Arcot in T.N. and Eastern part of Sikkim.

Reserve of Pb and Zn of India

Total Reserve of Pb & Zn ore = 160.48 MT

Reserve of Pb ore = 2.33 MT

Reserve of Zn ore = 7.49 MT

The most important lead ore deposit of economic value in India, is Zawar deposit in Udaipur, Rajasthan. There are 10 or 12 hills around Zawar, full of ancient workings for Pb or Zn of which Mochia Magra, Barai Magra and Zawar Mala hills contain most extensive and deep workings. Now Zawar

Table 18.2: Lead Distribution in India

States	Areas
1. T.N.	S-Arcot
2. A.P.	Chittoor, Kurnool (Chelima area), Guntur (Karempndi), Cuddapah, Nalgonda, Khamam
3. Jharkhand	Hazaribagh, Ranchi, Singhbhum, Palamu
4. Bihar	Bhagalpur, Monghyr
5. H.P.	Mahasu, Simla, Kangra, Kulu & Sirmur
6. J.K.	Baramula, Udhampur
7. M.P.	Sarguja, Durg, Balaghat, Bilaspur, Hoshangabad, Indore, Rewa, Jabalpur, Shivpuri and Gwaliar
8. Karnataka	Chitradurg, Mysore, Bellary, Chitaldurg
9. Punjab and Haryana	Kula Subdivision and at Tabo in Karngra
10. Maharashthra	Chanda
11. Rajasthan	Galena is found associated with Zincblende (Zn S) in Zawar Mines in Udaipur, Dungarpur, Banswara, Alwar and Bhilwara. Also at Bharatpur, Ajmer and Sawai Madhopur.
12. Gujarat	Baroda, Danta, Panch-mahal & Surat
13. Uttaranchal	Galena occurs in Almora, Tehri-Garhwal, Dehradun. A recoverable possible reserve of 0.99 MT has been estimated in Pithoragarh district. Metal content here is estimated to be 0.28,000't of Pb and 0.30,000't of Zn. Askot area of Pithoragarh
14. Assam	Khasi and Jaintia hills
15. W.B.	Jalpaiguri, Darjeeling, Midnapur, Purullia district.

mines are nationalized and worked by HZL and the development work has been extended to the adjoining Balaria hill. Principal rock types of Zawar area consist of phyllites, slates, mica-schists, dolomites and quartzites of Aravalli S.G. The ore body occurs as veins, fissure-filling or replacement lodes, the ore-horizons being confined to the highly sheared dolomitic limestone bounded by quarzites and phyllites. The ore minerals consist of argentiferous galena associated with sphalerite containing a small percentage of cadmium, pyrite and small amount of arsenopyrite and chalcopyrite. The ore contains 1.5-2.0 % lead and 4.5-5.0% zinc. The ore treated in 1967 was 6.25% (2.05% Pb and 4.20% Zn). The ore reserves at Zawar are about 9 MT of which about 1 MT are of good ore, containing on an average 7.16% Pb and 7.49% Zn. Fig. 18.1 shows the distribution of Lead in India.

Fig. 18.1: Distribution of Lead in India

At present, lead-zinc mine at Zawar is the chief mine in India, producing Pb-Zn ores. Pb-concentrates are sent to Pb smelter at Tundoo in JCF (Dhanbad, Jharkhand) where Pb and Ag are recovered. As there was no Zn smelter in the country, Zn concentrates were exported to Japan on condition that a particular percentage of the metallic content is returned to India. A zinc smelter has now been installed by HZL at Udaipur.

Zawar Mine

Zawar area is situated 39 Km south of Udaipur (Rajasthan). Mochia Magra hill in which Zawar mine is located has an average height of about 120 m. and extends EW for more than 3 km with a width of about 2 km in east and 1.5 km in west. The general strike of ore-body is EW and the dip is towards south varying from 75° to almost vertical at depth. The width of ore bodies varies from 45 m to as small as 3 m. The ore horizons are mainly confined to dolomite (Aravalli S.G.) which are highly sheared and form parallel sheets. The ore generally occupies these sheeted zones between two shear-zones, a tension fracture forms a link-lode. Southern flank of dolomite exposes older phyllitic rocks and northern flank of the quartzites. The ore consists of galena, sphalerite and pyrite; galena is generally concentrated in some particular portions of the deposit but sphalerite is more or less evenly distributed. The gangue is dolomite and quartz. The country rock as well as wall rocks are fairly hard and strong enough and require no supports. The depth of deposit can be reasonably assumed to be more than 350 m.

The deposit has been opened up for mining by a number of adits, and No. 6 adit at valley level is the main production adit and meets the ore-body at the second level. An auxiliary shaft having a depth of 147.5 m worked by 70 H.P. Koepe winder and furnished with rigid guides connect the 6th level. The main shaft sunk from the surface to a depth of 321 m and having a size of 5.2 m × 3.8 m is used as the main production shafts. The level interval below second level has been kept to 36 m. When a block of ore is formed by connecting two levels by two raises approximately 60 m apart, a preparatory drive is put at a height of 6 m from the floor level of the lower drive to connect the two raises. The drive is connected from the level below by a number of raises to dispose the muck. Again, from the preparatory level drive, attempt has been made to delineate the ore-body either by putting boreholes or by putting crosscuts, two at the ends and one in the middle. After that, stripping at preparatory level starts which is continued upto the economic limit of the ore-body.

The host rock being competent, method of stopping is done on the width of the ore-body. In case of narrow ore-bodies upto 2.4 m in width, Shrinkage method of stopping is done without preparatory level, and cross-cuts at an angle of nearly 50° are put from the trimming level drive. Shrinkage method of stopping has been adapted to a width of 10m. Beyond this, sub-level methods of stopping is adopted.

The beneficiation of ore is carried out by crushing, grinding and flotation. The run of ore has a size of 200 mm approximately. From the coarse ore hinds, the ore is fed to Jaw crushers to reduce its size to –30 microns and then into Gyratory crushers where the size is reduced from 30 to 10 microns. The size of the ore is further reduced in Symon Cone crusher to –0.5 microns and then the entire quantity of crushed ore is collected in these fine-ore bins. In order to liberate the particles of sphalerite and galena from the gangue, this ore is ground in three ball mills in a closed circuit to –300 microns. The entire pulp is fed into a conditioning tank where $ZnSO_4$ and NaCN are added for the purpose of depressing Zn and Fe minerals, and potassium ethyl xanthenes is used as collector for Pb mineral. The conditioned pulp is then fed into Dener Froth Flotation Cells where, by adding suitable reagents, the two metallic sulphide ore minerals (galena and sphalerite) are differentially floated. The concentrates so obtained contain 70-72% Pb metal and 54-55% Zn metal respectively. The entire tailing from the beneficiation plant is pumped to a pond for disposal and a good percentage of tailing water is recovered for re-use in the plant.

Indian production of lead is far behind the requirements of this metal, and large quantities of lead and lead alloys are imported every year to meet the demand for electrical, chemical, and other industries. The consumption of metal is estimated to increase in near future. Production of lead in India has been expanded by the expansion of mining and milling capacity of Zawar mine. G.S.I. proved 17 MT of Pb-Cu ore in Agnigundala area (Guntur district) of A.P. There is about 3 km long Pb-Zn-Cu belt containing estimated reserves of 5.5 MT with 9.2% Pb and Zn and 1.6% Cu in Ambamata-Devi area of Gujarat and Rajasthan and a 2,200 m long Pb-Zn-Cu belt with 4-5% metal content in Umpyrtha area of Meghalaya have been reported.

Reserve: Table 18.3 shows the reserve scenario of lead ore and metal.

Table 18.3: Reserve of Lead Ore and Metal

Category	*Total*	*Remaining*	*Resource*
Ore (MT)	522.58	125.75	396.83
Pb (000't)	7208	2591	4617

Uses

The most important properties which have given Pb a wide range of industrial uses are its softness, malleability, easy workability, high specific gravity, alloying properties, electric properties, low melting point, high boiling point, good corrosion resistance, adaptability to the manufacture of pigments, and impenetrability to short-wave radiation. Principal uses of Pb are for making plates for electric storage batteries; coverings for electrical cables; in pipes and sheets for buildings (water pipes and roofings) and chemical plants (as sulphuric acid plants) which are subjected to chemical corrosion; paint pigments like white lead (2 $PbCO_3$. $Pb(OH)_2$) litharge (PbO), red lead (Pb_2O_4) and sublimed white lead (PbO, Pb_2SO_4), used in rubber, pottery, enamel, glass and iron-steel industries and storage battery plates; alloys as solders (Pb-Sn), pewters (Pb-Sn-Cu-Sb), type metals (Pb-Sn-Sb), low-fusion alloys (Pb-Bi-Sn), anti-friction or bearing metals (Pb-Ca-Sr-Ba) and ammunition (Pb-Sb); collapsible lead rules and lead foils for wrappings; Terne plates (steel plates coated with Pb-Sn-alloy) used for packing dry goods and roofing; tetraethyl lead chemical added to automobile gasoline to improve the qualities; and as nuclear shield in atomic plants against certain types of dangerous radiation, specially gamma rays.

19

Zinc Deposits

Zinc is an important base metal which occurs in association with copper and lead. In India Zn production has ever been fallen short of meeting the demands. As a result the industry is all set to go in for massive expansion of capacities and capabilities encouraged by the good pick up in demand from domestic and industrial consumers, especially steel makers who need it for their galvanising operations.

Zn-Ore

Important ore minerals of Zn have been shown in Table 19.1.

Table 19.1: Ore Minerals of Zinc

	Ore	*Composition*	*Zn%*	*Nature of Minerals*
1.	Sphalerite (Zinc blende)	ZnS	67	Primary
2.	Smithsonite (Calamine- English name)	$ZnCO_3$	52	Secondary
3.	Hemimorphite (Calamine -American name)	$2ZnO.\ SiO_2.2H_2O$	54.2	

Smithsonite and Hemimorphite are regarded to be secondary minerals derived from zinc blende and are usually found in the upper oxidized portions of Zn deposits. Zinc blende usually occurs in intimate association with galena. Most of the lead deposits/ores of world are also Zn producer and nearly all Zn-deposits carry Pb-ores. Fig. 19.1 shows the distribution of zinc in India.

Occurrence: Zinc ore bodies are found as contact metasomatic deposits, fissure veins, cavity fillings and replacement loads formed by hot hydrothermal solutions and as deposits from descending surface waters. These deposits commonly occur in limestone and dolomites.

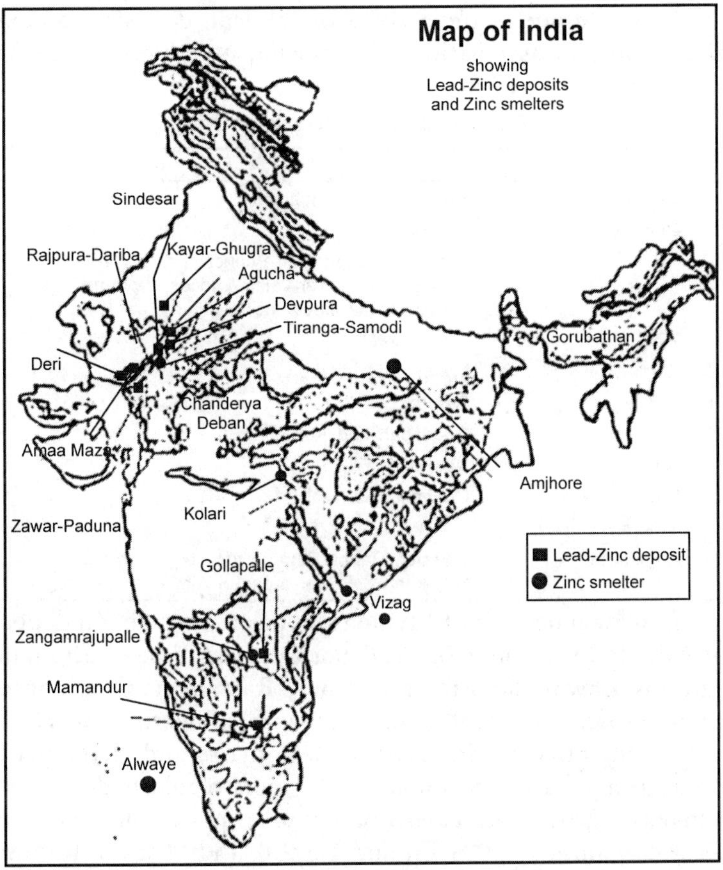

Fig. 19.1: Distribution of Zinc in India

Mineralogy

Zinc ore generally contain 2-12% Zn. Zinc blende often contains some Cd. Some complex Zn ores carry Au and Ag which are important byproducts of these deposits. Galena (PbS), chalcopyrite ($CuFeS_2$), Pyrite (FeS_2), Pyrrohtite are the ore minerals commonly associated with sphalerite. Common gangue minerals present are quartz, calcite, dolomite etc. Before the treatment of Zn-ores for extraction of Zn-metal, they are concentrated to remove other minerals and to reduce the proportion of Pb minerals. Zn ore which is sent to smelter usually contains 45 to 60% Zn.

Distribution: In India some of the lead deposits contains zinc blende in association with galena. Notable deposits are shown in Table 19.2.

Table 19.2: Zinc Deposits of India

	State	*District*
1.	Jharkhand	Hazaribagh, Singhbhum
2.	Gujarat	Baroda
3.	J.K.	Riasi, Udhampur and Poonch district (ZnS occurs in lenticular veins & lodes in permo-carboniferous limestone in Riasi)
4.	T.N.	South Arcot
5.	Karnataka	Mysore, Chitaldurg
6.	H.P.	Sirmur
7.	A.P.	Cuddapah and Kurnool
8.	Maharashtra	Nagpur
9.	Uttaranchal	Chamoli (Garhwal)
10.	Sikkim	–
11.	Rajasthan	Udaipur, Dariba-Rajpura area of Alwar and Lohakhan & Sewar areas of Ajmer.

Bawdwin deposit of Mynmar is the greatest Pb-Zn deposit of Asia. In India, only Pb-Zn deposit proved to be of economic value is Zawar deposit of Udaipur, Rajasthan. Ores consist of argentiferous galena associated with sphalerite containing minute quantity of Cd. Pyrite, arsenopyrite and chalcopyrite occur in small amounts. Sphalerite is more regularly distributed whereas argentiferous galena occurs in large and small pockets. Ores contain 4.5-5.0% Zn and 1.5-2.0% Pb. Dariba-Rajpura belt is a continuation of Pb-Zn-Ag belt of Zawar. In Dariba-Rajpura belt, Zn is predominantly followed by Pb and Cu. Here Cd, As and Sb ores occur in association with Zn-Pb ores. Rajasthan and Gujarat have estimated reserves of 12 MT base metal with the ore content of Cu-Pb-Zn at around 7.6%. Among prominent Zn producers of India are Hindustan Zinc Ltd (HZL) and Binnani Zinc Ltd. HZL with its operation in Rajasthan, A.P., Orissa and Jharkhand is the largest producer of Zn and is competent enough to provide Zn for another 30 years (Table 19.3). Pb and Zinc smelters are:

Table 19.3: Statistics of HZL (2001-02)

1. Capacity (in lakh t)	
(a) Present	1.69
(b) Ultimate (projected)	4.00
2. Production (in lakh t)	
(a) in 2000-01	1.48
(b) in 2001-02	1.76
3. Mines	
(a) No. of Mines	6
(b) Combined Capacity (in lakh t)	3.49
4. No. of Smelters	4

(a) Pb smelter - at Tundoo in Jharia coal field, Jharkhand (earlier used)

(b) Zn smelter - at Devri near Udaipur, Rajasthan

(c) Zn smelter - near Cochin in Kerala (use Zn-Concentrates imported from Canada, Australia & S America)

(d) Zn smelter - at Visakhapattanam in A.P. (for imported Zn-Concentrates)

These smelters produce electrolytic Zn, H_2SO_4, Cd, Pb and Ag.

Ore reserve (Indicated and inferred): Estimates for Jawar deposits are 116 MT of all grades ranging from 3-12.5% combined metals, the ore being generally richer in Zn than Pb (as per GSI estimates). Reserve, production and resources of Pb in India are shown in Table 19.4-19.6.

Table 19.4: Zn-Pb Reserves of India (MT)

Reserve	*Year* 1998-99	2001-02
Recoverable	179.13	176
Restricted	203.87	—
Probable	150.00	—
Estimated	350.00	—
Zn Content	**10.00**	**9.80**

Table 19.5: Zn Production in India ('000 t)

Year	*Production*
1990-91	137
1998-99	350
1999-2000	360
2000-01	365
2001-02	399

Table 19.6: Zinc Resources

Category	*Total*	*Reserve*	*Remaining*
a-Ore(MT)	522.58	125.75	396.83
b-Metal (000't)	24260	11093	13167

Uses: Zinc is an important industrial non-ferrous metal owing to its malleability, ductility, resistance to corrosion, adaptability to manufacture of pigments and its ability to form useful alloys with other metals. The largest single use of Zn is for galvanizing iron and steel products (such as corrugated roofing sheets of iron) to protect them from atmospheric corrosion. Zn is also used in making brass, bronze, German silver, anti-friction metal and other alloys with copper, in manufacture of batteries for telegraphic and other services and in zinc-alloy die-castings for automobile carburetors, petrol pumps, machine tools and electric appliances. Zn in the form of rolled sheets is used in manufacture of dry-cell battery cans, fruit-jar tops, collapsible tubes, photo-engraving and lithographic plates and boiler plates. Zn dust and Zn-shavings are used to precipitate gold from cyanide solutions in the treatment of gold ores. Important quantities of this metal are used in production of white pigments such as zinc oxide and lithophone (precipitated ZnS and $BaSO_4$ which, in addition to their use in paints, are also employed as fillers or pigments in manufacture of rubber and linoleum and in plastics. Zinc oxide is also used in pharmacy and zinc salts such as $ZnCl_2$ and $ZnSO_4$ are used in textile industry, timber preservation and in agriculture (anti-fungicide spray).

Zn-Recycling

This metal is widely recovered due to its features like complete recyclability without any loss in physical or chemical properties. Over 80% Zn is presently recovered and 30% of world supply (approx 3 MT) comes from recycling. About 6 lakh tons of Zn is recovered from brass recycling alone and Zn-Coated steel scrap is estimated to increase by more than 50% by 2010.

Consumption Pattern: Developing countries in general and India in particular use comparatively more Zn for protection purposes in steel industry. About 75% of Zn consumed in India is used by galvanising sector for producing steel sheets, pipes, structures, wire rods and fasteners. Prominent consumers of Zn are Zindal Iron Industry (Zisco) with an annual intake of 30,000t and Tata Steel and Ispat Industries.

Principal reason for global increase in consumption of Zn-ingredient products is their maintenance-free durability for more than 100 years which make it a valuable and sustainable resource for future generations. Hence, consumpiton as well as production of Zn in world, is expanding at an unprecedented rate. Presently, production at global level is outstriping consumption (Table 19.7).

Table 19.7: Production and Consumption of Zinc in World (ILZDA, 2000-2001)

Country/Region	*Production (MT)*	*Consumption (MT)*
Asia	3.96	3.71
Europe	2.88	2.74
America	1.68	1.90
Oceania	0.56	0.24
Africa	0.14	0.30
India	0.40	
World Total	**9.21**	
World Surplus	**0.46**	

The table shows that if, on one hand some countries like America are producing lesser in comparison to production, on the other hand, there are countries producing more, consuming lesser leaving a global surplus of 0.46 MT. India is deficit in production of Pb and Zn but has enough potential in consumption.

20

Bauxite Deposits

Aluminium (Al) is a very important non-ferrous metal which is used in modern industry. Growth of aluminium industry in recent years is remarkable. There has been a large scale use of this metal and its alloys in construction of airplanes, automobiles, trains, machinery, building materials and electrical appliances. This metal is of great industrial value due to its lightness compared to its weight, relatively high electrical and thermal conductivity, corrosion resistance, silvery lustre, high heat and light reflectivity and also its characteristics of rolled and drawn into any form. Aluminium is capable of forming light alloys which possess good strength and could easily be of replacing iron and steel in transportation industry; Cu in electrical industry; Sn in containers, packing foils and collapsible tubes; Fe, Cu, brass and bronze in use as cooking utensils; and iron, steel, wood and roofing in building industry. Powdered Al can be used as a powerful reducing agent in the production of several metals, in welding of heavy parts of machinery and structural materials, and also as a pigment. Main competitors of this metal in modern industry are plastics, stainless steel, Mg and Ti, but considering the resources, costs of production and qualities, Al has a bright future and thus its requirements will keep on growing.

Aluminium Ore: Quarrying for bauxite started in Katni (M.P.) in 1908, Khaira (Gujarat) in 1920 and on a large and continuous scale in Ranchi (Jharkhand) in 1946. Al-industry is of only recent origin and till then bauxite produced was used for refining kerosene, manufacturing alum and making refractory bricks. Al-metal was first produced in India in 1943 by Aluminium Production Co (a British-Canadian Co. named later as Indian Aluminium Co.) in their smelter set up at Alwaye (Kerala) due to availability of cheap hydroelectric power. The plant was first based on imported alumina till their own plant for manufacture of alumina from Bihar (present Jharkhand) bauxite was installed at Muri near Ranchi. Rolling mills of this firm are located at Belur near Kolkata. Another pioneering effort was made by Aluminium Corporation of India which in 1937, started an integrated plant at Jaykaynagar near Asansol (W.B.) and produced alumina for the first time from Jharkhand bauxite in 1942 and aluminium metal in 1944. Indian Aluminium Co. also established their second plant at Hirakund (Orissa) and the third at Belgaum (Karnataka). Two more aluminium companies started their works in 1970s—Hindustan Aluminium Corporation (Hindalco) at Renukot (Sonebhadra, U.P.) and Madras Aluminium Co. at Mettur (Salem, T.N.). Two public sector plants were stated in 1960s under Bharat Aluminium Co. (BALCO) one at Koyna (Ratnagiri, Maharashtra) and the second at Korba (M.P.). Gujarat has one plant, and Karnataka has one at Sharavathy. Though India has still to import annually appreciable Al-metal, it is expected that with the expansion of industry, it will not only gain self-sufficiency but occupy an important position as an aluminium producing country in world.

Bauxite: Global requirements of Al are met mainly by bauxite, although some plants have been constructed in bauxite, lacking countries to use other Al bearing minerals and rocks, like clays, anorthosite, syenite and nepheline syenite, leucite rocks, alunite, etc., as Al ore. Secondary Al recovered from the scrap is also an important source of the metal. Bauxite is an aluminous rock of which the main constituent is $Al(OH)_3$, with varying proportions of Fe-oxide, SiO_2 and TiO_2 $Al(OH)_3$ present in the bauxite ore are

(1) Diaspore and boehmite, Al_2O_3, H_2O (Al_2O_3, 85%, Al-45%).

(2) Gibbsite (hydrargillite) Al_2O_3 $3H_2O$ (Al_2O_3-65.4%, Al-34.6%).

(3) Bauxite (containing colloidal alumina hydrogel), Al_2O_3 $2H_2O$ (Al_2O_3-73.9%, Al-39.1%).

Fe in bauxite ore is present as hematite or goethite, SiO_2 occurs partly as clay and partly as free quartz while TiO_2 occurs as leucoxene or rutile.

Generally, it is difficult to distinguish in hand specimen between bauxite and clay. Pisolitic texture, if present, and granularity instead of plasticity, when rubbed between fingers, are two distinguishing properties of typical bauxite. Bauxite, when struck with a glancing blow by a hammer, will not show the marks of the hammer. It may, contain some clay as an impurity. Dried bauxite, if thrown on a hard floor, will make a distinct rattling sound while bauxite clay will not do so. If a dried sample is ground in an agate mortar for 30 seconds, good bauxite will be hard to grind, stick to the mortar and have to be scoured out of it while poor bauxite or bauxitic clay will grind easily and stick very little, whereas clay will grind much more easily and will not stick to the mortar. A high Al_2O_3 content in bauxite is generally indicated by the lighter shade of its colour and its greater density and compactness.

Occurrence: Most bauxite deposits have developed from the decay and weathering (mostly chemical) in warm, moist, tropical and subtropical climates, of rocks rich in Al-silicates and low in free quartz, like syenites, nepheline-syenites, basalts, feldspathic gneisses, schists and granulites, clayey limestones and clays. Al-bearing silicates are first kaolinised and then SiO_2 is also removed in solution. Bauxite deposits thus represent residual concentration of Al-hydrates after removal of all the other constituents of parent rock.

Bauxite deposits according to mode of occurrence:

(a) blanket deposits (at or near the surface, in the form of capping or flat or undulating sheets or lenses under some soil cover)

(b) Beds and lenses (interstratified with sedimentary rocks)

(c) Pockets or irregular masses (enclosed in rocks)

(d) Detrital deposits (derived from previous three types)

Bauxite deposits according to mode of formation:

(a) Lateritic (associated with the rock laterite, a product of tropical alternation of ferruginous aluminous rocks)

(b) Terra-rossa (underlain and overlain by terra rossa, a 'kankar'-like material derived from decomposition of clay-bearing limestones and dolomites).

Indian, African and American bauxites are of lateritic type. They are porous in nature and often contain more combined water than terra rossa type which is typically characteristic of France and other Mediterranean coast countries of Europe. There is no sharp distinction between laterite and bauxite and the former may gradually pass into the latter with decrease of Fe-oxide and increase of Al-oxide. Laterite containing Al_2O_3 50% or above is commercially called bauxite and that containing Al_2O_3 40% or above but below 50% is termed aluminous laterite.

Indian bauxite deposits occur mostly with high level laterite which caps high flat-topped hills or plateaus, and is derived from weathering of Deccan Trap basalts (M.P., Chhattisgarh, Maharashtra and Jharkhand), Archaean (feldspathic gneisses and granulites (T.N., Jharkhand and Orissa) or Eocene clays, clayey limestones, (J & K). Deposits are generally located on the concave portions of scraps at the edges of these hills and plateaus, or in the beds of streams on their summit. Inter-laminated layers of lithomarge (clay) and red or yellow ochres usually occur at the base of bauxite deposits. Bauxite deposits of coastal tracts of Saurashtra and Kutch occur with low level laterite almost at sea level below thin over-burden. Katni area bauxite also occurs with low lying laterite.

Distribution : Workable deposits occur in states shown in Table 20.1 and Fig. 20.1.

Table 20.1: Indian Distribution of Bauxite

States	*Areas/Districts*
A.P.	Vishakhapattanam, E and W-Godwari
Bihar	Monghyer
Chhattisgarh	Durg, Bilaspur, Sarguja, Raigarh
Gujarat	Jamnagar, Kaira, Broach, Surat, Kutch
J&K	Riasi, Poonch
Jharkhand	Ranchi, Palamu
Karnataka	Belgaum, Chikmanglore
Maharashtra	Kolaba, Thana, Kolhapur, Ratnagiri
M.P.	Jabalpur, Mandla, Shahdol, Balaghat
Orissa	Kalahandi, Koraput, Sambalpur
T.N.	Salem, Nilgiri

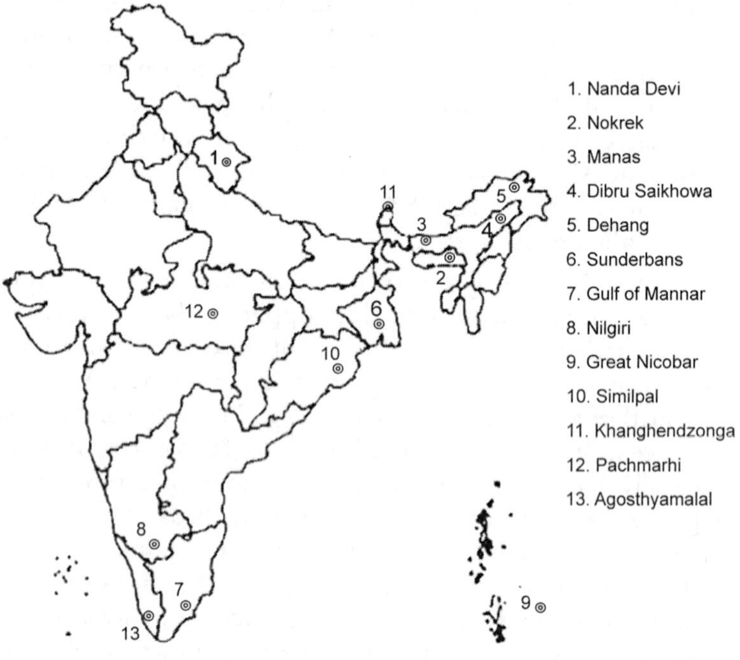

Fig. 20.1: Major Bauxite Mining Areas

(1) **Jharkhand and Bihar:** Bauxite of Ranchi and Palamu districts is of high grade, Al_2O_3 52-60% and SiO_2 2-5% (of which 1-2% is of reactive nature). It, however, contains a very high contents of TiO_2 and Fe_2O_3 (8-10% each) and is, therefore not suitable for chemical and refractory industries.

Most of the bauxite produced comes from the plateaus near Lohardaga (Ranchi) and is utilised in Al industry at Muri near Ranchi and Jaykaynagar near Asansol.

(2) **Gujarat:** Saurashtra bauxite is rich in Al_2O_3 content (usually 60%), and low in Fe_2O_3 and SiO_2 (about 2% each), and also in TiO_2 (1-3%). It is ideal for chemical and refractory industries and as this bauxite is rich in gibbsite, Al_2O_3 can be extracted from it more easily than with other bauxites. Kaira bauxite is high in Fe and is used in manufacture of cement and in petroleum-refining industry. Jamnagar deposits are unique and lie in a flat surface, almost at sea level.

(3) **J & K:** Jammu bauxite is exceptionally rich in Al_2O_3 (70-80%) as it contains mostly diaspore. The ore is, however, very refractory in nature, not easily digested and consumes large quantities of caustic soda in extraction of Al_2O_3. Further it is highly siliceous, though low in Fe_2O_3 and TiO_2. These deposits occur as extensive surface strata over nummulitic limestone along the coal belt of Jammu region.

(4) **M.P. and Chhattisgarh:** Bauxite of these states is more or less similar to that of Bihar and Jharkhand. Katni bauxite is somewhat richer in Al_2O_3 while SiO_2 and TiO_2 contents are somewhat lower (6-8% each) than those of the other districts. This bauxite is used in chemical, refractory, cement and abrasive industries. Amarkantak plateau bauxite is used in Al-industry.

(5) **Maharashtra and Karnataka:** Bauxites of Kolhapur and Belgaum districts, though high in Al_2O_3 content (50-58%) have high Fe_2O_3 and TiO_2 contents (6-10% each) and are low in SiO_2 (1-3% of quartz type and 1-2% of reactive nature). These bauxites are used for Al extraction. Kolaba bauxite is not so high in Fe_2O_3 or TiO_2 as of other districts.

(6) **Orissa:** Kalahandi bauxite is usually ferruginous in character, although SiO_2 and TiO_2 contents are usually low. It may be used for extraction of the metal. High grade bauxite is found in Kalahandi and Sambalpur districts.

(7) **T.N. :** Shevaroy hills bauxite of Salem district usually contains <55% Al_2O_3 and is usually ferruginous (Fe_2O_3,

8-10%), high in SiO_2 (15-20% of quartz type and 2-4% of reactive nature) but low in TiO_2. Bauxite is mostly gibbsite, very soluble in caustic soda and amenable to easy extraction of Al_2O_3. It is used for manufacture of artificial emery. It is also suitable for refractories, and may be used for manufacture of Al-metal. High grade bauxite is found in Salem district and Nilgiris.

Grades of Al-Ore: Indian Standards Institution (ISI), has Graded Al-ore in to 3 grades (Table 20.2).

Table 20.2: Grades of Al-Ore

		ISI Standard	
	Grade 1	*Grade 2*	*Grade 3*
Total Al_2O_3%	>51	48-51	45-48
SiO_2% (Max)	3.5	5	5
P_2O_5% (Max)	0.2	1	1
V_2O_5% (Max)	0.2	0.2	0.2
Fe_2O_3 + TiO_2% (Max)	30	30	30
L.O.I at 1100°C	20	20	20

According to GSI, bauxite reserves with >45% Al_2O_3 are distributed in various states as shown in Table 20.3. Major reserves are concentrated in E-Coast bauxite deposits of A.P. and Orissa.

Table 20.3: Reserves of Bauxite (in MT)

State	*Indicated and Inferred*		*Potential*	
	$A1_2O_3$ (45-50%)	*$A1_2O_3$ (50%)*	*$A1_2O_3$ (45-50%)*	*$A1_2O_3$ (50%)*
1. Bihar & Jharkhand	—	16.23	0.02	10.04
2. Gujarat	—	—	1.50	12.61
3. J & K	—	—	—	12.80
4. Kerala	0.10	0.09	—	0.70
5. M.P.	12.12	6.56	1.90	8.86
6. Maharashtra	3.97	44.12	4.72	29.39
7. Karnataka	—	—	4.88	0.67
8. Orissa	—	—	—	0.35
9. T.N.	—	—	2.2	—
10. U.P.	—	—	6.00	—

Industrial Uses and Specifications

Most of the bauxite produced in India is consumed by Al-industry. Indian Aluminium Company have their mines and alumina plant in Ranchi (Jharkhand) and their smelters at Alwaye (Kerala) and Hirakund (Orissa). Aluminium Corporation of India also have their mines in Ranchi and their smelter is at Asansol (W.B.). Hindustan Aluminium Corporation get their bauxite from Ranchi and Rewa (M.P.) and have their smelter at Rihand in Mirzapur (U.P.). More plants for manufacture of the metal were established, *e.g.*, one by Madras Aluminium Company at Mettur in Salem (T.N.), which uses Al-ore of Salem; the other in public sector at Ratnagiri (Maharashtra), which uses bauxite of Kolhapur; another in Karnataka to use Belgaum deposit and one other has been set up in public sector at Korba (M.P.) to utilise Amarkantak bauxite.

Commercial bauxites are classified according to their nature as:

(i) Oolitic or pisolitic

(ii) Spongy or porous

(iii) Amorphous or clayey

According to the relative amounts of the respective oxides present, bauxites are classified as :

(i) Normal

(ii) Siliceous

(iii) Titaniferous

(iv) Ferruginous

Most of Indian bauxite is titaniferous, Kashmir bauxite is siliceous and T.N. and Orissa bauxites are ferruginous. Industrial use of bauxite in different industries depends upon the limits of Fe oxide, SiO_2, TiO_2 and other impurities present. SiO_2 may be present in free state or as reactive silica (combined with Al or other metals). The latter silica largely affects use of bauxite for Al-industry.

Besides Al-industry, bauxite is used in chemical, refractory, abrasive, cement and petroleum refining industries.

Al-Industry: For the preparation of metal, bauxite (dried) should contain Al_2O_3 ≥50%; SiO_2 <6%; Fe_2O_3 <8%, and TiO_2 <4%. A high SiO_2 content (especially of a reactive type) diminishes recovery of alumina and increases consumption of caustic soda; and oxides of Fe & Ti remain only as insoluble residue (red mud) after extraction of Al_2O_3. Occasionally, ferruginous bauxites containing upto 20% Fe_2O_3, and Siliceous bauxites containing upto 15% SiO_2 have been used for extraction of aluminium. Main raw materials to manufacture Al are bauxite, electric power, cryolite, fluorite, petroleum or pitch coke and caustic soda. About 4t of bauxite are required to produce 2t of Al_2O_3 from which 1t of Al is recovered. About 20,000 KW hrs, 0.1t of cryolite, 0.01t of fluorite, 0.44t of claimed petroleum coke and 0.2t of caustic soda are needed to produce 1t of Al.

Chemical Industry: Here bauxite is used for manufacture of aluminium salts like alumina, alums, aluminium sulphate, aluminium chloride, etc., which are used for various purposes (tanning, dyeing, printing, purification of water, decolourising of sugar and treatment of sewage). Bauxite ore for this purpose should contain Al_2O_3 >52%; Fe_2O_3 <2.5%; TiO_2 <3% and SiO_2 upto 11%.

Refractory Industry: As refractory material, bauxite bricks are used in cement kilns and metal working furnaces. Ore for this purpose should have at least 58% Al_2O_3 and should be low in Fe_2O_3, TiO_2 and SiO_2; but for use as lining in basic open hearth steel furnaces, bauxite bricks should contain SiO_2 >12%.

Abrasive Industry: Various abrasives are produced by the fusion of calcined bauxite or prepared alumina in an electric furnace. Bauxite ore should contain >50% Al_2O_3; but for production of high grade artificial corundum, highest grade of bauxite with very low Fe_2O_3, SiO_2 and TiO_2 content is required. For production of artificial emery, percentage of Fe_2O_3 may be upto 15%.

Cement Industry: Bauxite cements set and attain strength more quickly than ordinary cement. They have great resistance against heat, chemical action, and sea-water action. Bauxite suitable for high alumina cement should have a minimum ratio of Al_2O_3/SiO_2 of about 10 : 1 and that of Al_2O_3/Fe_2O_3 of about 2 : 1.

Oil Industry: In petroluem-refining, activated (calcined) bauxite is used for decolourising and desulphurising petroleum products like kerosene. Bauxite should be of high grade and should contain a high percentage (about 24%) of combined water.

Mining of bauxite: Quarrying is the common method of bauxite mining. Deposits which occur as surface blankets or in beds, lenses or pockets, overlain by thin OB or interlayered with shallow unconsolidated sediments, are mined by OC method removing the OB mainly by manual labour, using picks, axes, shovels, wheel barrows and head loads. If OB is thicker and of hard lateritic layer, as is the case with some of the bauxite deposits of Lohardaga (Jharkhand) power shovels, mechanical scrappers, drag lines, and dumpers are used for its removal, and drilling by rotary drills is carried out. The upper surface of bauxite layer is thoroughly cleaned by means of picks and shovels. Holes are driven into ore-bed vertically starting at the surface and penetrating into the lower portion of bauxite layer. Generally such holes are drilled in a series parallel to the existing quarry faces. Depending on the thickness and topography of ore-bed, mining may be thus carried out in benches of upto 8-10m. Due to the need for larger outputs, mechanised mining of bauxite has also gained ground in India.

Much of the bauxite mined in India is manually broken to a size of 4–8". In larger mines, bauxite is crushed to a size of minus 4" by Jaw Crushers. Transport of crushed ore is done by road, rail or ropeway. Large lumps of mined bauxite, received at the consuming alumina works, are crushed in primary and secondary crushers like Jaw Crusher and Hammer Mill or Gyratory Crusher. Crushed product is ground to a fine powder by wet or dry grinding carried out in closed circuits. Product size after grinding varies widely depending upon the characteristics of bauxite ore. Extraction of alumina from ground bauxite is then carried out by Bayer process.

Reserve and Production

India has the fifth largest resources of bauxite of which a large proportion comprises gibbsitic ores (Table 20.4 and 20.5). Indian alumina producers are looking beyond meeting domestic smelting needs, by substantially increasing exports of metallurgical as well as speciality grade alumina. There is a great deal more of latent potential to tap India's Bauxite assets. Table 20.6 shows major bauxite producing countries.

Table 20.4: Reserves and Production of Bauxite in India (MT)

	Years			
	1998-99	*1999-2000*	*2000-01*	*2001-02*
Reserve				
1. *In situ*	3037.0	—	—	2953
2. Recoverable	2525.0	—	—	2462
3. Metallurgical grade (in %)	89	—	—	86
4. Conditional	0.50	—	—	0.75
5. Prospective	90	—	—	90
Production	6.6	7.0	7.9	8.5

Table 20.5: Bauxite Resource in MT

Category	*Resource*	*Reserve*	*Remaining*
Ore	3290	899	2391

Table 20.6: Major Bauxite Producing Countries

Country	*Production (MT)*
Australia	47
Brazil	12
Jamaica	12
India	8
Indonesia	8
China	6
Venezuela	3
Surinam	3

21

Mica Deposits

In early historic times, mica had been used in medicines by Indians and also by Greeks and Romans. But its importance increased significantly in 20th century (Electrical Age) in electrical industry due to excellent insulating properties and its high resisting capacity.

In India production of mica from Kodarma (Jharkhand) and Nellore (A.P.) is recorded prior to 1900. After the failure of early European efforts to work the mines, mining was well established under the direction of F.F. Chrestian in 1894. By 1899 his company was operating 110 mines out of a total of 170 mines in Kodarma Mica field. The process of splittings discovered during this period gave a great impetus to mica industry and the local workmen became experts in this manual process. Many countries in subsequent years began to export mica to India for fine splittings only. Modernisation of mica mining started in 1910. Several Indian companies started and after World War II the industry came completely into Indian hands. Though there was some production of mica from Rajasthan early since 1904, Rajasthan mica-field was extensively developed later and Rajasthan mica began to be sent to Jharkhand for splitting during World War-II when this strategic mineral had to be flown from India to USA for war-purposes and stock-piling. India has practically the world monopoly in production of sheet mica. Bihar-Jharkhand, A.P. and Rajasthan are the chief producing states for muscovite mica. Intermittent production of phlogopite is from Kerala. There were, in 1970s, 537 mica mines in India of which 287 were in and around present Jharkhand, 185 in Rajasthan and 60 in A.P. Demand for mica in the country is small and bulk of the mica produced is exported. Internal demand is, however, increasing with the development of heavy electrical industry. Mica is a mineral which has been regularly earning foreign exchange as an export commodity.

Mica group minerals are composed principally of orthosilicates of Al with K and H and generally Mg and Fe^{+2}, sometimes Fe^{+3}, Va, Na & Li and rarely Rb & Ce. One element may replace another in part and some may contain also F, Ti or Cr. Micas are characterised by a highly perfect basal cleavage due to which they can be easily split into thin sheets. Important varieties of mica from the commercial point of view are : (1) Muscovite (white mica or K-mica), $H_2KAl_3(SiO_4)_2$ (2) Phlogopite (amber mica or Mg-mica), $(K,H)_2.Mg_2Al(SiO_4)_3$ (3) Biotite (black mica or Mg-Fe mica), $(K,H)_2 (Mg, Fe)_2 Al_2(SiO_4)_3$; (4) Lepidolite (Li-mica), $(OH,F)_2 KLiAl_2Si_2O_{10}$; (5) Zinnwaldite (Li & Fe mica and (6) Roscoelite. (V-mica). Of these, lepidolite, zinnwaldite and roscoelite are rather rare varieties and are used in industry mainly due to their chemical contents as the source of elements Li or V. Lepidolite is of importance for India. The other micas of commercial value—muscovite and phlogopite and less frequently biotite owe their usefulness in industry mainly due to their physical properties. These minerals alongwith their commercial names and chief constituents are presented in Table 21.1.

Table 21.1: Chief Mica Minerals

Mineral Name	*Commercial Name*	*Chief Constituents in Addition to Al and SiO_2*
Muscovite	White mica	Potash
Phlogopite	Amber mica	Magnesia, potash
Biotite	Black mica	Magnesia, iron, potash
Vermiculite	Jeffersite	Magnesia, iron, potash
Lepidolite	Lithia mica	Lithium, fluorine, potash
Zinnwaldite	Lithia-iron mica	Lithium, iron, potash, fluorine
Roscoelite	Vanadium mica	Vanadium, magnesium, iron
Fuchsite	Chrome mica	Chrome

The first three are the only micas employed commercially as mica, and the first two are the only principal ones; lithium and vanadium micas are used as sources of those elements, and vermiculite is used chiefly for sound and heat insulation.

Physical Properties

Muscovite and phlogopite possess the unique properties of elasticity, toughness, flexibility, transparency, resistance to heat and sudden changes of temperature, high dielectric strength and the property of splitting into thin films (muscovite or phlogopite splittings are usually 1/1000 of an inch or one mil in thickness).

In thin films, muscovite is colourless or transparent when pure, but may be stained or tinted due to impurities. In thick crystals it is white, reddish as Jharkhand-Bihar's ruby mica or greenish as Nellore mica. Phlogopite is yellow to dark brown in colour, less transparent, more flexible but less elastic than muscovite. Biotite is usually dark brown to black in thick crystals, often opaque in thin sheets but thin laminae are transparent and brownish red or brownish green in transmitted light; it splits less readily; is much less heat resistant than muscovite or phlogopite; and its dielectric strength (i.e. the property of withstanding high voltages without puncture) is very low due to its high iron content. Hardness of muscovite and softer phlogopite varies from 2-2½, and harder phlogopite and biotite from 2½-3. Bihar-Jharkhand muscovite is usually harder than phlogopite. For all ordinary purposes, mica is unaffected by acids and is infusible. Muscovite can be used for electrical purposes where temperatures are not in excess of 700°C but phlogopite can withstand temperatures upto 1000°C.

Occurrence

Muscovite and biotite in small flakes form essential constituents of many of the igneous rocks, gneisses and schists. Muscovite is a common mineral of sands and sandstones, but biotite is not found in sedimentary rocks due to its being easily disintegrated by weathering agencies. Commercial biotite is usually obtained from biotite-schists. Workable deposits of micas, however, are found only in pegmatites (very coarse grained igneous rocks occurring as dykes or veins). Commercial deposits of muscovite (sometimes with biotite and rarely with lepidolite) occur in acid granite, pegmatites consisting essentially of quartz, feldspar and mica. Important commercial pegmatites are characterised by extreme coarseness of texture, notable uneven segregation of their constituent minerals and vast size (giant crystals) and extreme rarity of some of the minerals found in them. Minerals associated with mica in these pegmatites are quartz, feldspar, tourmaline, garnet, apatite and rare minerals like beryl, columbite-tantalite, monazite, pitchblende, lepidolite etc. Good workable deposits of mica generally occur in those pegmatite veins which traverse mica-schists or mica-gneisses and contain soda feldspar (albite or oligoclase) much more than potash feldspar (microcline or orthoclase) and in which hydrothermal replacement has played an important role in deposition of minerals. Many of the mica-bearing pegmatites are zoned, they have usually a core of quartz surrounded on both sides by feldspar, and mica concentrations are commonly on two sides of the vein (at the junction of vein and country rock) and also between quartz core and feldspar on the two sides. Phlogopite mica is restricted to basic pegmatites like pyroxene-pegmatites, composed chiefly of pyroxenes and essentially quartz-free and is also found in crystalline magnesian limestones or dolomites.

Grades

Value of mica depends upon its physical perfections. Mica should be free from flaws and cracks and should be in as large pieces as possible. Average mica content of workable portion of a pegmatite is about 6% of the total rock excavated. Individual blocks or crystals of crude mica (known as books or book mica) are carefully sorted from the broken rock in the mine. Mica books average 6-12" across and 3-4" thick, rarely they are 3" or more in width and 2" or more in thickness. The rough books of mica are first split into thick sheets and are then trimmed (i.e. cracked and flawed pieces around their edges removed) by sickle in Bihar-Jharkhand (sickle-trimmed mica), or cut by shears or scissors in Nellore (shear-trimmed mica). Shape of sickled-trimmed mica is of an irregular-rounded rectangle with bevelled edges, whereas that of shear-trimmed mica is roughly rectangular with its cut edges perpendicular to cleavage planes of mica. The trimmed mica is known as sheet mica or block mica. Minimum size of sheet mica covers an area of 2" × 2½" and thickness is 1/8". Percentage of block mica produced from crude mica varies from 20 to 40%, the rest going as scrap mica. Trimmed mica is then sorted according to colour, quality and size. According to the degree of inclusions and stains present, Bihar-Jharkhand ruby mica is commercially termed 'clear', 'slightly stained', 'heavily stained', 'black spotted', etc; the Andhra green mica is classified into 'clear green', 'spotted green'; 'yellowish', 'brown' etc. Physical imperfections in muscovite influencing its classification are, air inclusions, waviness, pinholes, flexibility in thin films, and wedge, ruled or other such structures which make it difficult to split mica crystal into thin films. Clear mica which is hard, of uniform colour, flat and free from all stains and other defects is the finest quality of mica. Quality of phlogopite mica is determined by its resistance to heat. The higher heat resisting phlogopite is known as spark plug quality.

Grading of Indian mica, according to size, is based on the limiting area of the rectangle of good material, which can be cut from a sheet of mica. These grades, generally recognised for muscovite and phlogopite, (if found in larger sizes) are shown in Table 21.2.

Table 21.2: Grades of Mica

No.	*Indian grading (usable area based on rectangular sizes) Square inches*	*USA Grading (Approx. corresponding sizes) Inches*
6	1—2.5	Punch
5½	2.5—3	Punch
5	3—6	2 × 2 and 1.5 × 2
4	6—10	2 × 3 and 3 × 3
3	10—15	3 × 4
2	15—24	3 × 5
1	24—36	4 × 6
A–1 or Special	36—48	5 × 8
Extra Special	48—64	6 × 8
Extra Extra Special	64—80	8 × 8
Over Extra Extra Special	80—100	8 × 10

In Bihar and Jharkhand a very small proportion (about 17%) of block mica obtained from a mine after cutting is of bigger size (No. 4 and above). Sheet mica of bigger size (No. 4 and above) is usually marketed as block mica. High quality mica of smaller sizes (No. 4 and below) is split into films (usually 0.0012 to 0.004' or 1¼ to 4 mils in thickness and used for condensers) and into mica-splittings (usually 0.001' to 0.0012' or 1 to 1¼ mils in thickness and used for micanite). Small size, low quality sheet mica can be subdivided further into punch, circle and washer according to the diameter of the disc obtained. Splitting is done by local expert workmen, by applying a long, sharp pointed knife along the bevelled edges of trimmed mica. Many countries export mica to India for fine splittings only.

Methods of grading and classification of muscovite mica blocks, films and splittings have been specified by Indian Standards Institution. Mica production is broadly graded into (i) sheet mica which includes blocks, films and splittings and (ii) waste or scrap mica comprising bulk of the output and is known in the trade as mine scrap and factory scrap.

Distribution

India occupies a unique position regarding mica production in world and supplies about 80% of world requirements of block mica and mica-splittings. Mica has been found practically in all parts of India (Table 21.3).

Amongst these occurrences, phlogopite is restricted to Quilon and Trivandrum (Kerala), Vishakhapattanam (A.P.) and Coimbatore (T.N.) only. Other occurrences are of muscovite mica.

India's production of muscovite comes mainly from mica-belts of Bihar, Jharkhand, Rajasthan and A.P.

(1) **Jharkhand-Bihar mica-belt (3%):** This belt is about 32 Km wide and runs in an ENE-WSW direction for about 145 Km, extending from Gaya district through Hazaribagh and Monghyr districts to Bhagalpur district. The important areas of production in this belt are Kodarma Reserve Forest,

Table 21.3: Indian Distribution of Mica

States	*Districts*
A.P.	Nellor, Krishna, Cuddapah, W-Godavari, Khamman and Vishakapattanam
W.B.	Bankura, Midnapur & Purullia
Jharkhand	Hazaribagh, Dhanbad, Palamu, Ranchi, Singhbhum
Bihar	Gaya, Bhagalpur, Monghyr
J & K	Somjan area, Doda
M.P. & Chhatisgarh	Balaghat, Bastar, Bilaspur, Chindwara, Gwalior, Jhabua, Rewa
T.N.	Coimbatore, Nilgiri, Tirichirapalli, Salem
Maharashtra	Bhandara, Ratnagiri
Karnataka	Hassan, Kodur, Mysore, Coorg
Kerala	Quilon, Cannanore, Trivandrum, Kottayam, Kozikode
Orissa	Ganjam, Koraput, Mayurbhanj, Sambalpur, Sundergarh
H.P.	Kangra
Punjab	Patiala
Rajasthan	Alwar, Ajmer, Bharatpur, Bhilwara, Jaipur, Jodhpur, Kishangarh, Shahpura, Tonk, Sirohi, Dungarpur & Udaipur

and areas NE of this forest, and areas near Gawan and near Parsabad Rly. Stn. Important mining centres in Hazaribagh district are Kodarma R.F., Chatkari, Dharkakola, Domchanch, Gawan, Tisri, Mahesari, Chakai etc. Bihar-Jharkhand possess the richest deposits of excellent quality muscovite mica in world and contribute >50% of India's total output. Bihar-Jharkhand mica is of superior ruby-red type, generally clear and free from spots, buckles and other flaws.

(2) **Rajasthan mica belt (51%):** Mica-bearing pegmatites of Rajasthan are distributed over an area some 100 Km wide, stretching from Jaipur in NE for >320 Km to Udaipur in SE. Rajasthan mica is mostly of stained ruby and green varieties and it is extensively mined in Bhilwara, Tonk, Ajmer, Jaipur and Udaipur districts.

(3) **A.P. mica belt (28%):** Nellore mica-belt of A.P. stretches for some 100 Km between Gudur and Sangam and has a width of about 16 Km. Gudur and Nellore are the most important mining centres in A.P., muscovite is mostly green and stained, but ruby mica is also obtained from few mines.

India is world's leading producer of sheet mica and accounts for about 60% of global mica trade. Recoverable reserves in A.P. are 44,837, Bihar and Jharkhand 13,554t and in Rajasthan 1589t (Table 21.4–21.5).

Table 21.4: Reserve and Production of Mica in India (in 000't)

Year	*Reserve*	*Production*
1990-91	—	4.0
1998-99	60	1.0
1999-2000	—	2.0
2000-01	—	1.08
2001-02	60	1.2

Table 21.5: Mica Resources (2005-06)

Category	*Total*	*Reserve*	*Remaining*
Pegmatite	393855	68570	325285

Industrial Uses

Sheet mica is indispensible as an insulating material in electrical industry. It is used in different forms in a great variety of electrical devices, *e.g.* commutators of dynamos and motors, condensors of magneto and wireless apparatus, resistances of electrical heating devices, in rheostat brush holders, fuse boxes, aircraft spark plugs, radio, radar and television, high voltage lamps, telephones, etc. Non-electrical uses of clear sheet mica are in construction of stove fronts, electrical light sockets, lamp chimeys, protective spectacles, windows in coning towers of warship and in ovens and furnaces, photograph diaphragms, toasters, hot plates, coffee pots, etc. Synthetic mica and other substitutes have been developed, and they are being used in certain fields but none has been able to displace sheet mica in many important electrical applications. Mica is getting increasing use in equipments that encounter very high temperatures, like rockets, missiles, jet engines, ignition system, etc.

Muscovite and phlogopite splittings are used in making built-up mica or micanite insulation products, such as mica cloth, mica tape and mica plate. The cloth and tape are made by bonding a thin layer of mica splittings to a backing of cotton, silk, rayon of fibre-glass cloth with a shellac or gyptal (an alkyl resin) binder. Mica plate is made by spreading alternate layers of mica splittings and binder until the required thickness is obtained and then drying under heat and pressure to secure uniformity of thickness and density. The greater part of built-up mica is used as an insulation between copper segments of electric generators and motors, as it wears down equally with copper; it is also being used as a substitute for sheet mica in manufacture of heating elements for electric household appliances.

Waste or Scrap mica

It is produced during the processing of crude mica and while punching mica. It is used for manufacture of light weight insulating mica-bricks (used in steel industry) and ground mica (prepared both by wet or dry grinding methods). Wet ground mica is used in paints (like road making, aluminium and fire resisting paints), decorative wall paper and as a filler in rubber and some plastics. Dry ground mica is used for coating roofing materials, coating fabrics to make them waterproof, in pipe and boiler-laggings, oil-well muds, lubricants and dusting powders. During the War, it was used to surface a composition material for aircraft landing strips. Abrak bhasm, an inorganic indigenous medicinal preparation prepared from a particular variety of biotite, is used in India as a tonic for certain chronic diseases. Muscovite sheets are extensively used for decorations and mica powder is sprinkled on clothes and toys.

Waste mica is also utilised in manufacturing of heat insulating bricks which are used in industrial furnaces to conserve heat. These bricks are better in heat insulation efficiency in camparison to imported vermiculite bricks (Table 21.6).

In addition to saving on fuel consumption to the extent of 35-40%, mica brick also help to keep heat radiation to the barest minimum.

Table 21.6: Utilization of Waste Mica

Property	*Mica bricks*	*Commercial Vermiculite bricks made from imported Vermiculite*
Bulk density (lbs/ft.^{2})	35-45	35-45
Thermal conductivity at avg temp, 500°C (B,Th.V/Ft.^{2})	0.8-1.3	0.95-1.35
Cold Crushing Strength (lbs/inch^2)	100-150	100-125
Safe operating temp under load of 1 ton/ft^2	975°C	860°C

(After Krishnaswamy, 1979)

Mica is a strategic mineral. Average annual output of crude mica in India is about 1,200t (IBM). India is the largest producing country for muscovite (block and sheet) mica in world and holds a monopoly on sheet mica (Table 21.7). As the development of electrical industry was limited, there had been in past only a small domestic consumption of mica in rubber, radio-valves, railways and small electrical appliances; and almost the entire production was being exported. Due to the recent development of Heavy Electrical Equipment industries and the increase in production of electrical goods in different parts of country, internal demand for mica has increased greatly, but at least for the next decade or so, India will still have to depend on external market to keep her mica-mining industry flourishing.

Table 21.7: Major Mica Producing Countries (1994-95)

Country	*Production ('000t)*
India	12.65
Brazil	3.20
Medagaskar	1.32
Canada	0.91

22

Limestone Deposits

Limestone (Chuna ka Pathar), as its name suggests, is the main source of lime. It is a rock composed mainly of $CaCO_3$ which may be present in a finely-divided noncrystalline state as in 'amorphous limestone', or a crystallised grains of mineral calcite as in 'crystalline limestone'. Limestones are sedimentary rocks of chemical, organic or mechanical origin and are seldom pure. They may be 'argillaceous' (clayey), 'arenaceous' (siliceous) 'carbonaceous' (containing bitumen or coaly matter), 'ferruginous' (containing iron minerals—siderite, limonite or hematite), or 'fossiliferous' (containing corals, crinoids and other fossils or their fragments). Pure limestones are readily soluble with effervescence in acids but impure limestones are not so readily soluble. Limestones often contain some $MgCO_3$, either as dolomite, $CaMg\ (CO_3)_2$, or magnesite, $MgCO_3$, mixed with calcite. They are then termed 'dolomitic' or 'magnesian' limestones if they contain 10 to 40% $MgCO_3$ Limestone altered by dynamic or contact metamorphism becomes coarsely crystalline and are referred to as 'marbles' and 'crystalline limestones'. Other common varieties of limestones are 'marlstone', impure limestone containing varying proportions of clay; 'oolite', (oolitic limestone) composed wholly ore partly of small spherical concretions of $CaCO_3$ built up in concentric layers; 'chalk', a white, extremely fine-grained, usually soft and friable variety of limestone, composed wholly or largely of microscopically small remains of foraminifers and broken shelly fragments; 'kankar', irregular nodules and concretions of impure $CaCO_3$ of all sizes, found in the older surface alluvium or soils; 'travertine' ('calc-tufa' or calcareous sinter), $CaCO_3$ rock deposited from solution in limestone regions by spring waters; and 'stalagmites' and 'stalactites', $CaCO_3$ deposits on the floors and roofs of caves in limestone regions formed by percolating waters. The two other varieties of limestone are 'hydraulic limestone' which when burnt gives lime or cement which sets or hardens under water; and 'lithographic limestone', a uniformly fine textured compact but sufficiently porous and soft limestone used in lithography in the form of rectangular slabs.

Distribution

Limestone deposits are widely distributed in India and are found in many of the geological formations, and are worked in almost all the States. Karnataka is the leading state followed by A.P., Gujarat, Rajasthan, Meghalaya, Chhattisgarh, M.P., Orissa, Maharashtra, Uttaranchal. The Dharwar, Cuddapah, Vindhyan and Tertiary formations contain the largest deposits, and the States of A.P., Assam, Bihar, Jharkhand, M.P. Chhattisgarh, Rajasthan, U.P. and Uttaranchal possess large reserves of limestone. In A.P., limestone mostly of Archaean, Cuddapah or Vindhyan formations, occurs in Adilabad, Anantapur, Cuddapah, Guntur, Hyderabad, Khammam, Krishna, Kurnool, Nellore, Srikakulam and Vishakhapatnam districts. Some deposits of Kurnool district are suitable for manufacture of bleaching powder and CaC_2 and for use in glass and sugar industries. In Assam good reserves of high grade Tertiary nummulitic limestone suitable for cement manufacture and for use as a flux are found in the Garo, Jaintia, Khasi and Mikir hills. In Bihar and Jharkhand which are the leading producers of limestone, crystalline limestone of Archaean age occurs in Hazaribagh, Ranchi Palamau and Singhbhum districts and the Vindhyan limestone in Shahabad district. Most of the limestone is used in cement works. Some of the limestone of Palamau district is suitable for glass and sugar industries, and some of the Vindhyan limestone of Shahabad and Palamau districts is of flux-grade. In W.B., limestone deposits occur in Darjeeling, Jalpaiguri and Purulia districts. Calc-tufa of Darjeeling and Jalpaiguri districts and 'kankar' concretions in the soil of Raniganj coalfield provide local material for lime-burning. In Gujarat, massive limestone occurs in Banaskantha, Subarkantha, Broach and Kaira districts, 'calc-tufa' in Broach district and 'coralline' and 'miliolitic' limestones in Halar, Sorath and Amreli districts. They are being used in cement, chalk, bleaching powder and alkali industries. In H.P. high grade limestone is reported from Kangra, Mahasu, Sirmur and Kulu districts. In J & K, limestone suitable for cement industry occurs in Anantnag, Baramula, Srinagar and Udhampur districts. In Kerala, 'lime-shell' deposits occur in Cannanore and Kottayam districts and 'kankar' deposits in Palaghat district. In M.P. and Chhattisgarh high grade limestone deposits of Dharwar, Cuddapah and Vindhyan age are found in many districts, the important ones being in Bilaspur, Durg, Raipur, Jabalpur, Rewa and Bhopal districts. Most of the limestone is used in cement works, some in sugar and paper mills, but limestones of Raipur, Durg and Jabalpur districts are also used as flux in iron and steel plants. In T.N. Archaean limestones are found in Coimbatore, Madurai, Ramanathapuram, Salem, Tiruchirapalli and Tirunelveli districts. Shelly, tufaceous and argillaceous limestones of Cretaceous age occur in Pondicherry and in parts of South Arcot and Tiruchirapalli districts. In Maharashtra, limestone occurs in Chanda and Yeotmal districts, some in Nagpur and Satara districts, and 'kankar' occurs in Ratnagiri district and 'calc-tufa' in Ahmadnagar districts. In Karnataka crystalline limestone is found in Chitaldrug, Mysore, Shimoga and Tumkur districts. The limestone of Shimoga district is used as a flux. Vindhyan limestone of Bhima series occurs in Bijapur district and is worked for cement. Shelly limestone is found in S Kanara district. In Orissa, the most important deposits of limestone (calcitic marble) associated with dolomite occur in Sundergarh district and belong to the Gangpur series of Dharwar Group and are extensively worked for use as a flux in iron and steel industry and for the manufacture of lime and cement. Limestone deposits of Cuddapah age are found in Sambalpur and Koraput districts. In Haryana, limestone suitable for cement occurs in Ambala and Mahendragarh districts. In Rajasthan, although the celebrated Makrana marble of Jodhpur and some crystalline limestones of other districts yield good lime when burnt, the most important deposits of limestone for lime and cement are of the Vindhyan S.G. and are found in Alwar, Bundi, Kota, Jodhpur, Tonk, Ajmer, Banswara, Bikaner, Dungarpur and Udaipur districts. Some of the limestones of Jodhpur and Alwar districts are suitable for manufacture of calcium carbide, bleaching powder, alkali and glass. In U.P. limestone suitable for cement manufacture and for use as a flux occurs in the Mirzapur district (in Vindhyan formations). Limestone and good calc-tufa depostis occur in Dehradun, Almora and Nainital districts of Uttaranchal. Kankar deposits used for lime-burning and road metal and as building stone, occur in older alluvium in many districts of western U.P., and 'marlstone' associated with clays of sub-Recent age occurs under a thin cover of alluvium in Lucknow and some other districts of eastern U.P. Table 22.1 presents state-wise distribution and reserve of limestone in India.

Table 22.1: Statewise Distribution and Reserves of Limestone in India

State	*District/Area*	*Reserve (in MT)*
1. A.P.	Godavari, Krishna, Guntur, Cuddapah, Kurnool, Warangal, Karimnagar, Adilakad, Khammamett, Anantpur, Mahboobnagar, Nalgonda, Nellore, Rangareddi and Vizianagram	11,264 (Portland Cement Grade)
2. Arunachal Pradesh	Lohit	105 (Portland Cement Grade)
3. Assam	Cachar and Mikir hills	731.8 (Portland Cement Grade, paper grade, chemical grade)
4. Bihar	Rohtas	—
5. Jharkhand	Hazaribag, Ranchi, Palamau, Singhbhum	273.9 (Portland Cement Grade)
6. Gujarat	Banaskantha, Junagarh, Khera, Junagarh, Sabarkantha, Panch-Mahals, Surat and Dang	—
7. Haryana	Ambala, Mahendragarh	77.3 (Portland Cement Grade)

8.	H.P.	Sirmur, Mahasu, Chamba, Bilaspur, Kangra, Kulu, Mandi, and Simla	143
9.	J&K	Anantnag, Sind valley, Riasi area in Jammu district, Baramula, Kathua, Poonch, Rajauri & Udampur	56.5 (Portland Cement Grade)
10.	Karnataka	Gulbarga, Shimoga, Chitradurga, Tumkur, Mysore	406.9 (Portland Cement Grade)
11.	Kerala	Cannanore, Kottayam, Palghat, Alleppey, Iduke (Iddiki), Kozhikode, Malappuram, Trichur	
12.	M.P.	Jabalpur, Rajgarh, Sagar, Damoh, Rewa	1784.79 (Portland Cement Grade)
13.	Chhattisgarh	Bastar, Raipur, Durg	302.10 (BF and Cement Mixed grade)
14.	Manipur	Manipur East	0.472 (Portland Cement Grade)
15.	Meghalaya	Garo, Khasi and Jaintia hills	34.75 (Portland Cement Grade)
16.	Nagaland	Phek and Tuensang	0.7 (Cement Grade)
17.	Orissa	Sambalpur, Koraput	45.2 (Portland Cement Grade)
18.	Pondicherry	——	2.6 (Portland Cement Grade)
19.	Punjab and Adjoining areas	Foothills on the NE border of the state, Kangra, Hoshiarpur	26.4
20.	Rajasthan	Ajmer, Banaswara, Bikaner, Dungarpur, Jodhpur, Kota, Tonk, Udaipur, Bundi, Sawai-Madhopur, Also in Jhunjhun, Jaipur, Sirohi, Nagaur, Jaisalmer Alwar, Bhilwara, Chitaurgarh	(Portland Cement Grade)
21.	T.N.	Tiruchirapalli, Salem, Coimbatore, Madurai, Tirunelveli, Tanjore, N. and S. Arcot	182.2 (Portland Cement Grade)
22.	Tripura	N. Tripura	
23.	U.P.	Mirzapur, Barabanki, Lucknow, Rae Bareli, Unao	119.59 (Portland Cement Grade)
24.	Uttaranchal	Nainital, Dehradun, Tehri Garhwal	
25.	W.B.	Purulia, Darjeeling, Jalpaiguri, Burdwan, Bankura	

Industrial uses

Limestone either as such or after calcinations as lime, has numerous industrial uses, principally in building, chemical and agricultural industries. The important uses are : crushed stone for ballast and as filler in concrete and asphalt; as a building stone; as a lithographic stone; in the manufacture of limes and cements; as a flux in metallurgical processes; in glass, ceramic, paper, textile and tanning industries; in the manufacture of calcium carbide (CaC_2) alkali and bleaching powder; for water purification and sugar-refining; as a fertilizer (calcium-ammonium nitrate) and soil conditioning agent in agriculture; and as whiting (chalk and precipitated limestone) used as a filler in rubber, oil, cloth, paint, cosmetic, toothpaste, shoe polish and other industries. Limestone is also used in mine dusting to prevent explosions and also for the manufacture of rock or mineral wool, a heat insulating material. The manufacture of metallic calcium is one of the newer uses of lime. Ca is used in reducing organic compounds, desulphurising petroleum, debismuthizing Pb, production of hard Pb-alloys and Ca-Si alloys and in the manufacture of calcium hydride which is used as an efficient hydrogen carrier.

Specifications

Lime is prepared by heating limestone in kiln upto 1000°C when CO_2 is driven off and 'quicklime' (CaO) remains in the form and hard white lumps, which when slaked with water and mixed with sand, forms mortar or plaster. 100 parts of pure limestone give 56 parts of lime (quicklime). Commonly the commercial lime is prepared as a dry hydrated lime $Ca(OH)_2$ by adding to quicklime the water (18 parts to 56 parts of CaO). The value of lime for most purposes depends upon its CaO (or CaO+MgO) content. The chief impurities SiO_2, Al_2O_3 and Fe_2O_3 present in most limestones are not detrimental for producing lime to be used in construction. When the percentages of SiO_2 and Al_2O_3 are large and silica is present as a silicate rather than as quartz, the limestone may be hydraulic limestone, and the product produced from it is 'hydraulic lime', which sets in water and is preferred to ordinary mortar or plaster.

For some uses of lime, iron impurities in limestone due to their staining effects and chemical activity are objectionable.

Limestone for the manufacture of ordinary Portland cement usually contains 45% and above CaO and <3% MgO. Limestone is usually mixed with clay in varying proportions (commonly 1 part in 3 parts of limestone) so that the mixture may be composed of 75% $CaCO_3$; 20% SiO_2, Al_2O_3, Fe_2O_3; and 5% other oxides (MgO, alkalies, etc.). The percentage of SiO_2 in the mixture should be about equal to that of Al_2O_3 + Fe_2O_3. The presence of S and P is harmful. Gypsum (about 0.05t to 1t of clinker) is mixed at the end in order to regulate the setting time. About 1.5 to 1.6t of limestone are required to produce 1t of cement.

Limestone is used in iron and steel industry as a flux to lower the melting temperature and also to remove SiO_2 and Al_2O_3 of the iron ore as slag. For use as a flux, limestone should be dense, massive, fine-grained and compact and should contain >90% $CaCO_3$ and SiO_2 + Al_2O_3 should be low. For blast furnace, MgO may be high but SiO_2 should usually be <5%, but for basic open-hearth steel furnace, MgO should be <4% and SiO_2 <1%, and S and P as low as possible. For steel melting the insolubles in limestone should not exceed 4%. The consumption of limestone in the blast furnace is about 0.5 t/t of pig iron produced and is about 0.5 t/t of steel produced in steel melting. In addition, about 0.2 t of dolomite is also required in the blast furnace to keep the slag fluid over a wider temperature range.

Limestone when directly used (instead of lime) for the manufacture of calcium carbide (CaC_2) must be of high grade and be low in magnesia and phosphorus and poor in other impurities. About 2t of limestone are consumed for each tonne of carbide produced. For the manufacture of bleaching powder or for use in paper, textile and alkali industries, a high-calcium limestone containing only traces of Mg, Fe and MnO_2 and free from clay and silica is used. About 0.9t of limestone is required for production of 1t of bleaching powder and about 1.4 to 1.6t for the production of 1 t of soda ash or caustic soda. For glass industry, high calcium limestone is usually used and presence of Mg is not objectionable, but Fe-oxide and other colouring matter like carbon are undesirable (Fe_2O_3 ≤0.04% and free carbon ≤0.1% for colourless glass; for bottle glass Fe_2O_3 may be upto 0.5% and organic matter 0.3%). For ordinary glass, generally 0.2 to 0.25t of limestone is required per 1t of glass manufactured. For ceramic use (as a constituent of certain pottery glazes and enamels), limestone should contain ≥97% total carbonates with ≤0.3% Fe_2O_3, 2% SiO_2 and 0.1% SO_2. In sugar industry, lime is used either to precipitate impurities from the juice or to precipitate sugar from impure solutions; limestone used in this industry must be high in lime and low in SiO_2, Al_2O_3, MgO and Fe_2O_3. For the production of metallic calcium, limestone should be very high grade and should produce a high grade lime containing the smallest possible percentages of Mg, Ba and Sr oxides and alkalies; and SiO_2, Al_2O_3 and Fe_2O_3 must also be low. Table 22.2 presents specifications of limestone for commercial purposes.

Table 22.2: Specifications of Limestone for Various Purposes

1. For Metallurgical Purpose :		(%)	
(a) For blast Furnace	CaO	44.5 (minimum)	
	SiO_2	Usually not >7	
	MgO	<4	
	Al_2O_3	<1.5	
	S and P	as low as possible	
	S	<0.5	
	P	<1.5	
Only dense, fine grained compact and strong block of limestone which would be non-fritting on burning is suitable for the above purpose.			
(b) For Steel Melting Shop	CaO	47.5 minimum	
	SiO_2	<1, (but in India up to 4.4% is permitted.	
	S and P	as low as possible	
	Insolubles	4% (maximum)	
2. Cement	CaO	42 (minimum)	
	MgO	2.5 (maximum)	
	SiO_2	14 to 15 (maximum)	
	P_2O_5	<1	
3. Lime	For Mortar and Plaster	MgO	as low as possible
		SiO_2	10% (maximum)

(Contd...)

4. Sugar		%	
	CaO	>80	The limestone should be high in active lime and very low in impurities such as magnesia, iron, aluminium and silica.
	MgO	<1.0	
	$Fe_2O_3 + Al_2O_3$	<1.5	
	SiO2 + Insolubles	<4.0	
5. Glass : For colourless glass		%	
	CaO	55.2 (minimum)	
	Fe_2O_3	<0.04	
	Free carbon (Organic carbon)	<0.1	
	Mn, Pb, S and P	Each should be <0.1	
6. Paper	CaO	>45.0	
	MgO	<3.0	
7. Textile Industry	$CaCO_3$	>94.0	
	MgO	<3.0	
	$Fe_2O_3 + Al_2O_3$	<2.0	
	Insolubles + Silica	<2.5	

8. Chemicals	Calcium Carbide		Bleaching Powder	Soda Ash
	$CaCO_3$	95 to 97	>96	90 to 99
	SiO_2	1 to 3	Up to 1.5]	Nil to 3% (maximum)
	Fe_2O_3	1]	]	
	Al_2O_3	–]	<2]	
	MgO	0.5	<2	Nil preferably
Remarks:	Dense, non fritting limestones; Dense massive limestone required		There should be no clayey matter or Mn or Fe in limestone	Limestone should be free of clayey impurities

Mining of Limestone

The limestone deposits in India are being worked by quarrying or open-pit method, as most of them occur either as horizontal or only slightly dipping strata, usually without excessive OB requiring stripping operation. It is, however, advisable to start UG mining in cases where the limestone outcrops in the scrap face of a hill and the thickness of OB goes on increasing as the quarrying operation proceeds into the hill, or in cases where the limestone strata are steeply inclined. In USA, room-and-pillar, shrinkage and glory-hole methods of UG mining of limestone have proved to be economically successful under varying circumstances.

The present requirement of limestone in India is about >100 MT. In-exhaustible resources of limestone are available in India, especially for the cement industry and practically all the states of India mine limestone (Table 22.3). Total resources of limestone as per UNFC (2005-06) are estimated at 175345 MT of which 12715 MT are under reserve category and 162630 MT are under remaining resource category.

Table 22.3: Reserve and Production of Limestone (After IBM, 2002)

Years	*Reserve (MT)*		*Production (MT)*
	Recoverable	*Conditional*	
1990-91	—	—	70.1
1998-99	76446	360	111.0
1999-2000	—	—	128.0
2000-2001	—	—	126.1
2001-2002	75679	701	130.0

23

Phosphorite Deposits

Phosphorites are one among the very few naturally occurring raw materials which enjoy fundamental scientific significance and great socio-economic importance. Recently, there has been great scientific interest in phosphorites by virtue of their complex petrography, chemistry and mode of formation.

Occurrence

Life requires phosphorus for its existence. Primary origin of all phosphates (whether mineral or organic) is apatite (fluor-apatite, $3Ca_2P_2O_5$ CaF_2 or chlorapatite, $3Ca_2P_2O_5$ $CaCl_2$; both containing about 41-42% P_2O_5). Apatite is the most abundant crystalline phosphate mineral and is found as an accessory mineral (in the form of crystal grains) in practically all kinds of igneous rocks, but is concentrated sometimes in pegmatites, metallic veins and magmatic deposits. It also occurs in metamorphic rocks (crystalline limestones, gneisses and schists) and as a secondary mineral in phosphatic rocks of sedimentary origin. Colour of apatite is green, bluish green, reddish-brown or yellow. If forms large or small hexagonal prisms. It has hardness 5 and sp.gr. about 3.2. Primary phosphorus from apatite in igneous rocks is supplied to soil by weathering agencies or carried by water to rivers and then to sea, usually in the form of phosphoric acid and its salts. In sea, phosphorus is largely removed by living organisms. It is specially concentrated in fishes and supplied by them to sea-birds and land life or deposited as phosphatic nodules or beds. From soil it is absorbed by plants and then it passes to animals which in turn return it in form of excreta and bones to weathering agencies, to be again carried to sea for its living organisms or redeposited as beds.

Many large deposits of igneous apatite are mined today for production of fertilizers and chemicals. There are many other phosphate minerals, but they are not of economic importance as the source of phosphorus, except a few ones like monazite from which phosphoric acid or its salts may be obtained as a by-product of another industry. Phosphatic guano (deposits of droppings of sea-birds living primarily on fishes on certain sea-islands) is used direct in fertilizers when it is desired to incorporate organic matter and nitrogen in addition to phosphate. About 90% of world phosphate production is derived from phosphate rock (phosphorite) which is a natural rock containing one or more phosphatic minerals, usually calcium phosphate [$Ca_3(PO_4)_2$], of sufficient purity and quantity to permit its use directly or after concentration in manufacturing commercial products. The term includes phosphatised limestones, sandstones, shales and igneous rocks. Phosphorite occurs as nodular phosphates, residual weathered phosphatic limestones, vein phosphates and consolidated and unconsolidated phosphatic sediments.

Distribution

Apatite occurs in many places in India, especially in mica-pegmatites (and sometimes in the intruded mica schists) of Jharkhand, Bihar, Rajasthan and A.P. mica-fields; but the best known apatite deposit is in Singhbhum, Jharkhand, where an apatite-magnetite-chlorite rock forms a series of lenses or veins, arranged en echelon in the rocks of Iron Ore Series, parallel to their strike and to Cu-belt of Singhbhum. Deposits occur along a belt extending SE from Itagarh (W-of Tatanagar) to Khejurdari (S-of Mosabani), a distance of about 58 Km and then extends into district of Mayurbhanj, Orissa. Proportions of apatite, magnetite and chlorite in the veins or lenses vary greatly. Biotite and quartz are present in some places while in other places the associated iron ore is largely hematite. Some veins contain white-coloured apatite almost free from magnetite and chlorite. Material of inferior grade can be improved by magnetic separation. Reserves from three important areas—Nandrup Sector (from Itagarh to Rajdiha), Patharghara sector (from Rakha mine to Dhobani) and Sungri Sector (from Mosabani to Khejurdari) average 10-25% P_2O_5 upto a depth of 30m. $Ca_3(PO_4)_2$ content in the ore of this area can be increased by beneficiation increasing P_2O_5 by 30%. Mica-peridotites of Giridih, Jharia and Raniganj coalfields in Jharkhand and W.B. usually contain apatite as an essential constituent and $Ca_3(PO_4)_2$ content ranges between 4.5 to 12.5% of the rock, which is, however, too small to make the rocks of economic importance.

In Srikakulam and Vishakhapattnam districts, A.P., apatite is a universal constituent of Kodurite Series of Mn-bearing rocks and often occurs in the lithomerge resulting from decomposition of these rocks. It is an abundant mineral in some of the Mn-mines. A lenticular patch of apatite and some apatite-rich veins traversing the country rock are located near Sitarampuram in this area and another area of apatite deposit is Kashipatnam in the same district.

Phosphatic nodules of lime phosphates (containing 58-59% $Ca_2P_2O_5$,16-20% $CaCO_3$ and upto 14% SiO_2, Al_2O_3 and Fe_2O_3) occur in Cretaceous (Uttatur Stage) beds of Tiruchirapalli district, T.N. The nodules are light yellow to buff-coloured and contain 15-20% P_2O_5 (average 18-20%). Phosphatic deposits also occur in South Arcot district and Pondicherry in Creatceous beds, but they are associated with sandy shales. Phosphorite deposits have also been located in Jaisalmer area, Rajasthan. Here phosphatic beds occur in Birmania, 120 Kms from Barmer railway station. In rock phosphate deposits of Jhamar-kotra area near Udaipur the reserves have 20-30% P_2O_5. Extensive deposits in the form of nodules and layers of phosphate rock occur in limestone and black shale-chert rocks near Mussoorie in Uttaranchal and contain 65-76% $Ca_2P_2O_5$. They occur in workable quantities in Maldeota, Kinoi and other areas in beds upto 6 m thick. The indicated reserves at Maldeota are 5.1 MT (including a proved reserves of 3.1 MT). It has been reported that a good deposit of phosphorite has been also discovered in Palamu district, Jharkhand.

Some apatite deposits of Singhbhum have been worked by IISCO and used in their plants for manufacture of high-phosphorous pig iron and some of the deposits for manufacture of fertilizer at Sindri fertilizer plant. Apatite deposits of Visakhapatnam and rock phosphate deposits of Tiruchirapalli districts are consumed for manufacture of superphosphate (containing 16% water soluble P_2O_5) fertilizer. India is in short supply of mineral phosphate and major requirements of phosphate fertilizers are met by imports. The first recorded production of phosphate rock from India on a commercial basis has been in 1969 from Udaipur district and Rajasthan became major producer of rock phosphate in India. Some phosphate is recovered as a by product from monazite of Kerala and is transformed into trisodium phosphate and used in wool, textile, natural silk and rayon industries. Some quantity of phosphate may also be recoverable as a by-product from secondary phosphatic minerals of uranium like torbernite and autunite found in Singbhum district. A source of phosphorous for use as mineral fertilizer exists in the basic slags formed in manufacture of steel. Large quantities of this slag (P_2O_5-10%) are dumped at steel works for want of demand.

Phosphorite Investigation in India

Phosphate rock has been known for many years in some of the sedimentary basins of India. Several discoveries in 1960s and 1970s have proved of economic importance although the country continues to remain dependent upon imports for most of its phosphate requirements. Scope for further finds of phosphate rock have probably not been exhausted and a much more co-ordinated exploration program is needed. Proterozoic basins offer the best exploration potential and currently account for the bulk of known domestic phosphate rock resources. Nevertheless, geologically younger basins should not be discounted in terms of their phosphate potential.

Geological ages for main phosphorite sequences in India are not well established and there is no precise stratigraphic correlation with Proterozoic-Cambrian phosphorite provinces in other parts of world. Since the discovery of phosphorite beds of Palaeozoic age near Dehradun (Uttaranchal) and Jaisalmer (SE-Rajasthan), considerable progress has been made in exploration and exploitation of phosphate deposits in India. One notable, promising development in recent years has been research in India on beneficiation of low-grade and highly dolomitic phosphate rock.

Occurrences of phosphate nodules in some sedimentary basins of S-Peninsular and N-Extra-Peninsular regions of India have been known for over a century. In 1884, phosphatic nodules and shales were reported by GSI from near Mussoorie in Uttaranchal (King, 1884). Many other discoveries have been made since that time and geology of main deposits and occurrences has been well documented (Banerjee, 1986; Choudhuri, 1989; and others). In relation to the great size of sub-continent there are relatively limited resources of phosphate rock of either sedimentary or igneous origin. Thus, country's proven resources continue to be insufficient to meet growing domestic demand for phosphate rock and India remains heavily dependent on imports not only of rock, but also of phosphate in the form of phosphoric acid and high-analysis fertilizers such as diammonium phosphate.

Some of the initial discoveries during 1960s and 1970s proved of considerable commercial importance and raised hopes that country's phosphate rock resource base might be augmented substantially. Additional geological information available gave support to the view that further discoveries were very likely. Unfortunately, such expectations have not yet been realized. Indian phosphate rock resources currently amount to only 190 MT (Table 23.1). Nevertheless, phosphate investigations carried out in India during 1978-1988, notably under the aegis of Project 156, have been characterized by intense collaborative effort between GSI, Universities and other Governmental and Private agencies in India, mainly through National Working Group, and with many participants from overseas. As a result, much progress was made in understanding geology of Indian phosphorites in terms of lithofacial variations, petrography, geochemistry, sedimentary environment and regional tectonic setting. Useful comparisons between ancient and modern phosphorites have been made and detailed studies of the palaeogeographical aspects of phosphogenesis carried out. Discoveries such as those of Udaipur in Rajasthan and Mussoorie in Uttaranchal were based on the recognition of favourable lithological associations. Depositional models, subsequently developed for phosphorite deposits of these two areas, have been found to be applicable to other phosphogenic basins. In spite of significant progress, large tracts of India remain unexplored in detail in terms of their phosphate potential and occurrences will undoubtedly continue to be found as systematic phosphate exploration proceeds.

Significant Phosphorite Deposits of India

Several significant phosphorite discoveries of Late Proterzoic age have been made in N-India since systematic exploration was begun in 1966 by GSI and other organizations. The best known of these are situated in Udaipur, Rajasthan as well as the deposits investigated by GSI in Lalitpur and Bijawar basins of U.P & M.P. and near Jhabua in Chhattisgarh (Fig. 23.1). Proterzoic phosphorites have been recorded from many parts of India, but occurrences are generally too small to be of commercial interest.

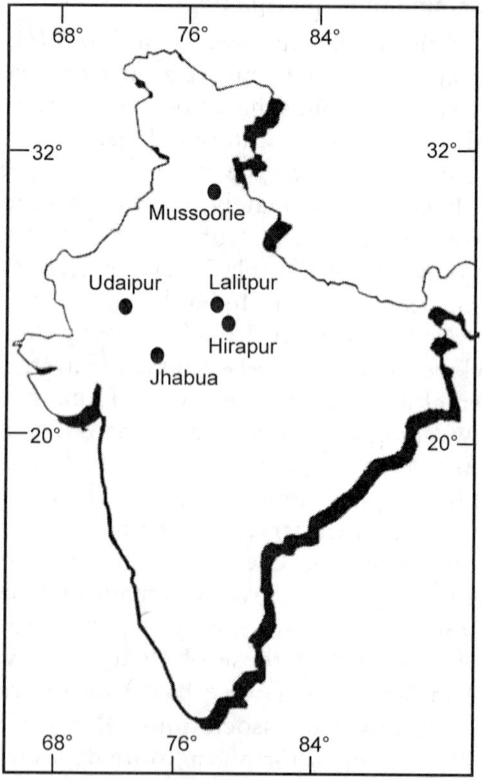

Fig. 23.1: Location of Major Phosphorite Deposits in India.

Table 23.1: Phosphate Resources of India

State/District	*Deposit*	*Grade* $\%P_2O_5$	*Reserves (MT)*	*Characteristics Value in %*
Rajasthan	Jhamarkotra	+30	16.8	High grade, associated with quartz (SiO_2, 8-12)
Udaipur	Jhamarkotra	25-30	5.0	High grade, associated with quartz (SiO_2, 12-20)
	Jhamarkotra	10-20	54.0	High dolomitic (MgO, 8-15, SiO_2, 3-5)
	Maton	+30	1.0	High grade, associated with quartz (SiO_2, 10-15)
	Maton	20-30	8.1	Siliceous phosphorite (SiO_2, 15-30)
	Kanpur	+30	0.8	High grade, associated with quartz (SiO_2, 10-15)
	Kanpur	20-30	1.0	Associated with quartz (SiO_2, 15-30), partly dolomitic (MgO, 2-5)
	Kanpur	11.6	6.2	Dolomitic phosphorite
	Kanpur	10-30	3.0	Dolomitic and siliceous
	Smaller deposits	6-9	1.0	Apatite in carbonatite
	Newania*			
Jaisalmer	Birmania	10-15	4.5	Calcitic phosphorite
Jaipur	Achraul	15-20	1.0	Siliceous, ferruginous phosphorite
Banswara	Sallopat	20-25	0.05	Dolomitie phosphorite
M.P.	Sagar-Chattarpur	-30	1.0	Siliceous phosphorite
Sagar, Chattarpur	Hirapur-Bassia	25-30	8.0	Siliceous-ferruginous phosphorite
	Mardeora-kachar	13-35	4.1	R_2O_3, 5-13
	Bassia	10-30	4.9	Siliceous
Chhattisgarh	Kelkua	10-30	3.2	Siliceous, dolomitic phosphorite
Jhabua	Khatamba	10-15	6.8	Dolomitic phosphorite
Uttaranchal	Mussoorie	+30	1.0	Siliceous, carbonate-rich phosphorite
Dehra Dun	Mussoorie	25-30	3.9	Siliceous, carbonate-rich phosphorite
	Mussoorie	10-30	26.3	Siliceous, carbonate-rich phosphorite
U.P.	Pisneri	10-30	23.9	Siliceous
Lalitpur				
A.P.	Kasipatnam*	15-35	3.0	Associated with magnetite
Visakhapatnam				
W.B.	Beldih*	15-30	1.5	Associated with magnetite
Purulia				
T.N.	Dharmapuri*	4-11	0.5	Apatite in carbonatite
Dharmapuri				

*Igneous deposit

Udaipur (Rajasthan)

By far the most important Proterozoic phosphorites in India are those forming Aravalli phosphogenic province, notably in Maton Fm or its stratigraphical equivalents. Maton Fm is best developed near Udaipur, Rajasthan, where phosphorites were first discovered by GSI in 1967.

Phosphorites, are essentially columnar, laminated, brecciated and fragmented stromatolites, intercolumnar portions of which being normally occupied by recrystallized dolomite. Stromatolites are characterized by alternate convex laminae of dark grey phosphorite and grey dolomitic limestone, all in a sheath of phosphorite. Total production of phosphate rock from Udaipur area is around 0.38 MTPA accounting for about 20% of India's total phosphate rock requirements.

At Jhamarkotra, SSE of Udaipur and the largest deposit of high-grade rock in India, phosphorite bed has an average thickness of about 15m and can be traced for a strike length of 16 Km, with a N-ward dip of 30-60°. Phosphorites around Udaipur are chemically similar (Banerjee *et al*, 1980), being characterized also by their crystalinity, grain sizes ranging from 10-50 mm. Resources are estimated at around 76 MT with >10% P_2O_5, of which some 17 MT is of high grade, 30% P_2O_5 or more (Choudhuri, 1989). A smaller open pit mine at Maton came into operation in 1972. The ore, which contains 20-28% and averages 25% P_2O_5, is beneficiated to obtain a flotation concentrate with 32-33% P_2O_5 and 8-12% SiO_2. Maton deposits are estimated to total about 9 MT over a strike length of 3.5 Km and to a depth of 200 m. Phosphorite horizon averages 4-7 m in thickness, with about 3 MT averaging 25% P_2O_5 being available to open pit mining to a depth of 100 m (Choudhuri, 1989). Reserves of high-grade and siliceous phosphorite amount to some 9 MT.

Jhamarkotra deposit has been divided aerially into 7-blocks, i.e. block A-G, from west to east. Block D is situated in the central part of outcrop. Open pit mining began in 1969 in block D, taking advantage of availability of high-grade ore suitable for fertilizer manufacture, the mine being capable of producing 1 MT of phosphate ore annually. High-grade ore has a relatively high silica content (5-10% SiO_2) and low magnesia (about 0.5% MgO). In contrast, low-grade Jhamarkotra rock generally contains <20% P_2O_5, with 2-5% SiO_2 and 7-14% MgO. To test this rock for eventual commercial use, a semi-industrial pilot plant came into operation in 1979. The most favoured beneficiation route is the froth flotation of carbonate and phosphate which has yielded a product with about 35% P_2O_5 and MgO content of 1.2-1.6%.

It is notable also that mining operations near Udaipur are confined to leached or weathered zones. These are highly variable in depth and result from pronounced and selective leaching of the more soluble carbonate fraction by meteoric waters circulating in structurally disturbed zones. Within these zones phosphatic stromatolitic columns have consequently been reduced to fragmented and powdery residual masses of high-grade ore. In the weathered zone, higher-grade rock generally contains higher SiO_2 and lower MgO, whereas unweathered rock is of lower grade, with low SiO_2 and very high MgO (Choudhuri & Balasubramanium, 1980). Thin encrustations and vein fillings of secondary phosphorite also occur locally in pockets underlying main phosphorite bed (Choudhuri, 1981).

Hirapur and Lalitpur (M.P. & U.P.)

Radioactive phosphorite was discovered by AMD in central India in 1976, within Proterozoic Bijawar Group (Banerjee *et al*, 1982). The group overlies Archaean Bundelkhand and consists of quartzites, shales and dolomites occurring in two isolated basins about 25 Km apart; Hirapur Basin in M.P. where Bijawar Group occurs over a strike length of about 80 Km. and Lalitpur Basin in southern part of U.P. where phosphorite was discovered in 1977. In Hirapur Basin, Sagar deposits have been worked since 1978 at the rate of about 22000 tpa. There are at least four distinct phosphorite horizons, each about 5 m in thickness and containing 15 to 35% P_2O_5 (Pant *et al*, 1989). Phosphorite deposits of Lalitpur Basin occur as lenticular and detached bodies ranging from a few metres to about 4 Km in length and up to 125 m in width. As in Hirapur Basin, there are four phosphorite horizons, the most important being the upper-most, generally brecciated horizon. Deposits contain 5-25% P_2O_5, averaging about 20% (Pant *et al*, 1979). However, in weathered and leached zones as much as 37% P_2O_5 may be present.

Jhabua (Chhattisgarh)

Pohosphorite occurring in a geological environment similar to that around Udaipur was first discovered in Jhabua, in 1973 (Banerjee & Basu, 1979). Economic potential of Jhabua deposits, which lie about 200 Km S-of Udaipur, has been investigated in detail. The area forms southern extension of the main belt of Aravalli rocks in Rajasthan, the rocks consisting mainly of phyllite, quartzite, and dolomitic limestone. Stromatolitic phosphorite is associated with dolomitic limestone as separate and elongated bodies up to 1500 m long. The phosphorite unit averages about 15 m. There are two distinct beds, the upper forming the main high-grade deposit of Jhabua district, averaging around 28% P_2O_5. Production of phosphate rock began in 1976, but there has been only limited output.

Marine, fossiliferous Palaeozoic rocks are well developed in Extra-Peninsular region, mainly in Kashmir, U.P. and H.P. Palaeozoic rocks occur also in W-Rajasthan in Jaisalmer Basin, a structure situated at NW margin of Indian Craton. Mesozoic and Tertiary sediments of the basin thicken to the west into Indus basin of Pakistan. The most extensive phosphate deposits are near Mussoorie, Uttaranchal about 20 Km NW of Dehradoon.

Mussoorie Syncline (Uttaranchal)

By far the most important Palaeozoic phosphate-bearing formation in India is the Lower Tal sequence in W-Kumaun Himalaya, Uttaranchal (Shanker, 1989). In Dehradoon and Tehri districts, near Mussoorie, beds of phosphorite have been mapped over a distance of 120 Km in Mussoorie Syncline, a doubly-plunging structure much dislocated by cross-faulting. Drilling, trenching and development of exploratory adits have led to delineation of 11 blocks which together have been estimated to contain about 45 MT of rock ranging from 16-18% P_2O_5 (GSI, 1981). Deposits in Maldeota and Durmala blocks near Dehradoon account for 10.64 MT averaging 19% P_2O_5 and 9.14 MT with 19.9% P_2O_5, respectively. Proved and probable reserves in the two blocks are 4.98 and 3.69 MT. Maldeota and Durmala deposits were brought into commercial production in 1970 and 1976, respectively. In recent years, production has exceeded 0.1 MTPA.

Distinctive lithology and easy recognition of contacts permit a division of Tal rocks into Lower Tal and Upper Tal Fm. Lower Tal, which is about 200 m thick, is further divisible into four members, depending on the dominance of chert, shale, sand/silt or carbonate in the sediments. It contains the main phosphate unit which is 10 m thick. Samples from phosphate unit contain averages ranging from about 8-19% P_2O_5 (by weight), but rock from weathered zones, such as those mined at Maldeota and Durmala, may analyse up to 33% P_2O_5, depending upon degree of alteration.

Beneficiation tests have shown that unweathered Mussoorie phosphorite, containing 17.8-19.5% P_2O_5, can be beneficiated by calcination to give a product with 31% P_2O_5 (Shanker, 1989). Similarly, a concentrate with 28-29% P_2O_5 has been obtained by flotation, but the content of 3% Fe_2O_3 is too high for its use in fertilizer manufacture. A major difficulty has been the liberation of phosphate grains from the predominantly carbonate matrix, with pyrites and organic matter hindering successful flotation. Currently, Mussoorie rock is finely ground and sold with 18-22% P_2O_5 for direct application to the soil as fertilizer.

Jaisalmer Basin (Rajasthan)

Mesozoic sequence in Jaisalmer Basin begins with continental Lathi Fm, which rests on Palaeozoic sandstones and is overlain by limestones, shales and sandstones of Jurassic and Cretaceous age, and a thin but relatively continuous shallow-water, marine sequence of Palaeocene and Eocene limestones and shales. Quarternary clastic rocks overlie middle Eocene rocks.

Among the earliest phosphate discoveries in this basin are those in Upper Palaeozoic Biramania Fm, an unfossiliferous formation exposed over an area of only 4Km^2 and comprising a folded sequence of limestone with sandstone quartzite sandstone, variegated shale, and dark shale believed to have been deposited during a Palaeozoic marine transgression over W-Rajasthan. Birmania rests, probably unconformably, on Randa Fm and is, in turn, overlain by Lathi Fm of Triassic-Jurassic age. Overlying quartzitic sandstone is a phosphorite series 0.56-9 m thick and averaging about 13% P_2O_5 that consists of thinly banded phosphatic calcareous shaly sandstones laterally grading into thinly banded phosphatic limestone.

Phosphate exploration and research in India over last 3 decades has provided valuable data on country's phosphate resources in terms of mode of occurrence, facies variations, structure, petrography, geochemistry, and palaeogeograpy. Most of the occurrences belong to late Proterozoic, the three main areas being those of near Udaipur, Jhabua and Hirapur-Lalitpur. There, phosphorite sections overlie Archaean basement rocks and comprise essentially phosphorite-carbonate sequences with basal quartzite units. These basins account for >80% of known domestic resources but others of Proterozoic age remain to be fully assessed for their phosphate potential, as do some of the geologically younger sedimentary basins. So, a systematic program of exploration and research is highly needed. For example, full potential of Cuddapah Basin in A.P. southern India, as a phosphogenic province has not yet been thoroughly evaluated. Within the basin, occurrence of a phosphatic quartzite horizon within Cumbum Fm of Cuddapah S.G. near Chelima, togther with the possible correlation of Cuddapah S.G. with phosphate-bearing, upper Proterozoic-Cambrian rocks elsewhere in India, suggests that further discoveries over 200 Km extent of the basin seem likely. Similarly, in Kutch Basin in Gujarat of NW-India, ferruginous sandstones and limestones of Bhuj Fm are reported to contain thin phosphatic concretions and nodules (Sant, 1980).

Unfortunately, in spite of considerable geological research on Proterozoic-lower Palaeozoic deposits in India, a major but unresovled problem centres on the absence of precise stratigraphical correlation, particularly with those deposits forming part of the vast Proterozoic-Cambrain Phosphogenic Province that embraces also Russian countries, China, Mangolia, and Australia. Udaipur phosphorites lie in the basal part of Aravalli geological event and have conventionally been considered to be of early Proterozoic age. However, study of stromatolite morphology has variously indicated a middle-upper Proterozoic or even post-late Proterozoic (Vendian) age (Choudhuri & Roy, 1986; Banerjee, 1986).

Similarly, age of phosphate-bearing Lower Tal continues to be controversial. Variously regarded as of Proterozoic, Permian, or Jurassic-Cretaceous in age, investigations by GSI have yielded microfaunas similar to Tommotian and Meishucun stages of Russian and Chinese platforms respectively, indicative of a lower most Cambrian age (Brasier, 1990). A lower Palaeozoic age is supported by well-documented Cambro-Ordovician conodonts from the phosphorite unit of Lower Tal (Azmi *et al*, 1981).

Economically one of the most notable developments in phosphorite research has been the successful beneficiation of low grade dolomitic phosphate ores. Proterozoic phosphorites are generally unreactive type and the dense and interlocking nature of constituent minerals, (including apatite, dolomite and quartz) usually renders such rocks difficult to beneficiate by physical methods. Recent agronomical investigation indicates that phosphorites are not too unreactive to be suitable for direct application as a fertilizer. However, considerable progress has been made in development of a commercial beneficiation process capable of providing good quality phosphate concentrates.

Types

Two types of phosphorite are normally distinguished :

(a) **Bedded, pelletal-oolitic-granular phosphorite:** This type occurs as continuous layers with phosphate pellets, grains, organic remains, pseudomorphs etc., bound in a cement which may be calcareous, sandy etc.

(b) **Nodular Phosphorite:** This variety comprises a wide array of phosphatic material ranging from tiny particules to pebble size, in a matrix of clay, quartz, carbonates etc. Some of the largest deposits of phosphorite of world fall in nodular category. They are considered to be formed mainly due to the reworking of older sediments.

Genesis

Complexity involved in the formation of phosphorite which represents a unique type of sediments has attracted the attention of geoscientists for a considerable period of time, and various theories based on one or more aspects of deposition and accumulation have been put forth. These can broadly be grouped into three main categories *viz.*, accumulation, replacement and precipitation.

The earliest theories attributed the source of phosphate to accumulation of coprolites (Sollas, 1873), planktons (Murray and Renard, 1891), fish remains following catastrophic deaths (Blackwelder, 1916) and residual accumulation at unconformities (Grabau, 1919).

The first departure from earlier biological source theories was by Mansfield (1920). He suggested that phosphorite was formed by progressive replacement of originally calcareous oolitic deposits, by the phosphate present in sea water in dissolved form. Emigh (1967) also put forward somewhat identical views. Replacement of calcareous deposits by phosphate in estuarine or shallow shelf environments was advocated by Pevear (1966).

Kazakov (1937), suggested a new theory of inorganic precipitation of phsophate from sea water. According to him, where cold phosphate-rich water from deeper levels of ocean rise to shallower levels by what is known as oceanic upwelling, phosphate will precipitate due to increase of pH and reduction in partial pressure of CO_2. Sequence of deposition under such an environment comprises calcareous sediments followed by a combined precipitation of $CaCO_3$ and $Ca_3\ (PO_4)_2$. He postulated a maximum precipitation of phosphate in pH range of 7.7 to 8.1 and at a depth range of 50 to 200 m.

Mckelvey *et. al* (1953) revised and expanded Kazakov's theory and suggested that independent precipitation of phosphate is possible and that it can be deposited prior to $CaCO_3$. They also envisaged a deeper range of deposition extending from 200 to 1000 m.

Basic concept of oceanic upwelling as the principal control of phosphate deposition has undergone considerable refinement over the years through intensive research and application of new ideas. These include : study of biological and physical parameters prevailing in the zones of upwelling, effect of varied intensities of biological activities in ocean, recognition of zones of dynamic upwelling and zones of upwelling caused by divergence identification and understanding of upwelling suite of rocks, etc. reconstruction of the whole palaeo-environment, clastic and chemical sedimentation, climate, wind and current directions etc.

Smirnov (1964) suggested that phosphate deposition was brought about not so much by internal changes in a cold phosphate-rich water mass rising to a different level in ocean by changes that occur as water masses from different sources meet and mix.

Sheldon (1964) on the basis of palaeomagnetic and palaeogeographic data envisaged the possible occurrence of ancient phosphorite between the equator and 40th palaeo-latitude. Some Russian geologists regarded climate responsible for depositions and accumulation of nodular phosphorite on continental shelf, under tropical climate on adjacent land mass and oolitic phosphorite in arid hinterland. Ronov and Korzhina (1960) indicated a broad cyclic increase and decrease in phosphate deposition through ages and suggested pronounced deposition in Silurian, Permian, Cretaceous and Tertiary periods.

Mansfield (1940) for the first time suggested that the source of phosphate in marine waters is submarine volcanism. Later, Miller (1964) suggested volcanic source for both, phosphate and fluorine.

There is considerable controversy in various views on the process of deposition, precipitation and fixation of phosphorite in sea water. For pre-Cambrian phosphorite of Rajasthan, Banerjee (1971) suggested that algal stromatolites played an important role in trapping and precipitating the dissolved inorganic phopshorus from sea water.

By and large Kazakov-McKelvey concept has been accepted by most and has been used successfully in locating many new deposits in different parts of world.

Prospecting Criterias

Ideas of genesis, environment of deposition, facies relationship etc., serve as useful guides in search for phosphorite. Considerable areas in various parts of world have been explored resulting in discovery of sizeable deposits. Deposits of Rajasthan can be quoted as an example of such a search in our country. Following aspects may be helpful in search of phosphorite occurrences in India.

Palaeogeography

Palaeogeographic considerations are useful for delineating possible phosphogenic areas. Most of the extensive deposits of phosphorite are formed in areas of upwelling by divergence. This takes place where surface waters are blown off-shore in the trade wind belts, and where off-shore current is augmented by Coriolis force. As surface waters move off-shore, cool, deep, PO_4^{-3} rich waters rise up to replace them. PO_4^{-3} is precipitated either inorganically or by a combination of organic and inorganic chemical precipitation. Thus, following belts can be considered most favourable for deposition of phosphorite (i) divergent upwelling areas along N, NW, S and SW coasts, between 0° and 40° palaeo-latitudes and (ii) divergent and dynamic upwelling areas along E and SE coasts between 0° and 40° palaeo-latitudes.

Sites of potential upwelling which can be identified from palaeogeographic maps of various periods and which have been found to contain large phopshorite deposits are N-coast of Africa (Miocene), northern part of India-Mussoorie area (Jurassic-Cretaceous) etc.

Tectonic environment

Study of tectonic environment of phosphorite deposition is also significant in the search. Tectonic elements can be identified from isopach and structural maps. Data now available throw light on the role of tectonic environment in determination of geographical distribution and character of phosphorite lithofacies. Most of the marine phosphorite deposits appear to have been formed in somewhat restricted basins, occupying bays or gulfs of pre-existing seas. Tectonically such areas formed either in neritic shelf areas of geosynclines or subsiding areas of relatively stable continental platforms.

Important phosphorite deposits of world in Karatau area of the then USSR, western part of USA and Mussoorie area of India are considered to have been formed in miogeosynclines. Sinian and Cambrian phosphorites of Siberia and Kazakhastan have been deposited in geosynclines while those of E-China on platform areas with mild relief. Most of the Mesozoic and Tertiary phosphorites of Russian platform are associated with depressions (synclines) in which favourable bethymetric conditions were created for deposition of phosphate (Bushinskii, 1969). In SE-USA, Ordovician, Carboniferous and Tertiary phosphorite are considered to be derived mainly from weathering and reworking of sandy and clayey phosphatic limestone, deposited on the flanks of a series of rising domes. Deposition of lower Eocene phosphorite in Algerian-Tunisian basin appears to have been greatly affected by development of a submarine ridge, south of which extensive phosphorite deposits occur on the flanks of synclines. Majority of Egyptian deposits are also associated with synclinal depressions or in shelter of a submarine anticline. In Nigev desert of Israel, the most favoured condition for formation of phosphorite was considered in the depressed areas of basin floors which formed before or simultaneously with the deposition.

Phosphate bearing formation of Udaipur area is considered to be related to facies changes and appears to have been formed in a somewhat restricted basin of deposition, occupying arms of pre-existing seas and developed in neritic shelf areas of Aravallies geosyncline (Banerjee, 1971). In Mussoorie area, (Uttaranchal) depth of water and basin topography are considered to be the main factors controlling the formation of phosphorite. Deposition probably took place in sheltered sea floor depressions or silled basin under shallow water conditions (Ravi Shanker, 1971).

Phosphorite lithofacies

One of the most effective aids to phosphate exploration is the identification of classic phosphorite-type phosphogenic province, which contains extensive marine phosphorite beds in the area of upwelling ocean currents. The upwelling suite of rocks with which phosphorite beds are commonly associated are carbonaceous shale, chert, limestone etc. With some variation, these lithoassociation characterises geosynclinal as well as platform deposition. In western part of the phosphate field of USA, phosphorite is associated with carbonaceous mudstone, cherty mudstone and chert, while calcareous rocks and sandy beds are generally absent. Mesozoic platform type phosphorite deposits of Russia are characterised by nodular and pebbly phosphorite with glauconite and arenaceous and clayey sediments. In Jordon, phosphorite is closely associated with limestone and marl. In pre-Cambrains of Udaipur area, phosphorite is associated with the sequence of grey phyllite, marble, jaspery quartzite and quartzite, while in Mussoorie area associated rocks comprise argillaceous limestone, chert, dark grey shale etc.

Phosphate bearing sediments are generally located in the zone of low rate of deposition in the basin. The resulting phosphorite lithofacies is therefore thin compared to the sedimentary sequence deposited elsewhere during the same interval. As such, a relatively thick sequence of limestone, shale, chert etc. containing low grade phosphorite bands may pass laterally into a much-reduced thickness of high grade phoshorite. Such condensed sequence of phosphorite have been noticed in USA, Morocco and Mussoorie belt, India, etc. Phosphorite lithofacies have also been found to grade laterally into other lithological types and on a regional scale show transgressive relationship.

One of the classic examples of the use of this criterion for phosphorite exploration is discovery of phosphogenic province in Australia (Russell, 1968).

Petrography

Phosphorite is not a distinctive rock, and as such, may be overlooked in hand specimens or outcrops. However, within the same phosphogenic province, its characters are quite constant and once identified as phosphorite can be recognized without difficulty.

Phosphorite from western USA, when fresh, is hard, dark brown to black in colour, comprising colloform carbonate fluorapatite cemented with detrital silicates, calcite, dolomite, chert, carbonaceous matter etc., while that of N-Africa is sandy, yellowish to light grey and friable, consisting of rounded to ovoid pellets of carbonate fluorapatite and bone fragments in a matrix of calcite, foraminite and clay minerals. Phosphorite from Mussoorie is dark grey, while that of Birmania light yellow to dark grey in colour. In Precambrians of Rajasthan and M.P. the phosphatic material is associated with bluish grey stromatolites.

Phosphorite in the field can be easily identified by its pelletal texture and a bluish grey coating or phosphate bloom on weathered surface. A drop of HNO_3 followed by ammonium molybdate gives a yellow precipitate with phosphatic rocks. A few drops of conc. HCl produce a white spot with a phosphatic rock. A common semi-quantitative method of Shapiro (1952) is also used in the field in which a standard solution of ammonium-vanado-molybdate in HNO_3 base is used.

Other criterias

Marine phosphate in many cases contain 0.005 to 0.02% U. Radioactivity can therefore be measured by means of G.M. Counter or Scintillometer. This method can however be applied only in the areas with little OB. Technique of γ-ray logging of drill holes has also been successfully used for discovery of new phosphorite deposits in Australia, Turkey and other Middle East countries.

In Florida (USA) the streams that drain the acid phosphatic terrain contain anomalous amounts of P_2O_5 and U in solution. This technique will be effective only in areas of acidic GW.

In N-Carolina, USA, the artesian water associated with phosphorites contains a mark concentration of I and Br. Such anomalies can also be used in locating individual buried phosphorite deposits.

Chronological Account of Indian Phosphorite Deposits

The earliest reference to the occurrence of phosphatic nodules made by W. King in 1885 is from the Lower Tal rocks of Mussoorie area, the next occurrence recorded was that of the phosphatic nodules in the Cretaceous rocks of Tiruchirapalli (T.N.). It is, however, the discovery of phosphorite in the Palaeozoic Birmania Formation of Rajasthan by Sheldon of the USGS during 1966 that has given a great impetus to the search of phosphorite in India. A country-wide search by GSI led to the discovery of large deposits of phosphorite in the Precambrians in Udaipur, Rajasthan.

A brief chronological account of various Indian phosphorite occurrences and deposits has been mentioned here.

Precambrian

The latest concept of phosphate geology was applied in the search for phosphorite in Aravalli region near Udaipur, Rajasthan. In this area Aravalli rocks rest unconformably over Banded Gneissic Complex (BGC) and comprise basal conglomerate and quartzite, followed by a sequence of carbonaceous phyllite, pure and impure dolomitic marble, jaspery quartzite, quartzite etc. which is subsequently overlain by a sequence of phyllite, greywackes and quartzite; thus broadly representing near shore littoral, unstable shelf and deeper water environment respectively, prevailing for a certain duration in the evolution of Aravalli geosyncline. The environmental and facies characters (upwelling suite of rocks) of the middle sequence were considered ideal for phosphate deposition. The search resulted in discovery of phosphorite near Kanpur, Maton, Kharbaria-ka-Gurha, and Dakan Kotra in Udaipur, Rajasthan, in 1967. This was followed by a discovery of phosphorite near Jhamarkotra and smaller deposits near Sisarma, Neemachmata, Badgaon, etc. in Udaipur near Sallopat, Ramka Munna and Jharmoti in Banswara, Rajasthan.

In all these places the phosphatic material is associated with bluish grey stromatolitic structures either in dolomitic marble or cherty quartzite.

The largest deposit of phosphorite in the country is located near Jhamarkotra, about 28 Km SE of Udaipur. It extends over a strike length of about 12 Km with width varying from a few metres to 30m. Although major part of this deposit consists of low to medium grade phosphorite, associated with dolomitic marble, a part of the deposit extending over a strike length of nearly 1500 m comprises high grade material containing over 31% P_2O_5. This is associated with jaspery quartzite. A reserve of 28.5 MT (indicated category) of grade 15 to 35% P_2O_5 was estimated from this deposit, and it is supplying ore to various fertilizer plants. Further explorations have also indicated large reserves.

The deposit near Maton is 8 Km SW of Super-Phosphate plant of Hindustan Zn Ltd. (HZL) near Debari. The phosphorite horizon is traceable over a strike length of 3.5 Km., with thickness varying from 0.5 to 20m. and the major part is associated with brecciated jaspery quartzite. Considerable exploration has been done by GSI and HZL, indicating reserves of about 9.4 MT with 20 to 30% P_2O_5. This mine has been developed for production by HZL.

The deposit near Kanpur is situated 10 Km east of Udaipur. Phosphorite is associated mainly with dolomitic marble, and as such is high in carbonate content. Thickness of phosphorite horizon varies from 2 to 6 m. Rocks of the area are folded into anticlines and synclines. The reserves estimated from western limbs of syncline were put at 4 MT of 10 to 12% P_2O_5. Besides this, several smaller deposits of phosphorite, are also located in Udaipur and Banswara, Rajasthan. But in most cases the reserves are small and the grade medium to low.

The search extended to south of Banswara, into Jhabua district resulted in discovery of phosphorite occurrences over a distance of 9 Km between Amalmal and Piplod. In this belt phosphorite occurs in the form of bands and lenses within stromatolitic dolomite found associated with chert of Aravalli age.

Occurrences of phosphatic stromatolites associated with Gangolihat dolomite formation of late pre-Cambrian Calc. Zone in Pithoragarh, Uttaranchal which were reported by Valdiya in 1969, are of no economic importance.

A search in lower units of Vindhyan Super Group, where marine conditions of deposition apparently existed has also revealed the presence of phosphatic horizons in Nimbahera shale of Chittorgarh, Rajasthan, in turbidite layers associated with limestone and glauconitic sandstone near Paisuni Dam site, Banda, U.P.; and calcareous shale, glauconitic sandstone sequence in Shahabad and Palamu, Jharkhand.

Palaeozoic

Phosphorite occurrences have been reported from palaeozoic rocks located near Birmania, Jaisalmer in folded unfossiliferous Birmania formation, comprising limestone, chert, shale, sandstone, etc. The phosphatic material is associated with the calcareous sandy phyllite as pellets and bands within limestone. Reserves of phosphorite from this area have been estimated at 4 MT averaging about 10% P_2O_5.

Jurassic

Phosphatic nodules and shales are known from Mussoorie area since 1885. The phosphorite zone is confined to bedded chert and shale of chert member (Lower Tal Fm) in Mussoorie Syncline. It is traceable for neary 120 Km, with thickness varying from mere streaks to >150 m. Phosphatic horizons are better in grade where Lower Tals are less developed. There are 10 blocks ranging from 300 to 6500 m in length, containing significant horizons, upto 15 m thick, with P_2O_5 content varying from 15 to 25%. The phosphatic material is of granular, pelletal, lenticular and nodular varieties, embedded in a calcareous, siliceous and cherty matrix.

Of the more promising deposits in this area, the one at Durmala consists of a 3 to 9 m thick phosphorite band, extending over a strike length of 1.6 Km with possible reserves estimated at 4.58 MT averaging 27.8% P_2O_5. In Maldeota W-block, the zone extends over a strike length of 1.35 Km with an average thickness of 5 m. Reserves for this block have been estimated at 5.13 MT averaging 18.6% P_2O_5 in Paritiba-Chamasari block, possible reserves, of 1 to 3.8 m thick band extending over a total length of 3.6 Km, are of the order of 2.79 MT with P_2O_5 content varying from 18 to 31%. Phosphorite horizons have also been traced in Bhusti-Jhamtialgaon, Jhlikhal, Mathiangaon Bagi, Kimoi, Nagani, Chaunpa-Kumali and other blocks, but the reserves are small. Total reserves from all the blocks have been worked out at 18.8 MT.

In a similar set up phosphatic zones have also been reported from Korgai and Nigalidhar synclines, in Sirmur district, H.P. They are associated with chert, cherty and carbonaceous shales of Lower Tal Fm. However, they have only limited length and thickness.

Minor phosphate bearing horizons also occur in ferruginous sandstone member of Bhuj Series near Godpar in Gujarat. Similarly the oolitic ferruginous limestone of Chari Series and Upper Patcham ferruginous gritty sandstones of Gujarat are also feebly phosphatic at places.

Cretaceous

Occurrence of phosphatic nodules in Cretaceous beds of Uttatur stage of Tiruchirapalli, T.N., is known since long. The phosphatic nodules occur down to a depth of 15 to 18 m, over an area of about 26 Km^2 between Varagupadi in north and Nambakkurichi in south. Probable reserves of phosphatic nodules from this area, down to a depth of 15m have been estimated at 0.12 MT, with P_2O_5 content varying from 21.1 to 26.6%.

Upper part of Ariyalur Stage also have phosphatised molluscan shells, in S-Arcot district, T.N. and in Pondicherry. The phosphatic zone is several metres thick and occupies an area of nearly 82 Km^2. These deposits are not of economic significance.

Tertiary

Lower portion of Fatehgarh Fm of upper Createceous to Eocene age, exposed near Fatehgarh in Jaisalmer, Rajasthan has a 0.5 to 2.5m, thick phosphatic horizon. The phosphatic material occurs as pellets, oolites, phosphatic clay etc. and as replacement of shells in ferruginous sandstone. Though the horizon extends over a strike length of over 20 Km the grade is poor (5 to 12% P_2O_5), with Fe content.

Upper Cretacoues-Tertiary sediments of Garo and Jaintia Hills, Meghalaya contain cherty phosphatic nodules and phosphatic marly limestone. Nodules occur in a shaly matrix in transitional rocks between the uppermost member of Sylhet limestone and Kopili Fm. The grade is poor and reserves are limited.

Phosphorite nodules have been reported from Eocene rocks, below Murrees, J&K and near Durla along Simla-Bilaspur road, H.P. Phosphorite horizon, varying in thickness from a few cm to over a metre, has also been recorded from the limestone of Subathu Fm, near Bakhlog, Mahasu (H.P.). All these occurrences are of minor importance.

Significance of Phosphorites

Phosphorites are significant specially in view of their great importance to fertilizer industry. Substantial increase in agricultural production is necessary to feed huge population as well as for successful implementation of plans for economic growth of the country. This problem can be overcome to some extent, with increasing use of different types of fertilizers.

Phosphorus, nitrogen and potash are three main plant nutrients. Phosphorus plays an important role in root development and in the synthesis of proteins, fats and carbohydrates. Plant nutrients are ubiquitous in most soils but not necessarily in the proportion and quantities required by the growing plants. The deficiency is either natural or is caused by intensive cultivation. It can be compensated by adding natural or manufactured fertilizers.

In earlier times the deficiency in phosphorus used to be compensated by adding bone-meal, the raw material being animal or human bones available from slaughter houses or battle fields. Raw bone-meal fertilizer is now completely replaced by chemically prepared phosphatic fertilizer for which major sources are rock phosphate.

The term rock phosphate includes both, (a) apatite from vein deposits, and (b) marine phosphorite sedimentary rocks. Phosphorite, contributes >90% of present world requirement of phosphatic fertilizer industry.

The resources of rock phosphate in world are immense. Morocco, Tunisia, Russia, USA and other countries make up most of the presently known resources. By comparison, our known indigenous resources are negligible, both as regards reserves and production. The first ever regular production of phosphorite started in 1969 from Jhamarkotra area, Udaipur (Rajasthan) meeting nearly 10% of country's requirement. But for this deposit and second deposit of Maton, also in Udaipur district, our requirements either in the form of rock phosphate or phosphatic fertilizers used to be met by imports. Consumption of phosphatic fertilizer in terms of P_2O_5 requirements during 5th and 6th Five Year Plan periods was 2.2 and 3.75 MT, respectively. As against this, reserves of all deposits of all grades during that period in India, was of the order of about 75 MT. In terms of P_2O_5 this was equivalent to about 15 MT only assuming an average 20% grade for all the deposits put together.

Except for a part of Jhamarkotra deposit in Rajasthan which is of high grade, ore from other deposits is of medium to low grade, requiring beneficiation before it can be utilized in fertilizer manufacture. It is, therefore, imperative that efforts must be intensified to discover more phosphorite deposits in different parts of the country for self-sufficiency. Some of the recent discoveries in India reveal that it may not be difficult to find new deposits by applying modern techniques and concept of phosphate geology. Therefore, we have to dwell upon some of the aspects, which may be helpful as guides in the search of new deposits of this important mineral.

The term phosphorite is commonly used for marine phosphate bearing rock. It, however, has no well-defined lower limit of P_2O_5 but should normally contain >5% by convention (Bushinkii, 1969).

Grading. There is no specific grading of phosphorite as it is interrelated with other factors. However, conventionally the phosphate bearing rocks can be classified into 3 grades (Table 23.2).

Table 23.2: Grades of Phosphate

P_2O_5%	*Grade*
>31	High
15-31	Medium
<15	Low

Phosphatic mineral in phosphorite is usually a carbonate Fluorapatite Ca_{10} $(PO_4\ CO_3)_6$, $F_{2\text{-}3}$, though locally Francolite $(Ca_2F)_2$ $(PC)_4$ $(O_3OH)_{24}$ $(CaC)_4$, Dahllite 5Ca $(PO_4)_2$ $CaCO_3$ etc. have been reported. The bluish phosphate bloom on phosphorite is possibly due to Wevellite Al_2 $(PO_4)_2$ $(OH, F)_3$ $5H_2O$ and Augelite Alg (PO_4) $(OH)_{12}$ which are common products of weathering.

Many interrelated factors contribute to economic exploitation of phosphorite deposit. These include their grade, tonnage, accessibility, physical characteristics etc. Medium to low grade deposits if amenable to economic beneficiation can also be worked.

Major part of phosphorite goes into manufacture of chemical fertilizers and to a lesser extent of elemental phosphorus. The acid grade phosphorite, used in manufacture of fertilizers should have specifications as shown in Table 23.3.

Table 23.3: Acid Grade Phosphorite

Constituents	*%*
P_2O_5	>31.5
SiO_2	4-6
P_2O_3	<4
CO_2	4
F	<4

The furnace grade material for elemental P has to confirm to the specifications as shown in Table 23.4.

Table 23.4: Furnace Grade Phosphorite

Constituents	*Ratio*
P_2O_5 : C	2.3-2.6 : 1
SiO_2 : CaO	0.8-1 : 1
Cao : P_2O_5	1.48 : 1

Most of the Indian phosphorites are of low to medium grade containing 10 to 25% P_2O_5, and can not be used without beneficiation for manufacture of various industrial products.

Beneficiation : These techniques depend upon the nature of objectionable impurities in them. There are two distinct types of Indian phosphorite, (i) with high siliceous gangue and (ii) with high carbonate gangue. Beneficiation techniques for the two are therefore quite different.

(i) Under the first type, typical example is the phosphorite from Maton deposit, Rajasthan, which has a P_2O_5 percentage varying from 22 to 28 and SiO_2 from 20 to 40. Beneficiation process of this rock is simple and comprises crushing to required size, washing and fatty acid flotation. This process can produce the concentrates assaying over 35% P_2O_5, 8% SiO_2 with a recovery of over 90%. HZL has erected a beneficiation plant at Maton. The concentrates are being used in its superphosphate plant at Debari.

(ii) High CO_3^{-2} phosphorite is available from a part of Jhamarkotra, Kanpur and several other deposits of Rajasthan and Mussoorie, Uttaranchal. Treatment for beneficiation in this case includes crushing to desirable size, roasting at 950° to 1000°C quenching in water, attrition, scrubbing, followed by classification or flotation. By this method, Rajasthan phosphorite is being beneficiated for 34 to 35% P_2O_5 content with about 75% recovery. Mussoorie rock, however, gives a marginal grade of 29 to 32% P_2O_5. Cost of beneficiation in the case of carbonate rich phosphorite thus works out to be comparatively more and recovery of P_2O_5 poor.

Reserve and Production

Phosphorite is the major resource of phosphate in India. These are located in M.P., Rajasthan, Uttaranchal, U.P., and

Gujarat. Apatite is also available, though in a very poor quantity. Apatite deposits of commercial importance are reported from Jharkhand, W.B., A.P., T.N., and Rajasthan. Total resources of apatite as per UNFC system (2005-06) in the country are placed at 26.86 MT, out of which 6 MT are reserves, 61% of which are located in W.B. Total resources of rock phosphates, are placed at 305 MT, out of which 53 MT are placed under reserves and 252 MT under remaining resources category. Bulk reserves are located in Rajasthan. Reserve and production of phosphorite and apatite in India are shown in Table 23.5.

Table 23.5: Reserve and Production of Phosphorite and Apatite

Years	*Phosphorite Reserve*	*Apatite Reserve*	*Production*
1990-91	—	—	0.6
1998-99	146.95	13.20	1.27
1999-2000	—	—	1.20
2000-2001	—	—	1.36
2001-2002	145.37	13.64	1.07

Industrial uses

Principal use of mineral phosphates is for manufacture of ordinary superphosphate, triple superphosphate and other phosphate compounds for fertilizers, or after grinding for direct application to soil or as fillers in mixed fertilizers. Superphosphate is the product made by mixing phosphate rock with sulphuric acid, whereas triple superphosphate is made by mixing phosphate rock with phosphoric acid. The remainder of mineral phosphates are used for making elemental phosphorus, phosphoric acid and many phosphorus compounds and in manufacture of high-phosphorus pig-iron and ferro-phosphorus. Elemental phosphorus is used for military smoke-screens, incendiary bombs, shells, grenades and tracer bullets; in manufacture of matches, vermin exterminators (pesticides); and in phosphorus alloys. Phosphoric acid is used in production of fertilizers, animal feed supplements, beverages, in ceramics as a catalyst, as an oil-refining agent, in photography and dental cements. Elemental phosphorus and phosphoric acid are used in textile industry, in baking powders, water-softening and cleaning products, rust-proofing and fire-proofing compounds, sugar-refining and in manufacture of glass.

Phosphate rock for manufacture of superphosphate is usually guaranteed to have a minimum content of 27% P_2O_5 or 63-65% of tricalcium phosphate ($Ca_3P_2O_5$); the most undesirable impurities are oxides of Fe and Al which together should be as low as possible and not more than 3%. Silica should also be as low as possible. Presence of $CaCO_3$ between 2 to 4.5% is desirable but large percentages (about 10%) are undesirable due to extra consumption of acid. For manufacture of phosphoric acid by wet process, phosphate rock should usually contain 30-32% P_2O_5.

Experiments by various agricultural laboratories in the country, to determine relative efficiency of low to medium grade phosphorite from Mussoorie for direct application in the field have been quite successful. Such application in combination with N_2-fertilizers can be best in lateritic, acidic and upland soils of Gangetic region. Weak phosphatic limestone of Palamu district, Jharkhand has also been found useful for direct application to acidic soils.

Probable Phosphogenic Regions

Various theories of phosphate formation, important criteria and guidelines for search and the success achieved in locating some new deposits in the country have been discussed earlier. The search should not be stopped but continued for locating more phosphate deposits of better grade and size.

Unfortunately updated relevant informations on palaeogeography of some of the geological era, for demarcation of possible phosphogenic provinces in the country are not available completely. However, making use of available data and applying various clues of phosphorite deposition, various geological formations in different parts of the country, suitable for localisation of phosphorite, can be demarcated.

Precambrians

Dharwars and their equivalents: Precambrian metasediments of Indian continent are Dharwars and their equivalents. Though no detailed palaeogeographical, palaeotectonic and other data are available regarding the mode and nature of deposition of metasediments, in some areas they show litho-association favourable for phosphorite, especially along possible palaeo-shore lines. Udaipur phosphorite deposits in Aravallis are the best examples of Precambrain phosphorite in the country. Following areas can be looked for some new and large deposits:

1. Basal rocks of Aravalli system in Rajasthan and their equivalent, Chapaner Series in Panchmahal (Gujarat) and Jhabua (Chhattisgarh).
2. Chlorite schist, argillite, limestone-dolomite, quartzite, etc. of Dharwar-Shimoga-Chitradurga and N-Kanara belts (Karnataka).
3. Mica schist, phyllite, carbonaceous phyllite and limestone of Gangpur Series, Sundergarh (Orissa).
4. Rocks of Kolhan Series, Singhbhum (Jharkhand).
5. Rocks of Daling Series, Darjeeling (W.B.).
6. Quartzite, conglomerate, phyllite, sericite-chlorite mica and hornblende schists of Shillong Series (Assam).
7. Carbonaceous slate, schist and carbonaceous limestone of Chail Series in Simla and Chaur areas (H.P.) and Salkhalas of Kashmir.

Upper Precambrian

(A) Delhi-Cuddapah and their equivalents: Reconstruction of Upper-Precambrian palaeogeography is rather difficult due to uncertain correlations of various rock formations. Broadly speaking one can reconstruct two entirely disconnected shallow seas in Indian Continent, (i) in south covering Bastar, (Chhattisgarh) and Cuddapah areas and (ii) in north covering Delhi basin in Rajasthan and its probable continuation in an E-W direction from Hazara (Kashmir) to as far as Nepal.

In Cuddapah and their equivalents such as Delhis etc., favourable litho-associations for phosphorite are:

(a) Rocks of Vempalle, Tadpatri and Cumbum stages in Cuddapah and Kurnool; cherty limestone, quartzites and micaceous argillites of Cuddapahs exposed to the east of Khammameth; and the rocks of Pakhal stage exposed in Warangal and Khammam (A.P.).

(b) Basal members of Kaladgis in Yadwad-Kaladgi-Bagal-Kot areas (Karnataka).

(c) Penganga beds in Yeotmal and Chanda (Maharashtra) and Adilabad (A.P.).

(d) Upper horizon of Cuddapahs in Nowrangpur and Jagdalpur areas, Koraput, (Orissa) and Bastar, (Chhattisgarh).

(e) Rocks of basal Raipur Series occupying southern part of Chhattisgarh basin.

(f) Basal Bijawar rocks in the area between Dudhi (U.P.) and east of Katni (M.P.) along Sone Valley.

(g) Bijawar rocks in Jabalpur-Katni, Barawah-Nimawar areas, and those fringing Bundelkhand Granite, east of Gwalior and Bijawar areas (M.P.).

(h) Ajabgarh rocks comprising phyllites, schists and impure limestone bordering Aravalli and banded gneissic craton extending from Ajmer (Rajasthan) to Khed Brahma (Gujarat).

(B) Vindhyans: Depositional characters of Lower Semri Series, Lower Vindhyan rocks indicate that they are marine deposits, formed in areas bordering a craton. This, coupled with suitable litho association, makes them a favourable unit for search of phosphates. Such associations are seen in following:

(a) Rocks of Lower Semri Series of M.P., U.P., Bihar and Rajasthan.

(b) Rocks of Bhima Series, Wadi-Shorapur area (Karnataka).

(c) Deoban limestone (often stromatolitic and oolitic in H.P. and U.P.).

Palaeozoic

During Cambrian-Silurian times, Himalayan region was the site of Tethys, which extended from Europe and Arctic, through Myanmar and China. There also appears to be an extension of this sea in Sindh area (Pakistan) and W-Rajasthan (India). During Devonian, there was regression of this sea, but during Permo-Carboniferous period, sea started transgressing and occupied an immense tract, now forming Himalayas, Tibet and part of China. Northern part of Gondwana land formed the shore region of this sea. Disposition of Indian subcontinent during this period was probably between 30°N and 60°N palaeo-latitude. During Permian period, marine conditions might have been favourable for phosphate deposition, on off-shore region of Gondwanaland. Findings of phosphorite near Abotabad in Pakistan, in Sirban Fm and existence of large deposits of phosphorite in Kazakstan, in rocks of Palaeozoic age, indicate a great possibility of existence of such deposits in India under similar set up.

Rock formation suitable for localisation of phosphate during various periods are :

Cambrian

(a) Fossiliferous shales, and dolomite of Upper Haimanta and dark slates of Lower Haimantas, Spiti (H.P.)

(b) Dark slates with quartzose parting, greywackes, thick bedded clays and sandy slates, with intercalated oolitic and pisolitic limestones and sandstones in Handwar area (J&K).

Ordovician-Silurian

Upper beds of Ordovician in Spiti (H.P.) and Lidar valley (J&K).

Devonian

Rocks of Muth Series and Spiti, lower shales of Po Series, in Kangra (H.P.) and Fenestella Series in Jhelum Valley (J&K).

Carboniferous-Permian

(a) Syringothyris limestone and black shales and quartzites exposed between Bundil and Banihal Pass, Anantnag area; Great Limestone of Riasi area; Productus shales horizon in Vihi and Gulabgarh pass areas (J&K).

(b) Deoban limestone and shales and limestone of Mandhali Series, limestone and slates of Blainis, dark shales and slates of Infra-Krol, (H.P.).

(c) Calcareous sandstone, and black often siliceous shale (productus shale) belonging to Kuling system of Spiti-Bashahr area (Punjab and H.P.).

(d) Extension areas of Birmania Fm comprising black shale, chert, calcareous shale, sandstone etc. in W-Rajasthan.

Mesozoic

During Triassic and most part of Jurassic times, Tethys sea occupied more or less the same area as during late Palaeozoic Era, but towards the closing stages of Jurassic, it started receding northwards, except in Garhwal area, where marine fossils of Jurassic age are found in Tal Series. In Peninsular India, there was at this time a sea occupying Kutch, Kathiawar and W-Rajasthan connecting with Tethys through Salt range.

During Cretaceous, the gulf that occupied Kutch are extended in land along Narmada valley, while on E-coast waters covering Bay of Bengal transgresses over Coromondal coast, and again during late Cretaceous these waters swept in N-E up to Shillong plateau. During Triassic, Jurassic and Cretaceous times, disposition of Indian continent was respectively between 25° and 40°N, 33° and 35°N and 28° and 52°N palaeo-latitude. Sedimentary characteristics of favourable suites of rocks found in Triassic, Jurassic and Cretaceous systems, indicate that they were of shallow water origin, deposited close to cratons. These favourable suites of rocks are:

Triassic

(a) Lower triassic rocks of spiti, (H.P.) and Garhwal (Uttaranchal).

(b) Black shales, grey to dark limestone, sandy limestones in Sona Marg-Gurias Belt and Vihi-Anantnag area (J&K).

Jurassic

(a) Limestones, sandstones, marls and shales, of Patcham Series; limestones and shales of Chari Series and sandstones, phosphatic grits and dark shales of Katrol series; Kutch (Gujarat).

(b) Rocks of Tal Series comprising carbonaceous shales, dark greywackes, black chert beds and dark limestone exposed in H.P. and Uttaranchal.

(c) Spiti shales in Spiti area (H.P.).

Cretaceous

(a) Glauconitic sandstone, limestones and shales of Chikkim Series in Balcha Dhura, Niti Pass, Chitichur, Sangeha Malla and Kiogar of Kumaon Himalayas, (Uttaranchal) and Spiti area (H.P).

(b) Cretaceous rocks of coastal area of T.N.

(c) Bagh beds in Rajpipla area (Gujarat); Barwah-Jhabua Alirajpur (Chhattisgarh). Also Cretaceous rocks of Wadhwan area (Gujarat).

(d) Shales and limestones of Mahadek stage and Langpur stage of Khasi and Jaintia hills (Meghalaya).

(e) Cretaceous rocks of Nicobar Islands and also rocks with diatomaceous and radiolarian cherts in parts of Andaman Islands.

(f) Abur shale and limestone of Jaisalmer (Rajasthan).

Tertiary

During upper Cretaceous-lower Eocene, as a result of Himalayan orogeny, Tethys Sea started shallowing. During the same period, in east northern part of Bay of Bengal got split into two gulfs separated by an archipelago (now Arakan Yoma), the eastern one extending into Myanmar and the western one into upper Assam. During middle Eocene, large W-Indian gulf, along northern flank of Himalayas, receeded from S-Tibet, but at the same time a small and shallower branch of this gulf made its way along the southern foot of Himalayan range, reaching at least as far as Nainital. At the end of Eocene, W-Indian gulf shortened considerably retreating from Himalayan region and northern Punjab, until its head lay to south of Bugti Hills in Sindh (Pakistan). During Upper

Oligocene, this gulf receeded giving rise to estuarine conditions, while in E-India the 2 gulfs of Bay of Bengal became more and more widely separated from each other, by the continued rise of Arakan yoma, W-gulf subsequently giving rise to Meghna valley. During Tertiary times, palaeogeographic disposition of Indian subcontinent was between 2°N and 30°N palaeo-latitude (Freas and Eckstrom, 1968). Lithological units having suitable association for localisation of phosphate under either lagoonal or platform conditions are following :

Eocene

(a) Calcareous sandstone, intercalated with limestone, in Rann of Kutch; Hower Eocene rocks in Kutch area and limestone marls and sandy beds of middle and upper Eocene, Kutch and Kathiawar area and clay beds with intercalations of limestone in Surat-Broach area (Gujarat).

(b) Sandstones, bentonite and fuller's earth in Barmer and the nummulitic limestone in Jaisalmer and Bikaner (Rajasthan).

(c) Limestone of Ranikot Series; pyritous and carbonaceous shales with nummulitic limestone and variegated shales of Chharat Series exposed in Uri, Punch, Pir Panjal and Riasi areas (J&K).

(d) Limestones with gypseous shales of Subathu Series exposed along the foot hills in H.P. and Uttaranchal.

(e) Rocks of Jaintia Series consisting of limestone shales of Sylhet stage; clays, shales, impure sandstone of Kopili stage; Garo Khasi-Jaintia hills (Meghalaya) and Mikir Hills (Assam).

Oligocene-Lower Miocene

(a) Shales, clays, limestones and marls of Nari and Gaj beds in Kutch and Dwarka beds, Jamnagar (Gujarat).

(b) Shale with occasional bands of green phosphatic shale of lower Murrees and sandstone and shale of upper Murrees in Punch area; calcareous sandstone and shale of lower Murrees in J&K; Murree rocks of Mandi area (H.P.).

(c) Calcareous and carbonaceous shales of lower Barail Series and shale-sandstone and mudstone of Surma Series of Garo and Khasi-Jaintia hills (Meghalaya).

(d) Limestones, pale green clays, thin bands of marl exposed in Baripada area, Mayurbhanj (Orissa).

In order to meet the considerable gap between demand and indigenous resources of phosphorite, location of new deposits is a must. Recent discoveries, on the basis of application of the new concept of phosphate geology, have given confidence that a diligent search on scientific lines can lead to discovery of new deposits.

Studies reveal that phosphorites can be found in formations of all ages ranging from Precambrian to Recent and as such their search cannot be confined to rocks of any particular age. However, location of large deposits in Aravallis of Rajasthan and M.P. has tempted to think that rocks of equivalent age and set up in other parts of country may lead to the location of new deposits. Thus, a high priority should be given to this group of rocks as compared to other formations in Himalayas where even if deposits are found, they may not be economical.

Most of our known deposits are low grade with a high carbonate content and require beneficiation. Efforts will have to be made for finding out cheaper methods by our metallurgists and ore dressing experts.